Markus Horn

Structure-Reactivity-Relationships of Substituted Tritylium Ions

Markus Horn

# Structure-Reactivity-Relationships of Substituted Tritylium Ions

A Kinetic Investigation

Südwestdeutscher Verlag für Hochschulschriften

**Impressum/Imprint (nur für Deutschland/only for Germany)**
Bibliografische Information der Deutschen Nationalbibliothek: Die Deutsche Nationalbibliothek verzeichnet diese Publikation in der Deutschen Nationalbibliografie; detaillierte bibliografische Daten sind im Internet über http://dnb.d-nb.de abrufbar.
Alle in diesem Buch genannten Marken und Produktnamen unterliegen warenzeichen-, marken- oder patentrechtlichem Schutz bzw. sind Warenzeichen oder eingetragene Warenzeichen der jeweiligen Inhaber. Die Wiedergabe von Marken, Produktnamen, Gebrauchsnamen, Handelsnamen, Warenbezeichnungen u.s.w. in diesem Werk berechtigt auch ohne besondere Kennzeichnung nicht zu der Annahme, dass solche Namen im Sinne der Warenzeichen- und Markenschutzgesetzgebung als frei zu betrachten wären und daher von jedermann benutzt werden dürften.

Verlag: Südwestdeutscher Verlag für Hochschulschriften GmbH & Co. KG
Dudweiler Landstr. 99, 66123 Saarbrücken, Deutschland
Telefon +49 681 37 20 271-1, Telefax +49 681 37 20 271-0
Email: info@svh-verlag.de

Approved by: München, LMU, Diss., 2011

Herstellung in Deutschland:
Schaltungsdienst Lange o.H.G., Berlin
Books on Demand GmbH, Norderstedt
Reha GmbH, Saarbrücken
Amazon Distribution GmbH, Leipzig
**ISBN: 978-3-8381-2936-5**

**Imprint (only for USA, GB)**
Bibliographic information published by the Deutsche Nationalbibliothek: The Deutsche Nationalbibliothek lists this publication in the Deutsche Nationalbibliografie; detailed bibliographic data are available in the Internet at http://dnb.d-nb.de.
Any brand names and product names mentioned in this book are subject to trademark, brand or patent protection and are trademarks or registered trademarks of their respective holders. The use of brand names, product names, common names, trade names, product descriptions etc. even without a particular marking in this works is in no way to be construed to mean that such names may be regarded as unrestricted in respect of trademark and brand protection legislation and could thus be used by anyone.

Publisher: Südwestdeutscher Verlag für Hochschulschriften GmbH & Co. KG
Dudweiler Landstr. 99, 66123 Saarbrücken, Germany
Phone +49 681 37 20 271-1, Fax +49 681 37 20 271-0
Email: info@svh-verlag.de

Printed in the U.S.A.
Printed in the U.K. by (see last page)
**ISBN: 978-3-8381-2936-5**

Copyright © 2011 by the author and Südwestdeutscher Verlag für Hochschulschriften GmbH & Co. KG and licensors
All rights reserved. Saarbrücken 2011

# Table of Contents

Summary .................................................................................................. I-XII

Introduction ............................................................................................. 1

**1. Stabilities of Trityl Protected Substrates:**
**The Wide Mechanistic Spectrum of Trityl Ester Hydrolyses** ........ 7
1.1. Introduction ................................................................................. 7
1.2. Kinetic Methods .......................................................................... 8
1.3. Results ......................................................................................... 9
1.4. Discussion ................................................................................. 17
    *1.4.1. Hammett Analysis* ............................................................ 17
    *1.4.2. Winstein-Grunwald Analysis* ........................................... 18
    *1.4.3. Rate-Equilibrium Relationships* ...................................... 21
1.5. Conclusion ................................................................................. 22
1.6. References ................................................................................. 24

**2. Electrophilicity versus Electrofugality of Tritylium Ions**
**in Aqueous Acetonitrile** ................................................................ 27
2.1. Introduction ............................................................................... 27
2.2. Results and Discussion ............................................................. 28
    *2.2.1. Kinetics* ............................................................................ 28
    *2.2.2. Quantum Chemical Calculations* .................................... 31
    *2.2.3. Linear Free Energy Relationships* .................................. 36
    *2.2.4. Hammett Analysis* ........................................................... 37
    *2.2.5. Electrophilicity Parameters of Tritylium Ions* ................ 38
    *2.2.6. Comparison of Electrofugality and Electrophilicity* ...... 39
    *2.2.7. Common Ion Return of Carboxylate Anions?* ................ 40
    *2.2.8. Complete Free Energy Diagrams for the*
        *Hydrolyses of Trityl Carboxylates* ................................... 43
2.3. Conclusion ................................................................................. 45
2.4. References ................................................................................. 46

## 3. Electrophilicities of Acceptor-Substituted Tritylium Ions ............ 48
3.1. Introduction ............ 48
3.2. Results ............ 48
   *3.2.1. Rates of Hydride Transfers* ............ 48
   *3.2.2. Rates of Reactions with Water* ............ 52
   *3.2.3. Theoretical Calculations* ............ 53
   *3.2.4. Product Study* ............ 54
3.3. Discussion ............ 54
3.4. Conclusion ............ 59
3.5. References ............ 59

## 4. Electrofugalities of Acceptor-Substituted Tritylium Ions ............ 61
4.1. Introduction ............ 61
4.2. Results ............ 62
4.3. Discussion ............ 69
   *4.3.1. Leaving Groups* ............ 69
   *4.3.2. Ion Recombination* ............ 70
   *4.3.3. Linear Free Energy Relationships* ............ 73
   *4.3.4. Winstein-Grunwald-Analysis* ............ 74
   *4.3.5. Hammett Analysis* ............ 75
   *4.3.6. Determination of Electrofugality Parameters $E_f$* ............ 77
4.4. Conclusion ............ 82
4.5. References ............ 84

## 5. Towards a General Hydride Donor Ability Scale ............ 86
5.1. Introduction ............ 86
5.2. Methodology and Results ............ 88
5.3. Discussion ............ 95
5.4. Conclusion ............ 108
5.5. References ............ 110

**6. Reduction Potentials of Substituted Tritylium Ions** ............................................. 112
  6.1. Introduction ............................................................................................. 112
  6.2. Results ..................................................................................................... 113
  6.3. Discussion ............................................................................................... 119
  6.4. References .............................................................................................. 125

**7. Miscellaneous Experiments** ................................................................................ 127
  7.1. Nucleophilicitiy Parameters for N-Heterocyclic Carbene Boranes ................ 127
  7.2. Hydride Transfers from Dihydropyridines to Tritylium Ions ........................ 129
  7.3. Reactivities of Tritylium Ions toward Imidazoles ........................................ 133
  7.4. References .............................................................................................. 137

**Appendix** .............................................................................................................. 138

**A. Carbocationic n-endo-trig Cyclizations** ............................................................. 139
  A.1. Introduction ........................................................................................... 139
  A.2. Kinetic Investigations ............................................................................. 139
  A.3. Product Studies ...................................................................................... 141
  A.4. Discussion .............................................................................................. 143
  A.5. Quantum Chemical Calculations ............................................................. 146
  A.6. Consequences for π-Participation in Solvolysis Reactions ......................... 150
  A.7. Effective Molarities ................................................................................ 152
  A.8. References ............................................................................................. 153

**B. Organocatalytic Activity of Cinchona Alkaloids:**
**Which Nitrogen is more Nucleophilic?** ............................................................. 155
  B.1. Introduction ........................................................................................... 155
  B.2. Product Identification ............................................................................. 155
  B.3. Kinetic Investigation .............................................................................. 157
  B.4. Discussion .............................................................................................. 160
  B.5. Computational Analysis ......................................................................... 161
  B.6. Intrinsic Barriers .................................................................................... 163
  B.7. Conclusion ............................................................................................. 166
  B.8. References ............................................................................................. 167

**Experimental Part** ........................................................................................................... 169
1. General Information ............................................................................................... 169
   1.1. Methods ............................................................................................................. 169
   1.2. Materials ........................................................................................................... 171
2. Synthetic Procedures ............................................................................................... 171
   2.1. Preparation of triarylmethanols ........................................................................ 171
   2.2. Preparation of tritylium tetrafluoroborates ...................................................... 179
   2.3. Preparation of triarylmethyl esters ................................................................... 184
   2.4. Preparation of triarylmethyl halides ................................................................. 188
   2.5. Preparation of nucleophiles .............................................................................. 198
3. Product Studies ........................................................................................................ 202
4. Kinetic Data ............................................................................................................. 207
   4.1. Solvolyses of triarylmethyl esters in aqueous acetonitrile ............................... 207
   4.2. Solvolyses of triphenylmethyl esters in aqueous acetone ................................ 213
   4.3. Ionizations of trianisylmethyl esters in aqueous acetonitrile
        in the presence of piperidine ............................................................................ 214
   4.4. Reactions of tritylium ions with water in aqueous acetonitrile ....................... 214
   4.5. Reactions of tritylium ions in aqueous acetonitrile
        in the presence of additives ............................................................................. 217
   4.6. Reactions of tritylium ions with hydride donors ............................................. 220
   4.7. Reactions of benzhydrylium ions with hydride donors ................................... 257
   4.9. Ionizations of trityl halides and esters .............................................................
5. Computational Data ................................................................................................. 300
   5.1. Methyl anion and hydroxide affinities of tritylium ions ................................. 300
   5.2. Organocatalytic Activity of Cinchona Alkaloids:
        Which Nitrogen is more Nucleophilic? ............................................................ 304
   5.3 Carbocationic n-endo-trig Cyclizations ........................................................... 311
6. References ............................................................................................................... 311

**Abbreviations and Symbols**

# Summary

## 1. Stabilities of Trityl Protected Substrates:
## The Wide Mechanistic Spectrum of Trityl Ester Hydrolyses

Ionization rates of *para*-substituted triphenylmethyl (trityl) acetates, benzoates, and *p*-nitrobenzoates have been determined in aqueous acetonitrile and aqueous acetone at 25 °C. Conventional and stopped-flow techniques have been used to evaluate rate constants spanning a range of 7 orders of magnitude by conductimetry and photospectrometry.

The varying stabilities of the differently substituted trilylium ions account for a gradual change of reaction mechanism. Poorly stabilized carbocations are generated slowly by ionization of their covalent precursors and trapped fast by water. Better stabilized carbocations are generated more rapidly and accumulate, so that ionization and trapping by water can be observed as separated steps in a single experiment. Finally, highly stabilized trilylium ions do not react with water, and only the rates of their formation could be measured.

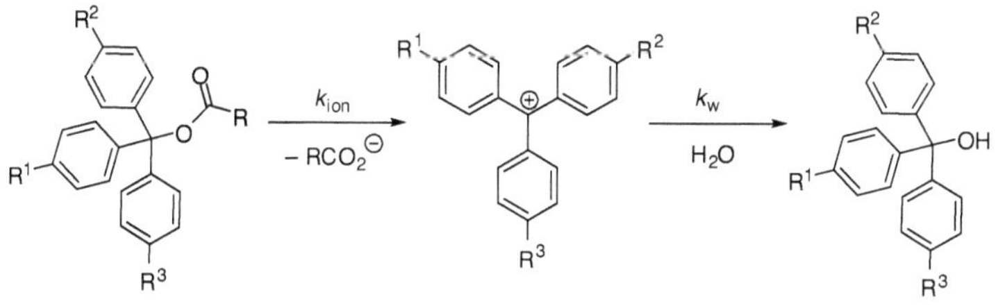

Scheme S1.

Ionization rate constants correlate linearly with Winstein's ionizing powers $Y$; the low slopes (0.17 < $m$ < 0.58) indicate non-carbocation like transition states. While the correlation between ionization rates and Hammett-Brown's $\sigma^+$ parameters is excellent for symmetrically substituted trilylium derivatives, deviations for unsymmetrically substituted systems are observed. The failing rate-equilibrium

| $R^1, R^2, R^3$ | Abbreviation |
|---|---|
| H, H, H | $Tr^+$ |
| Me, H, H | $MeTr^+$ |
| Me, Me, H | $Me_2Tr^+$ |
| Me, Me, Me | $Me_3Tr^+$ |
| MeO, H, H | $(MeO)Tr^+$ |
| MeO, MeO, H | $(MeO)_2Tr^+$ |
| MeO, MeO, MeO | $(MeO)_3Tr^+$ |
| $Me_2N$, H, H | $(Me_2N)Tr^+$ |
| $Me_2N$, MeO, H | $(Me_2N)(MeO)Tr^+$ |
| $Me_2N$, $Me_2N$, H | $(Me_2N)_2Tr^+$ |

relationship between the rates of ionizations (log $k_{ion}$) and the stabilities of the carbocations in aqueous solution (p$K_{R+}$) may be explained by the late development of resonance between a *para*-amino group and the carbocationic center of the tritylium ion during the ionization process (Figure S1).

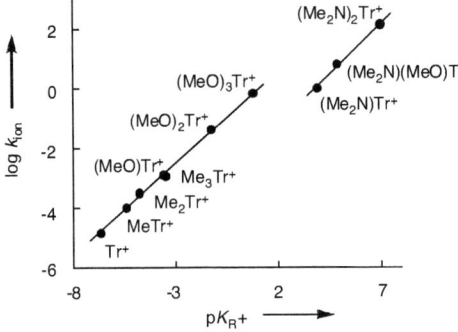

Figure S1. Plot of log $k_{ion}$ for trityl acetates in 90/10 (v/v) acetonitrile/water (25 °C) vs. p$K_{R^+}$.

## 2. Electrophilicity versus Electrofugality of Tritylium Ions in Aqueous Acetonitrile

First-order rate constants $k_w$ (Scheme S1) for the reactions of a series of donor-substituted triphenylmethylium (tritylium) ions with water in aqueous acetonitrile have been determined photometrically at 20 °C using stopped-flow and laser-flash techniques. The rate constants follow the linear free energy relationship log $k_w = s_N(N + E)$. Only the $k_w$ values of the methyl and methoxy substituted tritylium ions correlate linearly with the corresponding p$K_{R^+}$ values, the Leffler-Hammond coefficient α = δΔ$G^‡$/δΔ$G^0$ being 0.62. The amino substituted compounds react more slowly than expected from the correlation of the less stabilized systems.

Quantum chemical calculations of tritylium ions and the corresponding triarylmethanols and 1,1,1-triarylethanes have been performed on the MP2(FC)/6-31+G(2d,p)//B3LYP/6-31G(d,p) level of theory. The calculated gas phase hydroxide and methyl anion affinities of the tritylium ions correlate linearly with a slope of unity, indicating that the relative anion affinities do not depend on the nature of the anion. The p$K_{R^+}$ values of the methyl and methoxy substituted tritylium ions correlate linearly with the calculated gas phase hydroxide affinities, and the slope of this correlation shows that the differences in carbocation stabilities in the gas phase are attenuated to 66 % in aqueous solution. Mono- and bis(dimethylamino) substituted derivatives deviate from this correlation; their p$K_{R^+}$ values are higher than expected from their calculated gas phase hydroxide affinities. This is explained by the extraordinary solvation of unsymmetrically amino substituted tritylium ions. As a consequence, no general

# Summary III

linear correlation between electrofugalities (log $k_{ion}$) and electrophilicities (log $k_w$) does exist for the complete set of tritylium ions (Figure S2).

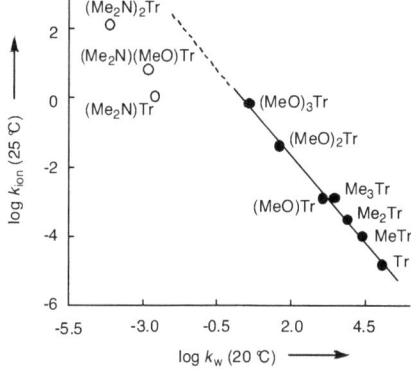

Figure S2. Plot of ionization rates log $k_{ion}$ for trityl acetates vs. rates of attack of water at tritylium ions log $k_w$, 90/10 (v/v) acetonitrile/water.

Complete free energy profiles for the solvolyses of substituted trityl benzoates have been constructed (Figure S3).

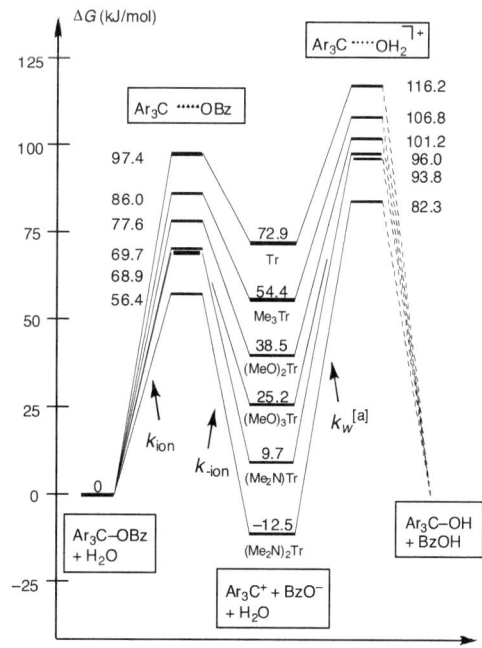

Figure S3. Free energy profiles for the hydrolyses of substituted trityl benzoates in 90/10 (v/v) acetonitrile/water, 25 °C.

## 3. Electrophilicities of Acceptor-Substituted Tritylium Ions

Rates of hydride transfers ($k$) from triphenylsilane to a series of substituted tritylium ions have been determined spectrophotometrically in dichloromethane solution at 20 °C (Scheme S2). The obtained kinetic data have been used to evaluate electrophilicity parameters $E$ for acceptor-substituted tritylium ions according to the linear free energy relationship log $k$ = $s_N(N + E)$, thus extending the previously established electrophilicity scale of differently substituted tritylium ions towards more reactive systems.

The rates of attack of water ($k_w$) at *meta*-fluoro substituted tritylium ions have been determined in aqueous acetonitrile solution using laser-flash techniques. Hydroxide and methyl anion affinities of fluoro-substituted tritylium ions have been calculated on the MP2(FC)/6-31+G(2d,p)//B3LYP/6-31G(d,p) level of theory. Rate-equilibrium relationships are discussed.

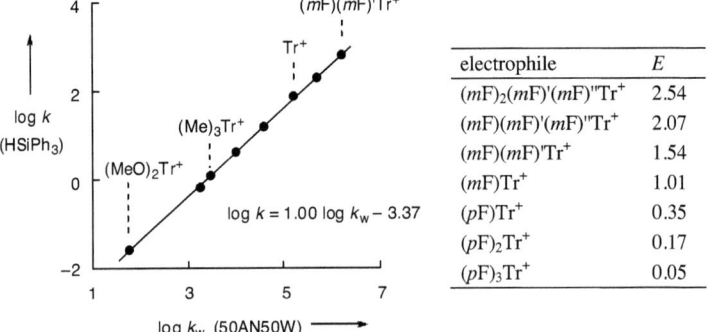

Figure S4. Plot of log $k$ for the reactions of triarylmethyl cations with HSiPh$_3$ (CH$_2$Cl$_2$, 20 °C) vs. log $k_w$ for the reactions with water (50 % aqueous acetonitrile, 20 °C), and empirical electrophilicity parameters $E$ of fluoro-substituted tritylium ions.

## 4. Electrofugalities of Acceptor-Substituted Tritylium Ions

Ionization rate constants ($k_{ion}$) of differently substituted trityl halides and carboxylates have been determined by means of conductimetry in aqueous acetonitrile and acetone at 25 °C (Scheme S3). Common ion return was suppressed by the addition of piperidine which traps the generated tritylium ions. The obtained rate constants have been subjected to Winstein-Grunwald and Hammett analyses. The solvolysis rate constants of trityl chlorides and bromides have been employed to derive electrofugality parameters $E_f$ of tritylium ions according to the linear free energy relationship $\log k_{ion} = s_f(E_f + N_f)$.

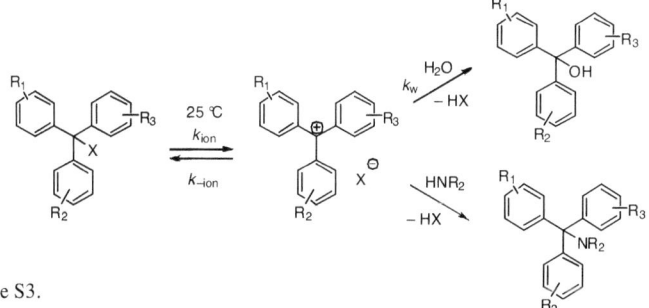

Scheme S3.

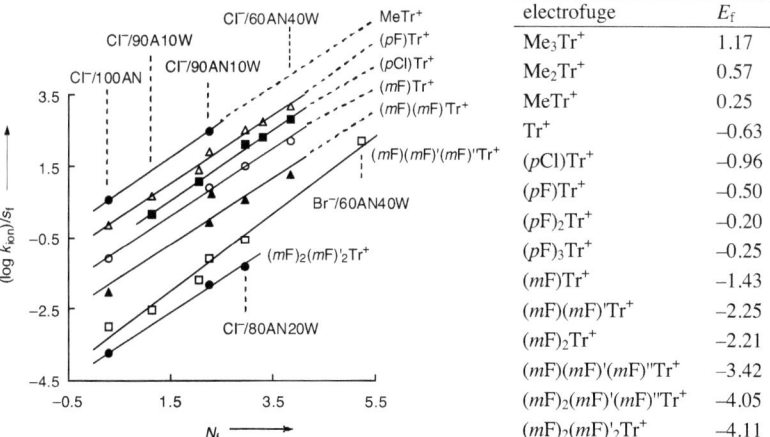

| electrofuge | $E_f$ |
|---|---|
| $Me_3Tr^+$ | 1.17 |
| $Me_2Tr^+$ | 0.57 |
| $MeTr^+$ | 0.25 |
| $Tr^+$ | −0.63 |
| $(pCl)Tr^+$ | −0.96 |
| $(pF)Tr^+$ | −0.50 |
| $(pF)_2Tr^+$ | −0.20 |
| $(pF)_3Tr^+$ | −0.25 |
| $(mF)Tr^+$ | −1.43 |
| $(mF)(mF)'Tr^+$ | −2.25 |
| $(mF)_2Tr^+$ | −2.21 |
| $(mF)(mF)'(mF)''Tr^+$ | −3.42 |
| $(mF)_2(mF)'(mF)''Tr^+$ | −4.05 |
| $(mF)_2(mF)'_2Tr^+$ | −4.11 |

Figure S5. Left: plot of $(\log k_{ion})/s_f$ vs. $N_f$ for the ionizations of substituted trityl chlorides and bromides in aqueous acetonitrile (AN/W) and aqueous acetone (A/W), 25 °C, solvents are given in vol%; right: empirical electrofugality parameters $E_f$ of substituted tritylium ions.

A simple scheme to estimate ionization rates of trityl derivatives is presented (Figure S6).

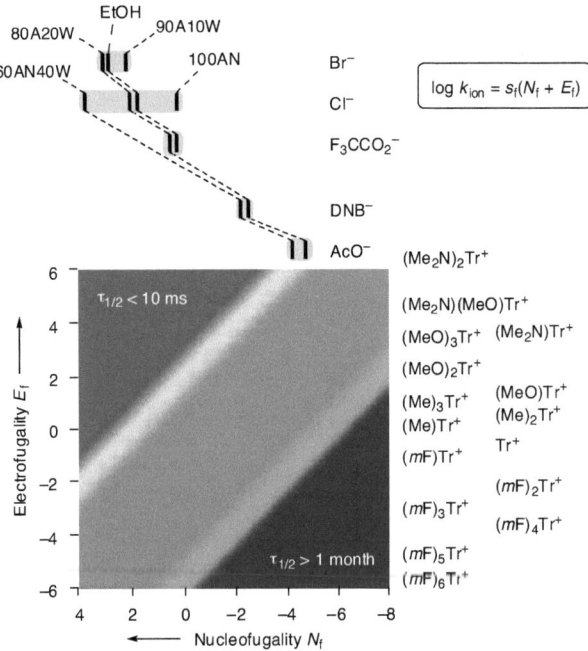

Figure S6. From high reactivity to inertness. A semiquantitative model for estimating half-lives of trityl derivatives in different solvents (given in vol%), AN = acetonitrile, A = acetone, W = water, EtOH = ethanol.

## 5. Towards a General Hydride Donor Ability Scale

Rate constants of hydride transfers from several hydride donors to benzhydrylium ions have been determined at 20 °C (Scheme S4). Empirical nucleophilicity parameters have been evaluated according to the linear free energy relationship $\log k = s_N(N + E)$. The experimental rate constants of the reactions of these hydride donors with tritylium ions agreed well with those calculated by this correlation (deviation factors of up to 39). The huge amount of published rate constants of hydride transfers to the triphenylcarbenium ion could, therefore, be used to incorporate a multitude of different hydride donors in our comprehensive nucleophilicity scale (Table S1).

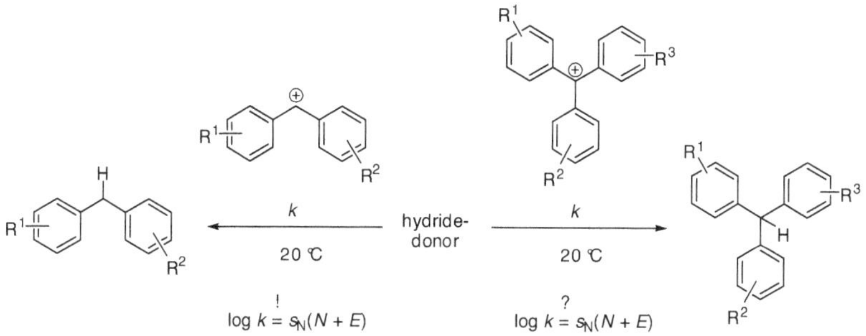

Summary VII

Scheme S4.

Table S1. Nucleophilicity parameters $N$ for several hydride donors derived from literature rate constants.

| | $N$ | | $N$ | | $N$ |
|---|---|---|---|---|---|
| HSiMe$_3$ | 2.6 | HW(CO)$_3$(C$_5$H$_4$CO$_2$Me) | −0.9 | Et$_4$Si | −6.5 |
| HSiMe$_2$Et | 2.3 | HMn(CO)$_5$ | 1.5 | BuSiMe$_3$ | −5.4 |
| HSiMeEt$_2$ | 2.3 | HCr(CO)$_3$Cp* | 1.6 | PhCH$_2$CH$_2$SiMe$_3$ | −3.5 |
| HSi$^n$Pr$_3$ | 2.4 | HW(CO)$_3$Cp | 1.7 | Ph$_2$CHCH$_2$SiEt$_2$Me | −6.4 |
| HSi$^n$Hex$_3$ | 2.6 | cis-HMn(PCy$_3$)(CO)$_4$ | 2.2 | Me$_3$Si(CH$_2$)$_6$SiMe$_3$ | −4.7 |
| HSiMe$_2$(CH$_2$Cl) | −0.6 | cis-HMn(PPh$_3$)(CO)$_4$ | 2.3 | Me$_3$Si(CH$_2$)$_3$SiMe$_3$ | 5.3 |
| HSiMePh(CH$_2^t$Bu) | 0.9 | HW(CO)$_3$(C$_5$H$_4$Me) | 2.4 | 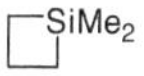 ⌐SiMe$_2$ (cyclobutyl) | −2.1 |
| HSiMe$_2$Bn | 1.9 | HMo(CO)$_3$Cp | 2.6 |  SiMe$_2$ (cyclopentyl) | −5.5 |
| HSiMe$_2$(m-ClBn) | 1.3 | HW(CO)$_3$Cp* | 3.5 | 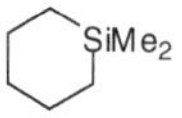 SiMe$_2$ (cyclohexyl) | −3.0 |
|  HMeSi (cyclobutyl) | 2.3 | HW(CO)$_3$(indenyl) | 3.5 |  SiMe$_2$ (cycloheptyl) | −3.1 |
| 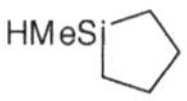 HMeSi (cyclopentyl) | 2.2 | HRe(CO)$_5$ | 3.5 | Et$_4$Ge | −4.7 |
| HSi(OEt)$_3$ | −1.8 | cis-HRe(PPh$_3$)(CO)$_4$ | 4.5 | Me$_3$Ge(CH$_2$)$_3$SiMe$_3$ | −4.8 |
| HSiMe$_2$(OTMS) | 2.5 | HW(NO)$_2$Cp | 4.8 | Me$_3$Ge(CH$_2$)$_3$GeMe$_3$ | −3.4 |
| HSiMe$_2$(OPr) | 2.4 | trans-HMo(CO)$_2$(PCy$_3$)Cp | 6.5 | Et$_4$Sn | −1.9 |
| HGeEt$_3$ | 4.0 | trans-HMo(CO)$_2$(PPh$_3$)Cp | 6.6 | Pr$_4$Sn | −0.5 |
| | | trans-HMo(CO)$_2$(PMe$_3$)Cp | 7.8 | $^i$Bu$_4$Sn | −0.1 |
| | | HMo(CO)$_3$Cp* | 4.5 | $^{sec}$Bu$_4$Sn | −1.1 |
| | | | | Me$_3$Sn(CH$_2$)$_3$SiMe$_3$ | −3.4 |
| | | | | Me$_3$Sn(CH$_2$)$_3$GeMe$_3$ | −2.4 |
| | | | | Me$_3$Sn(CH$_2$)$_3$SnMe$_3$ | −1.7 |
| | | | | Me$_3$SnCH$_2$Bn | −1.6 |
| | | | | Me$_3$Sn(CH$_2$)$_3$CMe$_3$ | −3.9 |
| | | | | Et$_4$Pb | 0.1 |
| | | | | Me$_3$Pb(CH$_2$)$_3$SiMe$_3$ | −1.2 |
| | | | | Me$_3$Pb(CH$_2$)$_3$GeMe$_3$ | −1.0 |
| | | | | Et$_2$Hg | −0.7 |

## 6. Reduction Potentials of Substituted Tritylium Ions

One-electron reduction potentials $E_{1/2}^{red}$ of a series of substituted tritylium ions have been determined in acetonitrile solution at 25 °C (Scheme S5). The silver/silver oxide (Ag/Ag$_2$O) electrode served as reference. Because ultramicroelectrodes were employed, no conductive salt was used.

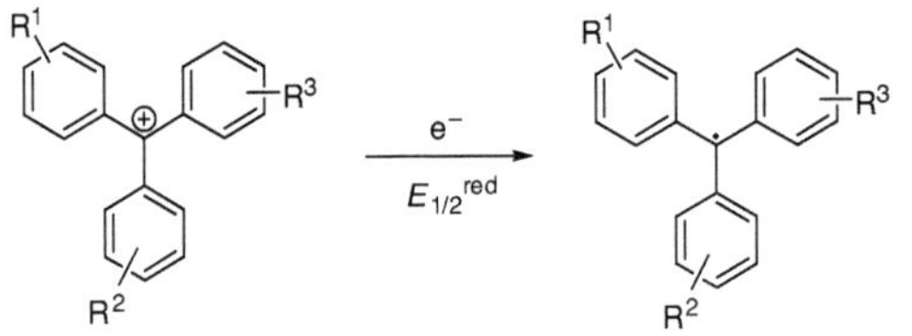

Scheme S5.

The obtained data were converted to the standard calomel electrode (SCE) as reference, thus enabling the direct comparison with substituted benzhydrylium ions. Two different correlation lines are found, when the electrophilicity parameters $E$ of tritylium and benzhydrylium ions are plotted against $E_{1/2}^{red}$ (Figure S7).

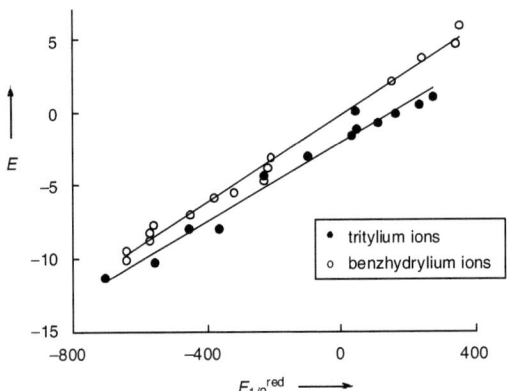

Figure S7. Plot of electrophilicity parameters $E$ of tritylium and benzhydrylium ions vs. $E_{1/2}^{red}$.

Hydride transfers from silanes to tritylium ions are shown to proceed via the polar, rather than the also conceivable stepwise mechanism, which consists of initial single electron transfer (SET) and subsequent hydron shift. Reaction free energies $\Delta G^0_{SET}$ for hypothetical single electron transfers, i.e., the first steps of the stepwise mechanism, have been calculated from reduction potentials of tritylium ions and oxidation potentials of silanes. They are compared with the free energies of activation $\Delta G^{\ddagger}_{obs}$, which have been obtained from experimentally determined rates of hydride transfers (Figure S8).

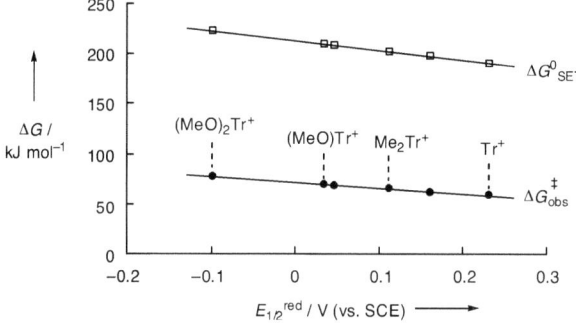

Figure S8. Reactions of substituted tritylium ions with dimethylphenylsilane. Plots of $\Delta G^0_{SET}$ (calculated with redox potentials) and $\Delta G^{\ddagger}_{obs}$ (from experimentally determined rate constants) against the reduction potentials of substituted tritylium ions.

## 7. Miscellaneous Experiments

The N-heterocyclic carbene boranes **1** and **2** have been used as hydride donors to reduce substituted benzhydrylium ions to the corresponding diarylmethanes. When the rate constants log $k$ were plotted against the empirical electrophilicity parameters $E$ of the benzhydrylium systems, linear correlations were obtained, from which the nucleophilicity parameters $N$ and $s_N$ for the boranes have been derived (Figure S9).

When 1,4-dihydropyridines, such as the Hantzsch esters **A** and **B**, were combined with tritylium ions in dichloromethane, the triarylmethanes and pyridium ions were formed exclusively, but for unknown reasons the reactions did not follow second-order kinetics.

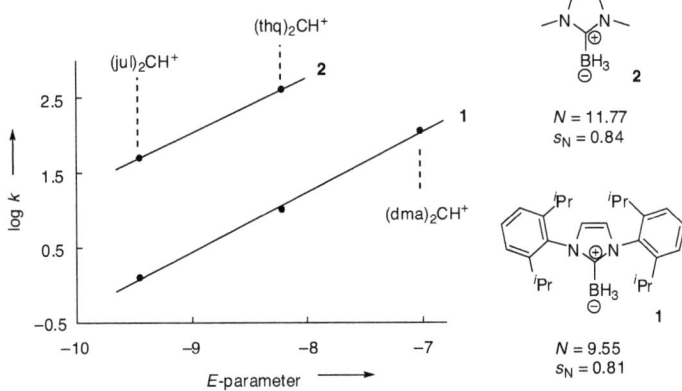

Figure S9. Plots of log $k$ of hydride transfers from carbene boranes to benzhydrylium ions against the corresponding electrophilicity parameters $E$ (CH$_2$Cl$_2$, 20 °C); dma = 4-(dimethylamino)phenyl; thq = 1,2,3,4-tetrahydroquinoline-6-yl; jul = julolidine-4-yl.

The rate constants $k$ of the reactions between substituted tritylium ions and imidazole in acetonitrile at 20 °C deviate only marginally from the predictions made by the free energy relationship log $k = s_N(N + E)$ (Table S2).

Table S2. Comparison of calculated and experimental rate constants for the reactions of substituted tritylium ions and imidazole (CH$_3$CN, 20 °C).

| electrophile | $k_{exp}$ [L mol$^{-1}$ s$^{-1}$] | $k_{calc}$ [L mol$^{-1}$ s$^{-1}$] | $k_{calc}/k_{exp}$ |
|---|---|---|---|
| (MeO)$_2$Tr$^+$ | 1.64 × 10$^5$ | 4.57 × 10$^6$ | 28 |
| (MeO)$_3$Tr$^+$ | 2.58 × 10$^4$ | 4.22 × 10$^5$ | 16 |
| (Me$_2$N)Tr$^+$ | 5.19 × 10$^1$ | 6.26 × 10$^2$ | 12 |

In contrast, the reactions of tritylium ions with 2-methylimidazole did not follow a second-order rate law.

## Appendix: A. Carbocationic n-endo-trig Cyclizations

Unsaturated benzyl cations $(4\text{-MeOC}_6\text{H}_4)\text{CH}^+\text{-(CH}_2)_n\text{-CH=CH}_2$ (**1c-e**) have been generated laser-flash photolytically in acetonitrile in the presence of enol ethers or 2-methylfuran.

| | | |
|---|---|---|
| MeO—⟨⟩—CH⁺—CH₃ | MeO—⟨⟩—CH⁺—(CH₂)₃—CH₃ | MeO—⟨⟩—CH⁺—(CH₂)ₙ—CH=CH₂ |
| **1a** | **1b** | n = 2  3  4 |
| | | **1c 1d 1e** |

The reactions of **1c,e** with these π-nucleophiles follow second-order rate laws with rate constants comparable to those of the analogous saturated species **1b**. Product studies show the absence of cyclization products.

In contrast, carbocation **1d** undergoes a highly reversible 6-endo-trig cyclization which is approximately $10^7$ times faster than the corresponding intermolecular reaction of **1b** with 1-hexene. This cyclization yields a highly electrophilic, partially bridged carbocation, which accounts for the finding that **1d** is consumed 10 times faster in the solvent trifluoroethanol than all other carbocations in this series.

Quantum chemical calculations (B3LYP/6-311G(d,p) and MP2/6-31+G(2d,p)) have been performed to elucidate the structures of the involved carbocations. Consequences of these findings on the role of π-participation in solvolysis reactions are discussed.

## Appendix: B. Organocatalytic Activity of Cinchona Alkaloids: Which Nitrogen is more Nucleophilic?

The cinchona alkaloids **1** react selectively at the quinuclidine ring with benzyl bromide and at the quinoline ring with benzhydrylium ions. The kinetics of these reactions have been determined photometrically or conductimetrically and are compared with analogous reactions of quinuclidine and quinoline derivatives.

# Summary

**1a**: R¹ = OMe; R² = H, quinine
**1c**: R¹ = OMe; R² = Ac
**1d**: R¹ = H; R² = H, cinchonidine
**1b**: quinidine
**1e**, **1f**, **1g**, **1h**

Quantum chemical calculations show that the products obtained by attack at the quinuclidine ring of quinine are thermodynamically more stable when small alkylating agents (primary alkyl) are used, while the products arising from attack at the quinoline ring are more stable for bulkier electrophiles ($Ar_2CH$, Figure S10).

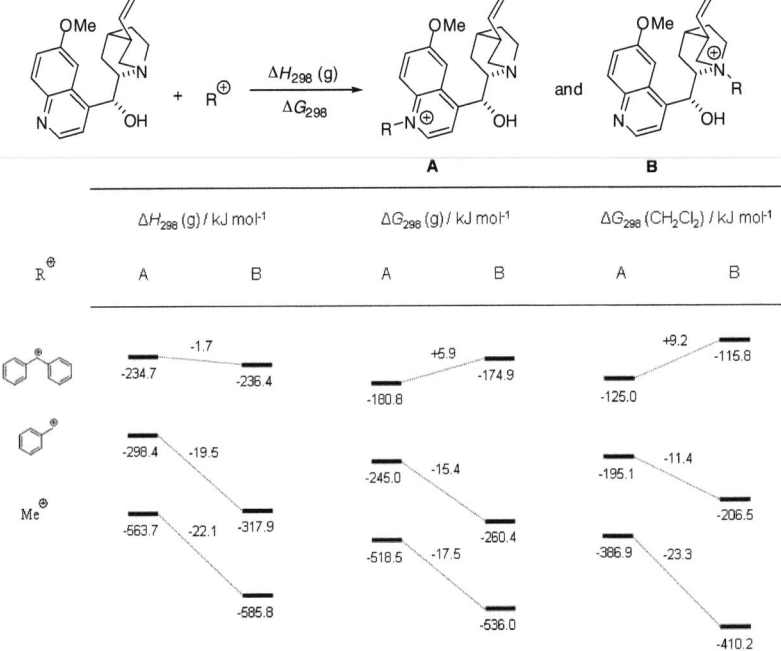

Figure S10. Benzhydryl, benzyl and methyl cation affinities of the different nitrogen atoms of quinine [MP2(FC)/6-31+G(2d,p)//B3LYP/6-31G(d)].

# Introduction

## 1. Reactivity Scales

For any chemical reaction, the questions of the reaction path ("What will happen?") and the thermodynamics ("To what extent will the reaction occur?") are joined by a third important aspect: the kinetics ("How fast does the reaction proceed?").

In 1929 Ingold introduced the terms "electrophile" and "nucleophile" for reactive organic species, characterized by a lack or a surplus of electrons, respectively.[1] In the 1930s he established the fundamental concept of an electronic theory concerning organic reaction mechanisms.[2]

Since then it has been a major objective of organic chemists to numerically quantify reactivities. One of the first systematic attempts to achieve this goal was undertaken by Swain and Scott in 1953.[3] In their equation (1), designed for $S_N2$ processes, $n$ represents a nucleophile-specific parameter, $s$ an electrophile-specific sensitivity parameter, $k$ is the bimolecular rate constant, and $k_w$ is the rate constant for the reaction of the electrophile with water. As a reference system, they chose methyl bromide ($s = 1.00$) in water ($n = 0.00$).

$$\log \frac{k}{k_w} = n \cdot s \qquad (1)$$

The terms "electrofuge" and "nucleofuge" are closely related to the ones mentioned above: While a bond is formed between an electrophile and a nucleophile during a combination reaction, the bond between an electrofuge and a nucleofuge is broken during a heterolysis reaction (Scheme 1).

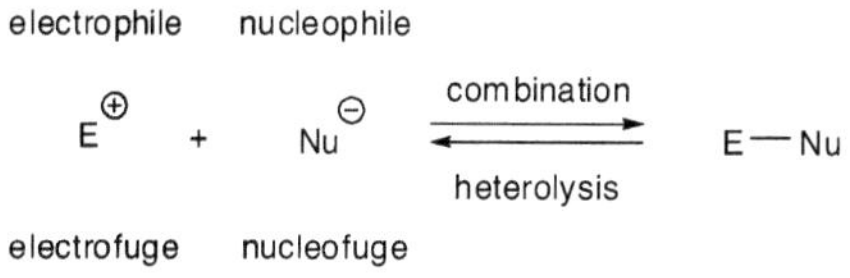

electrophile    nucleophile

$E^{\oplus}$ + $Nu^{\ominus}$ ⇌ E—Nu
(combination / heterolysis)

electrofuge    nucleofuge                                    Scheme 1.

Heterolysis reactions have been treated by the Winstein-Grunwald relationship (equation 2),[4] which aims at quantifying the ability of a solvent to ionize a specific substrate. Solvent ionizing powers $Y$ have been assigned to pure and mixed solvents, and were initially based on the reference system *tert*-butyl chloride ($m = 1.0$) in 80/20 (v/v) ethanol/water ($Y = 0.0$).

$$\log \frac{k}{k_0} = Y \cdot m \qquad (2)$$

In the so-called constant selectivity relationship (equation 3), Ritchie characterized nucleophiles by $N_+$.[5] He had noticed that the ratios of rate constants $k_{Nu1}/k_{Nu2}$ for two nucleophiles reacting with several triarylmethyl cations, aryldiazonium ions, and aryltropylium cations were constant, i.e., independent of the reactivities of the electrophiles.

$$\log \frac{k}{k_0} = N_+ \qquad (3)$$

Later, when the compounds under consideration were extended to other classes, deviations from the two-parameter relationship (3) were found.[6]

In equation (4) – developed by Mayr in 1994 – three parameters were used to deal with electrophile-nucleophile combinations.[7] The former are characterized by $E$, while the latter are described by $N$ and $s_N$.

$$\log k_2 = s_N(E + N) \qquad (4)$$

The analogous equation (5), which was designed for heterolysis reactions, was suggested a few years later.[8] Nucleofuges are characterized by $N_f$ and $s_f$, and an $E_f$ parameter is assigned to electrofuges.

$$\log k_{ion} = s_f(E_f + N_f) \qquad (5)$$

For both equations, (4) and (5), a series of substituted benzhydrylium cations (Scheme 2) is used as reference electrophiles and electrofuges, respectively. This class of carbocations provides the advantage to gradually adjust the electron density at the reactive site by suitable substitution patterns in the *para*- and *meta*-positions of the rings. Hereby, a large range of reactivity can be covered without changing the steric conditions at the central carbon. Quinone methides have been introduced as particularly unreactive electrophiles. Although they are neutral, the resonance structure in Scheme 2 demonstrates their similarity to benzhydrylium ions. Because the remote substituents do not interact with incoming nucleophiles or leaving nucleofuges, direct comparison of the nucleophilicities and nucleofugalities of species

Introduction 3

widely differing in structure becomes possible by purely electronic means. The bis(4-methoxyphenyl)methyl cation was defined as the origin of both reactivity scales ($E = E_f = 0.00$).

The efforts which have been made in order to develop the equations (1) to (5) demonstrate the desire of chemists for reactivity scales. The latter do not only provide an improved feeling and intuition for the reactivities of chemical species in general, but also allow the practical design of new synthetic strategies.

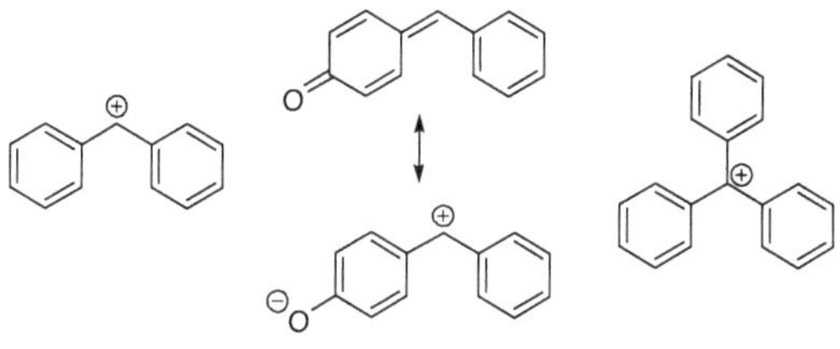

benzhydrylium    quinone methide    tritylium        Scheme 2.

## 2. Triarylmethyl Compounds

Historically, the triphenylmethyl (trityl, Scheme 2) radical was the first organic radical recognized (1900). In his attempt to prepare hexaphenylethane by combining triphenylchloromethane and zinc, Gomberg generated a very reactive material, which he considered as the triphenylmethyl radical.[9] Although it was found later, that this radical exists in equilibrium with its dimer, the constitution of this dimer remained an open question. The subsequent dispute about its nature, in which some of the most renowned chemists at that time were engaged, and which did not find an end until 1968, when the first NMR experiments revealed the truth, accessed the literature as the "hexaphenylethane riddle" (Scheme 3).[10]

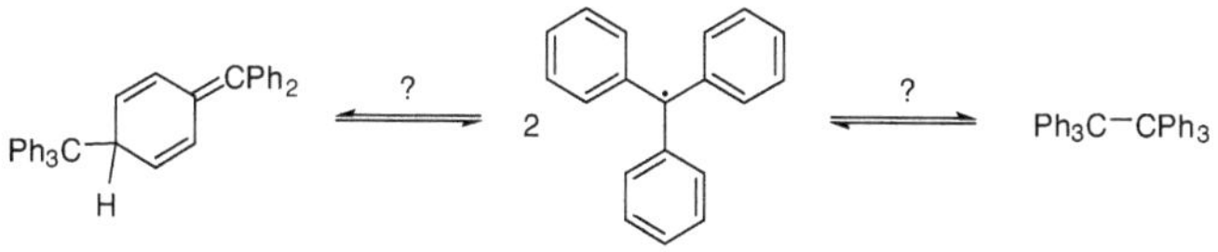

colorless               yellow              colorless
verified by NMR in 1968                    (fictitious)

Scheme 3. The hexaphenylethane riddle.

The aspect of color played an important role in this question, as the trityl radical is yellow, while its dimer is colorless. Because the electronic theory of molecules was not yet established at the beginning of the 20th century, concepts of color were continuously modified, and the triphenylmethyl dimerization began to trigger experiments which aimed at the explanation of the colors of triphenylmethane dyes in general.[11]

Dyes were one of the most important products of the chemical industry in the second half of the 19th century, and among the most important dyes were the triphenylmethane dyes.[12] They were characterized by a superior brilliancy, and the intense research in this field led to the synthesis of phenolphthalein and fluorescein (Scheme 4) by Adolf von Baeyer in 1871. Although compounds like crystal violet and malachite green, which were extensively used to dye textiles and fabrics, were soon replaced by other dyes of improved light-fastness, the triphenylmethane dyes did not lose their importance.

Today, compounds like rhodamine B and fluorescein serve as markers in biochemical staining techniques, phenolphthalein is a standard pH-indicator in every chemical laboratory, while patent blue V is an often used blue food dye (E 131). Crystal violet has relevance in medical applications, and eosin Y (tetrabromofluorescein) gives red ink its color.

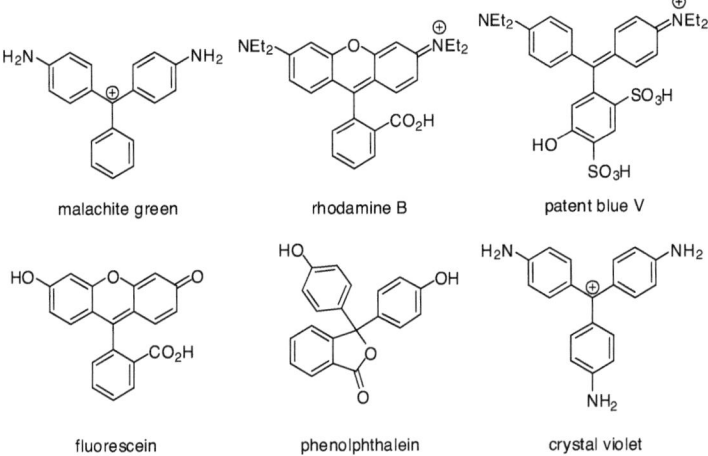

Scheme 4. Some triphenylmethane dyes.

Furthermore, tritylium ions find applications throughout the whole field of organic chemistry. They are employed as protecting groups,[13] hydride acceptors,[14] and initiators in cationic polymerizations.[15] All these examples demonstrate the abundance of trityl compounds, and underline their utility.

## 3. Goal of this Thesis

As tritylium ions were involved in the development of many basic concepts of organic chemistry, it was of interest whether this class of carbocations can be treated by equations (4) and (5), i.e., whether tritylium ions can be characterized by electrophilicity and electrofugality parameters $E$ and $E_f$. Compared with benzhydrylium ions and quinone methides, the additional aryl ring in tritylium ions accounts for pronounced steric hindrance in electrophilic additions. Nonetheless, a preliminary evaluation of electrophilicity parameters $E$ of tritylium ions has previously been undertaken.[16]

The goal of this work was to elucidate the applicability of equations (4) and (5) to tritylium ions as substance class, and to relate their electrophilic to their electrophobic properties in a classical structure-reactivity analysis. Therefore, electrophile-nucleophile combinations as well as heterolysis reactions of trityl derivatives have been investigated kinetically. Because electron-withdrawing as well as electron-donating ring-substituents have been studied, a wide range of reactivity could be covered.

As parts of this thesis have already been published, individual introductions are given at the beginning of each chapter.

## 4. References

[1]   C. K. Ingold, *Recl. Trav. Chim. Pays-Bas* **1929**, *42*, 797-812.
[2]   C. K. Ingold, *Chem. Rev.* **1934**, *15*, 225-274.
[3]   C. G. Swain, C. B. Scott, *J. Am. Chem. Soc.* **1953**, *75*, 141-147.
[4]   E. Grunwald, S. Winstein, *J. Am. Chem. Soc.* **1948**, *70*, 846-854.
[5]   a) C. D. Ritchie, P. O. I. Virtanen, *J. Am. Chem. Soc.* **1972**, *94*, 4966-4971; b) C. D. Ritchie, *Acc. Chem. Res.* **1972**, *5*, 348-354.
[6]   C. D. Ritchie, *Can. J. Chem.* **1986**, *64*, 2239-2350.

[7]   H. Mayr, M. Patz, *Angew. Chem.* **1994**, *106*, 990-1010; *Angew. Chem. Int. Ed.* **1994**, *33*, 938-957.

[8]   B. Denegri, A. Streiter, S. Juric, A. R. Ofial, O. Kronja, H. Mayr, *Chem. Eur. J.* **2006**, *12*, 1648-1656; *Chem. Eur. J.* **2006**, *12*, 5415; B. Denegri, A. R. Ofial, S. Juric, A. Streiter, O. Kronja, H. Mayr, *Chem. Eur. J.* **2006**, *12*, 1657-1666.

[9]   a) M. Gomberg, *J. Am. Chem. Soc.* **1900**, *22*, 757-771; b) M. Gomberg, *Chem. Ber.* **1900**, *33*, 3150-3163.

[10]  J. M. McBride, *Tetrahedron* **1974**, *30*, 2009-2022.

[11]  a) F. Kehrmann, F. Wentzel, *Chem. Ber.* **1901**, *34*, 3815-3819; b) A. Baeyer, V. Villiger, *Chem. Ber.* **1902**, *35*, 1189-1201; c) M. Gomberg, *Chem. Ber.* **1902**, *35*, 2397-2408; d) A. Baeyer, V. Villiger, *Chem. Ber.* **1902**, *35*, 3013-3033; e) L. C. Anderson, *J. Am. Chem. Soc.* **1933**, *55*, 809-812.

[12]  T. Gessner, U. Mayer, "Triarylmethane and Diarylmethane Dyes" in *Ullmann's Encyclopedia of Industrial Chemistry*, Wiley-VCH, Weinheim, **2002**.

[13]  a) M. Smith, D. H. Rammler, I. H. Goldberg, H. G. Khorana, *J. Am. Chem. Soc.* **1962**, *84*, 430-440; b) H. Schaller, G. Weimann, B. Lerch, H. G. Khorana, *J. Am. Chem. Soc.* **1963**, *85*, 3821-3827; c) C. Bleasdale, S. B. Ellwood, B. T. Golding, *J. Chem. Soc. Perkin Trans. 1* **1990**, 803-805; d) M. Sekine, T. Mori, T. Wada, *Tetrahedron Lett.* **1993**, *34*, 8289-8292; e) M. Sekine, T. Hata, *J. Org. Chem.* **1987**, *52*, 946-948; f) A. P. Henderson, J. Riseborough, C. Bleasdale, W. Clegg, M. R. J. Elsegood, B. T. Golding, *J. Chem. Soc. Perkin Trans. 1* **1997**, 3407-3414; g) P. G. M. Wuts, T. W. Greene, *Greene's Protective Groups in Organic Synthesis*, 4$^{th}$ Ed., Wiley-Interscience, Hoboken, **2007**.

[14]  a) H. J. Dauben, Jr., F. A. Gadecki, K. M. Harmon, D. L. Pearson, *J. Am. Chem. Soc.* **1957**, *79*, 4557-4558; b) H. J. Dauben, Jr., D. J. Bertelli, *J. Am. Chem. Soc.* **1961**, *83*, 497-498; c) P. Huszthy, K. Lempert, *J. Chem. Soc., Perkin Trans. 2* **1982**, 1671-1674; d) M. Green, S. Greenfield, M. Kersting, *J. Chem. Soc., Chem. Commun.* **1985**, 18-20; e) D. Mandon, L. Toupet, D. Astruc, *J. Am. Chem. Soc.* **1986**, *108*, 1320-1322; f) L. M. McDonough, Ph.D. Thesis, Washington, **1960**; f) G. Karabatsos, M. Tornaritis, *Tetrahedron Lett.* **1989**, *30*, 5733-5736.

[15]  a) S. D. Pask, P. H. Plesch, *Eur. Polym. J.* **1982**, *18*, 839-846; b) C. Schade, H. Mayr, *Macromol. Chem., Rapid Commun.* **1988**, *9*, 477-482.

[16]  S. Minegishi, H. Mayr, *J. Am. Chem. Soc.* **2003**, *125*, 286-295.

# 1. Stabilities of Trityl Protected Substrates: The Wide Mechanistic Spectrum of Trityl Ester Hydrolyses[‡]

## 1.1. Introduction

The triphenylmethyl (trityl) cation was the first carbocation ever recognized.[1] Since then, the trityl group has found numerous applications in organic chemistry. It has been used as a hydride acceptor[2] and as a catalyst in Lewis acid initiated reactions.[3] Stabilized tritylium ions, covalently bound to nucleosides and peptides, served as mass tags for improving the sensitivity in (MA)LDI-TOF mass spectrometry.[4] Most important in synthetic chemistry, however, is their use as protecting groups for OH and NH functionalities.[5-8] The unsubstituted parent residue has long been employed to protect alcohols, carboxylic acids, and amines.[5] Because the cleavage of trityl ethers often requires conditions which also lead to deprotection of other acid-labile groups, like glycosides, Khorana introduced the 4-methoxy- and the 4,4'-dimethoxytrityl derivatives which are more acid-labile.[6] 4-Methoxy substituted tritylium systems have been studied extensively by Maskill,[7] who determined rate constants of acidic detritylations of amines[7a] and trifluoromethanol.[7b] Destabilized trityl compounds, like the heptafluorotrityl system, were designed to serve as acid-stable protecting groups, e.g., for the γ-carboxy group of glutamic acid.[8]

While the qualitative trend, that the cleavage of trityl esters is facilitated by electron-donating substituents and impeded by acceptor groups is well-known, there are only few kinetic studies on the solvolyses of donor-substituted tritylium derivatives, probably because many of these reactions proceed very fast and require special experimental techniques, which were not generally available in the period when most kinetic investigations of solvolysis reactions were performed.

Swain and co-workers determined solvolysis rates of trityl acetate in aqueous alcohols and acetone at 25 °C.[9] Bunton proved that the alkyl-oxygen bond (rather than the acyl-oxygen bond) is cleft in the rate-determining step during the hydrolysis of trityl acetate.[10] Hammond reported rates of solvolysis of trityl benzoate in ethanolic ethyl methyl ketone at 55 °C and found the exclusive formation of the ethyl trityl ether.[11] Smith has studied the decomposition of tri-*p*-tolylmethyl benzoate in ethanolic methylene chloride and detected strong common ion

---

[‡] This part has been published in: M. Horn, H. Mayr, *Chem. Eur. J.* **2010**, *16*, 7469-7477.

rate depression.[12] In the course of their work on ion pairs, Swain[13] and Winstein[14] investigated the rate of equilibration of carbonyl-$^{18}$O-labeled trityl benzoate in pure acetone.

In view of the thousands of kinetic investigations of $S_N1$ reactions yielding less stabilized carbenium ions, it is surprising that no systematic investigation concerning the heterolyses of donor-substituted trityl esters has yet been undertaken. Because such information is essential for selecting trityl derivatives with tailor-made stabilities, ionization rates of trityl carboxylates, including dimethylamino and methoxy substituted derivatives which are of particular importance as protecting groups, were studied (Table 1.1).

Table 1.1. The tritylium systems studied in this work with corresponding $pK_{R^+}$ values.

| $R^1, R^2, R^{3\,[a]}$ | Abbreviation | $pK_{R^+}$ [b] |
|---|---|---|
| H, H, H | $Tr^+$ | –6.63 |
| Me, H, H | $MeTr^+$ | –5.41 |
| Me, Me, H | $Me_2Tr^+$ | –4.71 |
| Me, Me, Me | $Me_3Tr^+$ | –3.56 |
| MeO, H, H | $(MeO)Tr^+$ | –3.40 |
| MeO, MeO, H | $(MeO)_2Tr^+$ | –1.24 |
| MeO, MeO, MeO | $(MeO)_3Tr^+$ | 0.82 |
| $Me_2N$, H, H | $(Me_2N)Tr^+$ | 3.88 [c] |
| $Me_2N$, MeO, H | $(Me_2N)(MeO)Tr^+$ | 4.86 [d] |
| $Me_2N$, $Me_2N$, H | $(Me_2N)_2Tr^+$ | 6.94 [d] |

[a] For the location of the substituents see Scheme 1.1; [b] from ref. [15]; [c] from ref. [16]; [d] from ref. [17].

## 1.2. Kinetic Methods

If acyl cleavage is excluded, the hydrolyses of trityl esters follow Scheme 1.1, which includes 4 scenarios (A to D) depending on the relative magnitude of $k_{ion}$ and $k_w$. As discussed later, common ion return $k_{-ion}$ does not occur in the concentration range of interest and will therefore be neglected in the following discussion.

(A) $k_{ion} \ll k_w$. In the classical $S_N1$ reaction, carbocations are generated as short-lived intermediates, which are trapped immediately by the solvent. The progress of the reactions can conveniently be followed by conductimetry.

# 1. Stabilities of Trityl Protected Substrates

(B) $k_{ion} \approx k_w$. When the formation and consumption of the intermediates proceed with comparable rates, small concentrations of the tritylium ions are detectable, and the course of the reaction can be followed by conductimetry or photospectrometry.

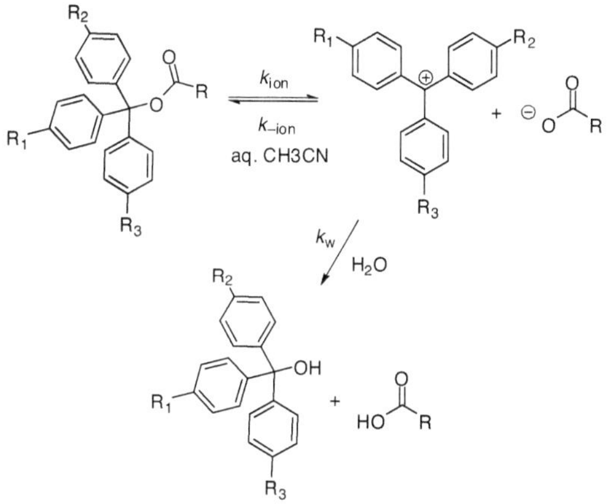

Scheme 1.1. Hydrolysis of a trityl acetate (R = Me), benzoate (R = Ph), or *p*-nitrobenzoate (R = *p*- NO$_2$C$_6$H$_4$).

(C) $k_{ion} \gg k_w$. In the so-called S$_N$2C$^+$ mechanism, which was first proposed by Ingold and co-workers,[18] carbocations are formed in a fast ionization process and trapped in a slow subsequent reaction. Because the intermediate carbocations are generated almost quantitatively before they are trapped by water, both steps of the reaction sequence can easily be followed by photospectrometry.

(D) $k_{ion}$ fast, $k_w$ not detectable. The solvolysis scheme is reduced to the ionization step, because the generated tritylium ions are so stable, that they do not react with water under the reaction conditions. Photospectrometry is a convenient method for monitoring the ionization.

## 1.3. Results

Only triphenylmethyl acetate, benzoate, and *p*-nitrobenzoate, i.e., the unsubstituted parent compounds have been isolated as pure substances. The esters of the donor-substituted tritylium systems were generated in acetonitrile solution directly before the kinetic measurements by mixing the colored tritylium tetrafluoroborates with tetra-$n$-butylammonium acetate or benzoate. Usually one equivalent of the ammonium carboxylate was sufficient to decolorize the solution. As mentioned above, different kinetic methods have been employed depending on the relative magnitude of $k_{ion}$ and $k_w$.

*Scenario (A)*: $k_{ion} \ll k_w$. Because $CH_3CO_2H$, $PhCO_2H$, and $p$-nitrobenzoic acid, which were generated in the hydrolyses, are weak acids, tertiary amines – usually triethylamine – were added to increase the sensitivity of the measurements by forming ionic ammonium carboxylates. Only relative conductivities $\kappa_{rel}$ were needed for the evaluation of the kinetic experiments, and we have not calibrated the conductivity cell for determining absolute values of $\kappa$. In order to examine the relationship between reaction progress and conductivity of the solution, a stock solution of (MeO)Tr-OAc in acetonitrile has been added portionwise to 50/50 (v/v) acetonitrile/water containing triethylamine ($\rightarrow$ Et$_3$NH$^+$AcO$^-$). The solvolysis was complete within a few seconds after each addition, and the resulting conductivities were plotted against the concentrations of the dissolved trityl ester. The linearity in the relevant concentration range of the kinetic runs (Figure 1.1) allowed for directly relating the conductivity with the progress of the hydrolysis.

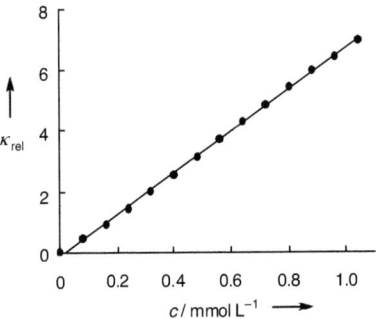

Figure 1.1. Plot of relative conductivities $\kappa_{rel}$ vs. the concentration of hydrolyzed (MeO)Tr-OAc; 50/50 (v/v) acetonitrile/water, 25 °C, [Et$_3$N] = 4.62 × 10$^{-3}$ mol L$^{-1}$.

Figure 1.2 shows a typical exponential increase of conductivity due to the rate-determining ionization of a trityl ester. Analogous first-order kinetics were observed for the hydrolyses of all other trityl carboxylates, implying that common ion return did not take place.[19] First-order rate constants $k_{ion}$ were evaluated by least-squares fitting of the curves to the monoexponential function $\kappa_t = \kappa_0(1 - e^{-kt}) + C$.

## 1. Stabilities of Trityl Protected Substrates

*Scenario (B)*: $k_{ion} \approx k_w$. The similarities of the rates of ionization and quenching of the carbocation during the hydrolyses of $(MeO)_3Tr\text{-}OAc$ and $(MeO)_3Tr\text{-}OBz$ gave rise to substantial transient concentrations of $(MeO)_3Tr^+$ which could be visualized by photospectrometry. No amine has been added in the experiments of Figure 1.3.

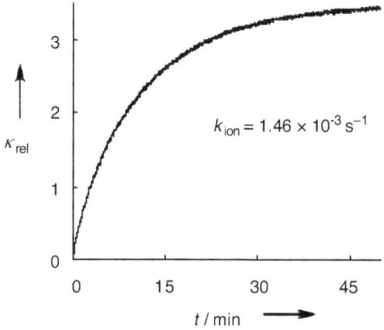

Figure 1.2. Conductivity $\kappa_{rel}$ vs. time $t$ for the solvolysis of MeTr-OAc ($c_0 = 1.06 \times 10^{-3}$ mol L$^{-1}$) in 60/40 (v/v) acetonitrile/water at 25 °C, 5 equiv. of NEt$_3$.

Numerical treatment of the time-dependent concentrations of $(MeO)_3Tr^+$ with the help of the computer program GEPASI[20] delivered individual rate constants for both steps, $k_{ion}$ and $k_w$ (Tables 1.2 and 1.3).

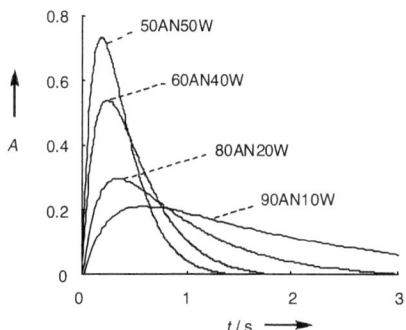

Figure 1.3. Time-dependent absorption of $(MeO)_3Tr^+$ during the solvolyses of $(MeO)_3Tr\text{-}OAc$ ($c_0 = 6.66 \times 10^{-5}$ mol L$^{-1}$) in aqueous acetonitrile at 25 °C; stopped-flow photospectrometry, $\lambda = 484$ nm.

At the maxima of the curves, ionization and trapping proceeded with the same rate. While the ionization of $(MeO)_3Tr\text{-}OAc$ was accelerated by a factor of eight when going from 90AN10W to 50AN50W, the trapping rate only increased by a factor of 1.4. This trend leads to a higher as well as earlier maximum absorbance in the case of 50AN50W, where the maximum of the carbocation concentration (25 % of initial substrate) was reached after 0.22 s. In

90AN10W the maximum of the carbocation concentration was reached after 0.63 s and corresponded to 7.2 % of the initial substrate.

Table 1.2. Rate constants for the solvolyses of $(MeO)_3Tr$-OAc in aqueous acetonitrile at 25 °C; determined by conductimetry and photometric monitoring of the intermediate carbocation.

| solvent [a] | from conductivity in the presence of piperidine | from Gepasi-Fit of $(MeO)_3Tr^+$ absorbance | |
|---|---|---|---|
| | $k_{ion}$ / s$^{-1}$ | $k_{ion}$ / s$^{-1}$ | $k_w$ / s$^{-1}$ |
| 90AN10W | $6.80 \times 10^{-1}$ | $5.15 \times 10^{-1}$ | 5.01 |
| 80AN20W | 1.58 | 1.13 | 7.13 |
| 60AN40W | 3.86 | 2.50 | 7.12 |
| 50AN50W | 5.56 | 4.09 | 7.22 |

[a] 90AN10W = 90/10 (v/v) acetonitrile/water, etc.

Table 1.3. Rate constants for the solvolyses of $(MeO)_3Tr$-OBz in aqueous acetonitrile at 25 °C; determined by conductimetry and photometric monitoring of the intermediate carbocation.

| solvent [a] | from conductivity in the presence of piperidine | from Gepasi-Fit of $(MeO)_3Tr^+$ absorbance | |
|---|---|---|---|
| | $k_{ion}$ / s$^{-1}$ | $k_{ion}$ / s$^{-1}$ | $k_w$ / s$^{-1}$ |
| 90AN10W | 3.79 | 2.34 | 5.08 |
| 80AN20W | 6.45 | 3.94 | 6.99 |
| 60AN40W | $1.12 \times 10^{1}$ | 6.18 | 7.46 |
| 50AN50W | $1.38 \times 10^{1}$ | 9.01 | 8.01 |

[a] 90AN10W = 90/10 (v/v) acetonitrile/water, etc.

When the solvolyses of $(MeO)_3Tr$-OAc and $(MeO)_3Tr$-OBz were carried out in the presence of piperidine, the intermediate carbocations were trapped immediately by the amine;[21] a transient absorption was not detectable. From the mono-exponential increase of conductivity, the first-order rate constants listed in the second columns of Tables 1.2 and 1.3 were obtained. The independence of $k_{obs}$ of the concentration of piperidine (Table 1.4) proved the ionization step to be rate-determining. Tables 1.2 and 1.3 show that the conductimetrically determined ionization rate constants are generally 1.3 to 1.6 times bigger than those derived from the absorbance of the intermediate carbocations.

# 1. Stabilities of Trityl Protected Substrates

*Scenario (C)*: $k_{ion} \gg k_w$. The esters of $(Me_2N)Tr$ ionized very rapidly compared to the reaction of the carbenium ion with water, as shown in Figure 1.4. Both, increase and decrease of the absorbance, were separated by a relatively large time gap, which allowed us to evaluate ionization rate constants by fitting the first parts of the curves according to the mono-exponential function $A_t = A_0(1 - e^{-kt}) + C$.

Table 1.4. Ionization rate constants of $(MeO)_3Tr\text{-}OAc$ ($c_0 = 8.33 \times 10^{-4}$ mol L$^{-1}$) in the presence of variable amounts of piperidine (stopped-flow conductimetry, 90/10 (v/v) acetonitrile/water, 25 °C).

| [piperidine] / mol L$^{-1}$ | $k_{obs}$ / s$^{-1}$ |
|---|---|
| $5.18 \times 10^{-3}$ | $6.74 \times 10^{-1}$ |
| $1.04 \times 10^{-2}$ | $6.79 \times 10^{-1}$ |
| $1.55 \times 10^{-2}$ | $6.65 \times 10^{-1}$ |
| $2.07 \times 10^{-2}$ | $6.80 \times 10^{-1}$ |

However, conversion to the alcohol only went to completion when an excess of carboxylate was used to quench the generated protons and thus suppress the reionization of the triarylcarbinol. Table 1.5 shows that an excess of AcO$^-$ did not affect the ionization rates.

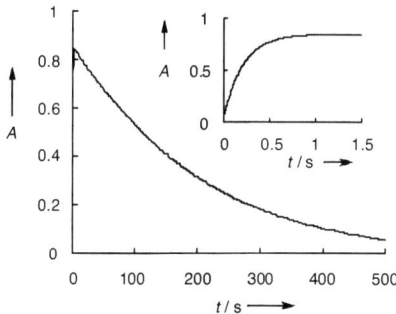

Figure 1.4. Solvolysis of $(Me_2N)Tr\text{-}OAc$ ($c_0 = 2.79 \times 10^{-5}$ mol L$^{-1}$) in 60/40 (v/v) acetonitrile/water, stopped-flow photospectrometry, $\lambda = 461$ nm, 25 °C; $[(n\text{-}Bu)_4N^+OAc^-] = 3.18 \times 10^{-4}$ M. The insert zooms the first stage of the reaction.

*Scenario (D)*: $k_{ion}$ fast, $k_w$ not detectable. As illustrated in Figure 1.5, the ionizations of $(Me_2N)(MeO)Tr\text{-}OAc$ in aqueous acetonitrile of different composition gave solutions of the tritylium ion according to the rate-law $A_t = A_0(1 - e^{-kt}) + C$. While the resulting solutions of the carbocations are fairly stable in the absence of base, a slow reaction with water takes place in the presence of carboxylate ions, converting this system to a case of scenario (C).

Table 1.5. Ionization rates of $(Me_2N)Tr\text{-}OAc$ ($c_0 = 6.00 \times 10^{-5}$ mol L$^{-1}$) in aqueous acetonitrile in the presence of variable amounts of $(n\text{-}Bu)_4N^+AcO^-$ (stopped-flow photospectrometry, $\lambda = 461$ nm, 25 °C).

| [$(n\text{-}Bu)_4N^+AcO^-$] / mol L$^{-1}$ | $k_{ion}$ / s$^{-1}$ | |
|---|---|---|
| | 50AN50W | 90AN10W |
| $1.32 \times 10^{-4}$ | 7.40 | 1.08 |
| $3.21 \times 10^{-4}$ | 7.41 | 1.07 |
| $1.87 \times 10^{-3}$ | 7.29 | 1.02 |
| $3.73 \times 10^{-3}$ | 7.34 | 1.04 |
| $6.78 \times 10^{-3}$ | 7.32 | 1.03 |
| $1.01 \times 10^{-2}$ | 7.22 | 1.05 |
| $1.35 \times 10^{-2}$ | 7.22 | 1.04 |

For the synthesis of $(Me_2N)_2Tr\text{-}OAc$, an ester of malachite green, a large excess of $(n\text{-}Bu)_4N^+AcO^-$ was needed. Its ionization proceeded so rapidly, however, that the kinetics of these reactions could only be followed by a stopped-flow device in solvents of low ionizing power, i.e., in 90AN10W and 80AN20W, but not in 60AN40W. According to a p$K_{R+}$ value of 6.94, the subsequent reaction with water would have needed strongly basic conditions and could not be observed even in the presence of large carboxylate concentrations.

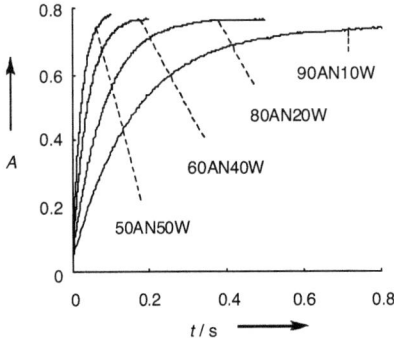

Figure 1.5. Time-dependent absorbance $A$ for the ionization of $(Me_2N)(MeO)Tr\text{-}OAc$ ($c_0 = 5.03 \times 10^{-5}$ mol L$^{-1}$) in aqueous acetonitrile (stopped-flow photospectrometry, $\lambda = 506$ nm, 25 °C).

*Salt effects.* When the trityl esters used for the kinetic studies were generated by combining tritylium tetrafluoroborates with tetra-$n$-butylammonium carboxylates, $(n\text{-}Bu)_4N^+BF_4^-$ was produced as a by-product. In order to examine the influence of this additional salt,[22] a series of experiments with variable concentration of $(n\text{-}Bu)_4N^+BF_4^-$ has been performed. Table 1.6 shows that even large amounts of this salt did not affect the rates of ionization. We, therefore, conclude that tetraalkylammonium salts do not exert a positive salt effect. This finding is in

1. Stabilities of Trityl Protected Substrates

line with Hojo's investigations of the solvolyses of adamantyl halides in 50/50 (v/v) sulfolane/water, showing that small concentrations of tetraalkylammonium salts hardly affected the ionization rates, while larger concentrations (up to 1 mol $L^{-1}$) caused slight retardations of the solvolyses.[23]

Table 1.6. Ionization rate constants of $(Me_2N)(MeO)Tr$-OAc ($c_0$ = 5.03 × $10^{-5}$ mol $L^{-1}$) in the presence of variable amounts of $(n$-Bu$)_4N^+BF_4^-$ (photospectrometry, $\lambda$ = 506 nm, 90/10 (v/v) acetonitrile/water, 25 °C).

| $[(n\text{-Bu})_4N^+BF_4^-)]$ / mol $L^{-1}$ | $k_{ion}$ / $s^{-1}$ |
|---|---|
| 0 | 6.23 |
| 6.07 × $10^{-4}$ | 6.37 |
| 1.75 × $10^{-3}$ | 6.13 |
| 3.17 × $10^{-3}$ | 6.27 |
| 4.53 × $10^{-3}$ | 6.19 |

*Summary of rate constants.* While a variety of differently substituted trityl acetates (Table 1.7) and benzoates (Table 1.8) have been investigated in aqueous acetonitrile, triphenylmethyl *p*-nitrobenzoate was the only system where the leaving group ability of *p* nitrobenzoate has been studied kinetically (Table 1.9).

Table 1.7. Ionization rate constants of trityl acetates in aqueous acetonitrile (25 °C).

| | $k_{ion}$ / $s^{-1}$ | | | | Scenario |
|---|---|---|---|---|---|
| | 90AN10W | 80AN20W | 60AN40W | 50AN50W | |
| Tr | 1.47 × $10^{-5}$ | 5.88 × $10^{-5}$ | 2.70 × $10^{-4}$ | 5.57 × $10^{-4}$ | A |
| MeTr | 1.03 × $10^{-4}$ | 3.59 × $10^{-4}$ | 1.46 × $10^{-3}$ | 3.01 × $10^{-3}$ | A |
| $Me_2Tr$ | 3.23 × $10^{-4}$ | 1.21 × $10^{-3}$ | 5.62 × $10^{-3}$ | 9.59 × $10^{-3}$ | A |
| $Me_3Tr$ | 1.30 × $10^{-3}$ | 4.98 × $10^{-3}$ | 1.77 × $10^{-2}$ | 3.33 × $10^{-2}$ | A |
| (MeO)Tr | 1.20 × $10^{-3}$ | 4.53 × $10^{-3}$ | 1.50 × $10^{-2}$ | 2.40 × $10^{-2}$ | A |
| $(MeO)_2Tr$ | 4.04 × $10^{-2}$ | 1.15 × $10^{-1}$ | 3.06 × $10^{-1}$ | 4.41 × $10^{-1}$ | A |
| $(MeO)_3Tr$ | 6.80 × $10^{-1}$ | 1.58 | 3.86 | 5.56 | B |
| $(Me_2N)Tr$ | 1.08 | 2.00 | 4.51 | 7.40 | C |
| $(Me_2N)(MeO)Tr$ | 6.23 | 1.22 × $10^1$ | 2.49 × $10^1$ | 3.93 × $10^1$ | D |
| $(Me_2N)_2Tr$ | 1.28 × $10^2$ | 2.15 × $10^2$ [a] | – [b] | – [b] | D |

[a] very fast reaction, approximate value; [b] reaction too fast to be measured.

# 1. Stabilities of Trityl Protected Substrates

Table 1.8. Ionization rate constants of trityl benzoates in aqueous acetonitrile (25 °C).

| | $k_{ion}$ / s$^{-1}$ | | | | Scenario |
|---|---|---|---|---|---|
| | 90AN10W | 80AN20W | 60AN40W | 50AN50W | |
| Tr | $5.34 \times 10^{-5}$ | $1.67 \times 10^{-4}$ | $5.14 \times 10^{-4}$ | $9.99 \times 10^{-4}$ | A |
| MeTr | $2.56 \times 10^{-4}$ | $8.08 \times 10^{-4}$ | $2.78 \times 10^{-3}$ | $5.01 \times 10^{-3}$ | A |
| Me$_2$Tr | $1.26 \times 10^{-3}$ | $3.55 \times 10^{-3}$ | $1.05 \times 10^{-2}$ | $1.71 \times 10^{-2}$ | A |
| Me$_3$Tr | $5.43 \times 10^{-3}$ | $1.51 \times 10^{-2}$ | $4.55 \times 10^{-2}$ | $6.97 \times 10^{-2}$ | A |
| (MeO)Tr | $4.45 \times 10^{-3}$ | $1.30 \times 10^{-2}$ | $3.86 \times 10^{-2}$ | $5.56 \times 10^{-2}$ | A |
| (MeO)$_2$Tr | $1.61 \times 10^{-1}$ | $3.34 \times 10^{-1}$ | $6.67 \times 10^{-1}$ | $9.30 \times 10^{-1}$ | A |
| (MeO)$_3$Tr | 3.79 | 6.45 | $1.12 \times 10^{1}$ | $1.38 \times 10^{1}$ | B |
| (Me$_2$N)Tr | 5.37 | 8.35 | $1.40 \times 10^{1}$ | $2.04 \times 10^{1}$ | C |
| (Me$_2$N)(MeO)Tr | $3.36 \times 10^{1}$ | $4.70 \times 10^{1}$ | $6.95 \times 10^{1}$ | $1.02 \times 10^{2}$ | D |

Table 1.9. Ionization rate constants of trityl $p$-nitrobenzoate in aqueous acetonitrile (25 °C).

| solvent | $k_{ion}$ / s$^{-1}$ |
|---|---|
| 90AN10W | $1.57 \times 10^{-3}$ |
| 80AN20W | $4.19 \times 10^{-3}$ |
| 60AN40W | $9.68 \times 10^{-3}$ |
| 50AN50W | $1.82 \times 10^{-2}$ |

Figure 1.6 illustrates the consistency of the different series of rate constants and shows, that the ionization rates of differently substituted trityl acetates in 50AN50W and 90AN10W correlate linearly with the ionization rates of the corresponding trityl benzoates.

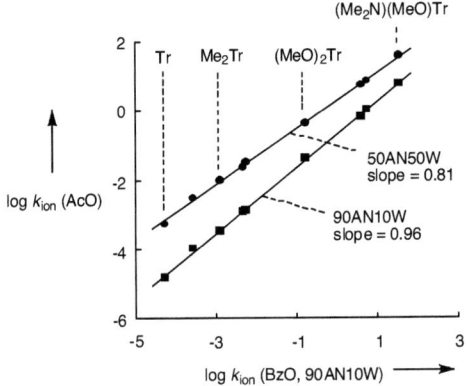

Figure 1.6. Correlation of ionization rates of trityl acetates in different solvents with those of trityl benzoates in 90AN10W, 25 °C.

# 1. Stabilities of Trityl Protected Substrates

The converging correlation lines indicate that the difference in rates between the two solvents decreases as one goes to better stabilized carbocations (Hammond effect).

Solvolytic studies in aqueous acetone were only performed with esters of the unsubstituted triphenylmethanol (Table 1.10).

Table 1.10. Ionization rate constants of several trityl esters in aqueous acetone[a] (25 °C).

|  | $k_{ion}$ / s$^{-1}$ | | | |
| --- | --- | --- | --- | --- |
|  | 90A10W | 80A20W | 60A40W | 50A50W |
| Tr–OAc | - | $1.38 \times 10^{-5}$ [c] | $1.99 \times 10^{-4}$ | $6.40 \times 10^{-4}$ |
| Tr–OBz | $9.31 \times 10^{-6}$ | $3.50 \times 10^{-5}$ [d] | $2.87 \times 10^{-4}$ | $6.95 \times 10^{-4}$ |
| Tr–PNB[b] | $3.63 \times 10^{-4}$ | $1.49 \times 10^{-3}$ | $1.08 \times 10^{-2}$ | $3.25 \times 10^{-2}$ |

[a] 90A10W = 90/10 (v/v) acetone/water, etc; [b] PNB = *p*-nitrobenzoate; [c] a rate constant of $1.45 \times 10^{-5}$ s$^{-1}$ has been reported in ref. [9]; [d] a rate constant of $3.33 \times 10^{-5}$ s$^{-1}$ has been reported in ref. [24].

## 1.4. Discussion

As benzoate is the better leaving group,[25] trityl benzoates always ionized faster than the corresponding acetates. The high polarity of water caused an increase of ionization rates with increasing amounts of water in all solvent mixtures. While Tr–OAc, Tr–OBz and Tr–PNB solvolyzed more slowly in aqueous acetone than in aqueous acetonitrile when the water portion was low (10 vol%), the opposite reactivity order was found in solvents with a high fraction of water (50 vol%). In line with the larger dependence of Winstein-Grunwald's ionizing power $Y$ (see later) on the percentage of water in aqueous acetone than in aqueous acetonitrile, the ionization rates of the trityl esters depend more strongly on the composition of acetone/water than of acetonitrile/water mixtures.

### 1.4.1. Hammett Analysis

Hammett-Brown parameters $\sigma_p^+$ were designed for reactions with a positively charged center developing at a position which is in conjugation to the substituents under consideration.[26] Figure 1.7 shows that the three symmetrical systems Tr, Me$_3$Tr, and (MeO)$_3$Tr, correlate perfectly linear with $\Sigma\sigma_p^+$, and deviations from the correlation line are most significant for systems which are unsymmetrically substituted with strong electron donating groups. The small value of the reaction constant ($\rho = -1.99$) is due to the propeller-like arrangement of the phenyl rings which inhibits full conjugation of the carbocationic center with all three aryl rings. Especially tritylium ions containing one or two dimethylamino groups deviate positively from the correlation line. The problem of additivities of $\sigma^+$ parameters in di- and triarylcarbenium ions has previously been discussed in detail.[27,28]

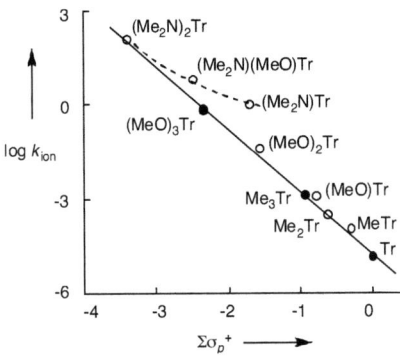

Figure 1.7. Plot of log $k_{ion}$ vs. $\Sigma\sigma_p^+$ for ionizations of trityl acetates in 90AN 10W, 25 °C; the line is drawn through filled circles, slope = –1.99; $\sigma_p^+ = 0$ (H), –0.31 (p-Me), –0.78 (p-OMe) and –1.70 (p-NMe$_2$) from ref. [29].

### 1.4.2. Winstein-Grunwald Analysis

The ionizing power $Y$ of a solvent was introduced by Winstein and Grunwald.[30] In equation (1.1) the parameter $m$ is a measure of the sensitivity of the rate of solvolysis to a change of the solvent; it has often been used as a criterion to determine the mechanism of a solvolysis reaction. Values below 0.5 were considered as evidence for $S_N2$ reactions, whereas values close to 1 are usually found for typical $S_N1$ reactions.[31]

$$\log k_{ion} = \log k_0 + mY \qquad (1.1)$$

In Figures 1.8 and 1.9, ionization rate constants of trityl acetates and benzoates are plotted against the ionizing powers $Y_{t\text{-BuCl}}$ of aqueous acetonitrile. The slopes of the linear correlations represent the $m$-parameters as defined by equation (1.1). It can easily be seen that $m$ decreases with increasing solvolysis rates, i.e., increasing stabilization of the carbocation. This trend may be explained by a Hammond shift[32] towards reactants as the exothermicity of the reactions is increased. The remarkably small $m$-parameters, particularly in the case of

## 1. Stabilities of Trityl Protected Substrates

donor-substituted systems, indicate non-carbocation like transition states. A similar behavior has recently been found for the ionizations of benzhydryl carboxylates.[25] The fact that the ionization rates of trityl benzoates are generally less sensitive to solvent polarity than those of trityl acetates may also be attributed to earlier transition states of the benzoate hydrolyses.

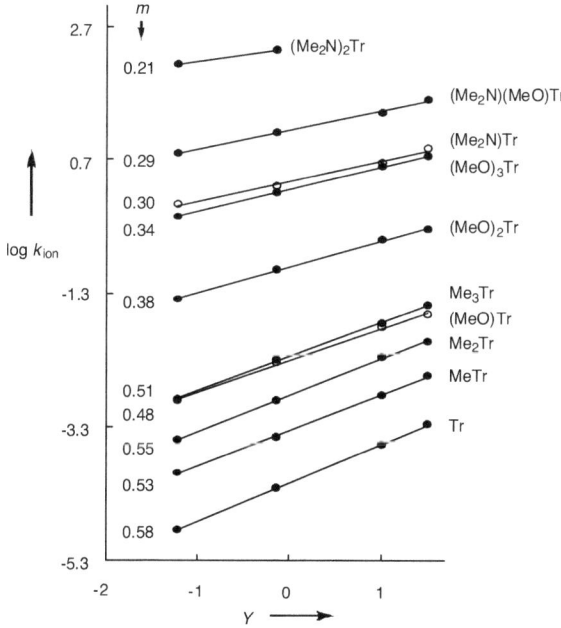

Figure 1.8. Plot of log $k_{ion}$ of trityl acetates vs. $Y_{t\text{-BuCl}}$ of aqueous acetonitrile, 25 °C; $Y = -1.23$ (90AN10W), $-0.14$ (80AN20W), 1.00 (60AN40W), 1.50 (50AN50W) from ref. [33].

Figure 1.10 shows that the $m$-value of triphenylmethyl $p$-nitrobenzoate is considerably smaller than that of the corresponding benzoate, which seems to exclude the rationalization of the different $m$-parameters by steric arguments, but supports an explanation by electronic effects. However, because trityl $p$-nitrobenzoates and benzoates have similar $m$-values in aqueous acetone, much smaller than trityl acetates (Figure 1.11), an unambiguous rationalization for the absolute magnitude of the $m$-values appears not to be possible.

# 1. Stabilities of Trityl Protected Substrates

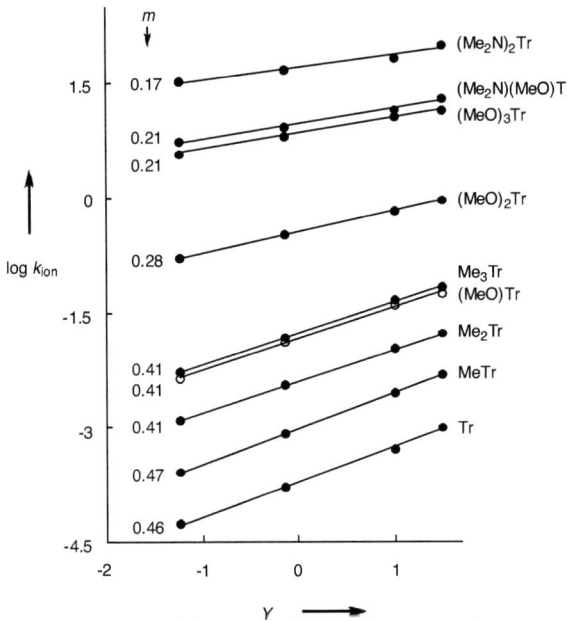

Figure 1.9. Plot of log $k_{ion}$ of trityl benzoates vs. $Y$ of aqueous acetonitrile, 25 °C.

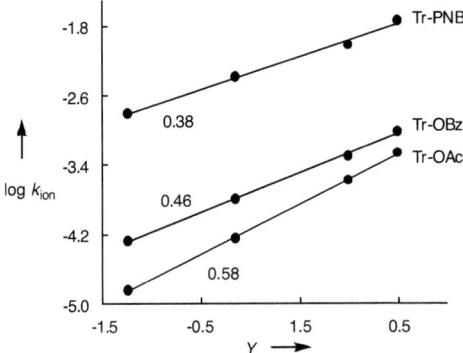

Figure 1.10. Plot of log $k_{ion}$ of trityl esters vs. $Y$ of aqueous acetonitrile, 25 °C.

## 1. Stabilities of Trityl Protected Substrates 21

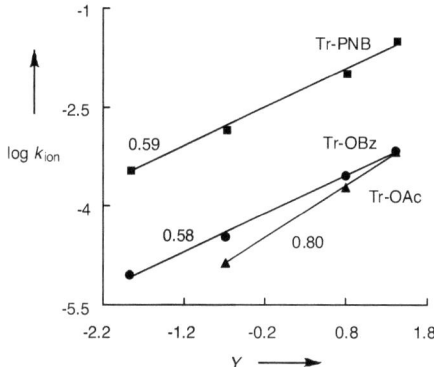

Figure 1.11. Plot of log $k_{ion}$ of trityl esters vs. $Y$ of aqueous acetone, 25 °C; $Y = -1.856$ (90A10W), $-0.673$ (80A20W), 0.796 (60A40W), 1.398 (50A50W), from ref. [34].

### 1.4.2. Rate-Equilibrium Relationships

Intuitively, one would expect a good correlation between the ionization rates of trityl esters and the stabilities of the corresponding tritylium ions in aqueous solution, $pK_{R+}$. Such a correlation has been reported for solvolyses of benzhydryl chlorides by Deno.[35]

As can be seen in Figure 1.12, there are two separate correlation lines between log $k_{ion}$ and $pK_{R+}$, one for the methyl and methoxy substituted compounds and one for the dimethylamino substituted systems. From the slope of the correlation for the methyl and methoxy substituted compounds (0.62) one might derive that more than half of the carbocation character is developed in the solvolysis transition states. Consideration of the full data set shows, however, that this conclusion is too simplistic. Although $(MeO)_3Tr^+$ and $(Me_2N)Tr^+$ differ by a factor of $1.1 \times 10^3$ in their thermodynamic stabilities in water (from $pK_{R+}$), the ionization rates of the corresponding carboxylates are almost identical. The unexpected low reactivities of the dimethylamino substituted trityl derivatives indicate that the product-stabilizing resonance of the amino group develops late on the reaction coordinate and contributes only slightly to the stabilization of the transition state. This is another example of Bernasconi's "Principle of Non-Perfect Synchronization",[36] and emphasizes the importance of intrinsic barriers for these reactions.

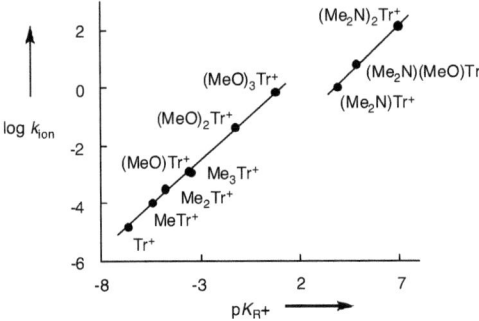

Figure 1.12. Plot of log $k_{ion}$ for trityl acetates in 90AN10W (25 °C) vs. p$K_{R+}$.

## 1.5. Conclusion

Hydrolyses of donor-substituted tritylium carboxylates in aqueous acetonitrile and acetone cover a wide mechanistic spectrum: From conventional $S_N1$ reactions with slow ionization and rapid trapping of the intermediate carbocations by water, over reactions where the carbocations can be observed by the appearance and disappearance of color (carbocation watching),[37] to reactions where the carbocations are formed as persistent species. From the small Winstein $m$-values one can derive non-carbocation like transition states despite the fact that we are dealing with well-defined ionization processes.

Most important for the use of tritylium ions as tailor-made protecting groups in organic syntheses is the finding that the well-known thermodynamic stability parameters p$K_{R+}$ cannot be used to predict kinetic stability and the ease of deprotection of strongly donor-substituted derivatives. While the relative hydrolysis rates of methyl and methoxy substituted tritylium esters correlate well with p$K_{R+}$, the expected further acceleration from the trimethoxy to the dimethylamino substituted system is not observed. The $p$-(dimethylamino)trityl protecting group is released by almost the same rate as the tri-$p$-methoxytrityl group, although the latter tritylium ion is less stabilized by a factor of $1.1 \times 10^3$ in aqueous solution.

According to Scheme 1.2, a correlation exists between the reaction times $\tau_{99}$ for 99 % ionization of trityl acetates and the semiquantitative reaction times reported for complete detritylations of 5'-O-trityluridines ($\tau_{uridine}$).[6a]

## 1. Stabilities of Trityl Protected Substrates 23

Because Tables 1.7 and 1.8 provide a quantitative comparison of the ionization rates of all $p$-methyl and methoxy substituted trityl systems, it is now possible to fine-tune the stabilities of trityl-protected OH and NH functionalities

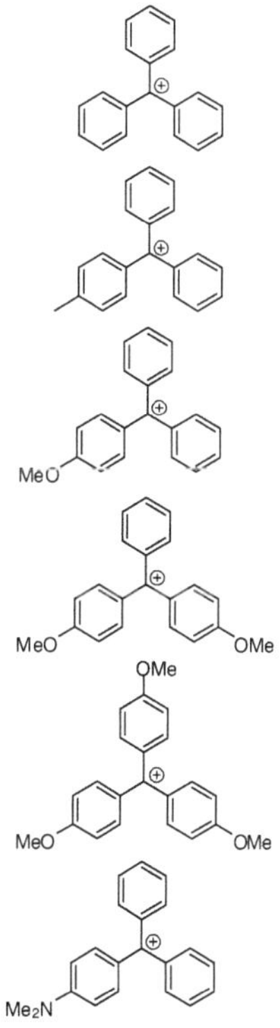

| structure | p$K_{R+}$ | $k_{rel}$[a] | $\tau_{99}$[b] | $\tau_{uridine}$[c] |
|---|---|---|---|---|
| Ph$_3$C$^+$ | -6.63 | 1.0 | 2.3 h | 48 h |
| (p-Me)Ph$_2$C$^+$ | -5.41 | 5.4 | 25 min | - |
| (p-MeO)Ph$_2$C$^+$ | -3.40 | 43 | 3.2 min | 2 h |
| (p-MeO)$_2$PhC$^+$ | -1.24 | 7.9 × 10$^2$ | 10 s | 15 min |
| (p-MeO)$_3$C$^+$ | 0.82 | 1.0 × 10$^4$ | 0.8 s | 1 min |
| (p-Me$_2$N)Ph$_2$C$^+$ | 3.88 | 1.3 × 10$^4$ | 0.6 s | - |

Scheme 1.2. Detritylation times for acetates and uridines. [a] Relative rate constants for the ionizations of acetates in 50AN50W, 25 °C; [b] time for 99 % ionization of trityl acetate in 50AN50W at 25 °C; [c] time for "complete" hydrolysis of 5'-protected uridine derivatives in 80 % AcOH at r.t., from ref. [6a].

## 1.6. References

[1] Historic reviews: a) G. A. Olah, in *Carbocation Chemistry* (Eds: G. A. Olah, G. K. S. Prakash), Wiley-Interscience, Hoboken, **2004**, pp. 7-41; b) C. N. Nenitzescu, in *Carbonium Ions* (Eds: G. A. Olah, P. v. R. Schleyer) Wiley-Interscience, New York, **1973**, Vol. 1, Chapter 1; c) H. H. Freedman, in *Carbonium Ions* (Eds: G. A. Olah, P. v. R. Schleyer) Wiley-Interscience, New York, **1973**, Vol. 4, Chapter 28.

[2] a) H. Volz, *Angew. Chem.* **1963**, *75*, 921. b) H. J. Dauben, Jr., F. A. Gadecki, K. M. Harmon, D. L. Pearson, *J. Am. Chem. Soc.* **1957**, *79*, 4557-4558; c) D. Mandon, L. Toupet, D. Astruc, *J. Am. Chem. Soc.* **1986**, *108*, 1320-1322; d) C. I. F. Watt, *Adv. Phys. Org. Chem.* **1988**, *24*, 57-112; e) T.-Y. Cheng, R. M. Bullock, *Organometallics* **1995**, *14*, 4031-4033; f) J. Chojnowski, L. Wilczek, W. Fortuniak, *J. Organomet. Chem.* **1977**, *135*, 13-22; g) J. Chojnowski, W. Fortuniak, W. Stańczyk, *J. Am. Chem. Soc.* **1987**, *109*, 7776-7781; h) C. A. Mullen, M. R. Gagné, *J. Am. Chem. Soc.* **2007**, *129*, 11880-11881.

[3] a) J. P. Kennedy, E. Markhal, in *Carbocationic Polymerization*, Wiley, New York, **1982**, p. 94; C. Schade, H. Mayr, *Makromol. Chem. Rapid Comm.* **1988**, *9*, 477-482; b) T. Mukaiyama, II. Iwakari, *Chem. Lett.* **1985**, 1363 1366.

[4] a) A. V. Ustinov, V. V. Shmanai, K, Patel, I. A. Stepanova, I. A. Prokhorenko, I. V. Astakhova, A. D. Malakhov, M. V. Skorobogatyi, P. L. Bernad Jr., S. Khan, M. Shahgholi, E. M. Southern, V. A. Korshun, M. S. Shchepinov, *Org. Biomol. Chem.* **2008**, *6*, 4593-4608; b) M. S. Shchepinov, R. Chalk, E. M. Southern, *Tetrahedron* **2000**, *56*, 2713-2724.

[5] P. G. M. Wuts, T. W. Greene, *Greene's Protective Groups in Organic Synthesis*, 4th Ed., Wiley-Interscience, Hoboken, **2007**.

[6] a) M. Smith, D. H. Rammler, I. H. Goldberg, H. G. Khorana, *J. Am. Chem. Soc.* **1962**, *84*, 430-440; b) H. Schaller, G. Weimann, B. Lerch, H. G. Khorana, *J. Am. Chem. Soc.* **1963**, *85*, 3821-3827; c) C. Bleasdale, S. B. Ellwood, B. T. Golding, *J. Chem. Soc. Perkin Trans. 1* **1990**, 803-805; d) M. Sekine, T. Mori, T. Wada, *Tetrahedron Lett.* **1993**, *34*, 8289-8292; e) M. Sekine, T. Hata, *J. Org. Chem.* **1987**, *52*, 946-948; f) A. P. Henderson, J. Riseborough, C. Bleasdale, W. Clegg, M. R. J. Elsegood, B. T. Golding, *J. Chem. Soc. Perkin Trans. 1* **1997**, 3407-3414.

[7] a) M. C. López, I. Demirtas, H. Maskill, *J. Chem. Soc. Perkin Trans. 2* **2001**, 1748-1752; b) M. C. López, I. Demirtas, H. Maskill, M. Mishima, *J. Phys. Org. Chem.*

## 1. Stabilities of Trityl Protected Substrates 25

2008, *21*, 614-621; c) M. C. López, W. Clegg, I. Demirtas, M. R. J. Elsegood, J. Haider, H. Maskill, P. C. Miatt, *J. Chem. Soc., Perkin Trans. 2* **2001**, 1742-1747; d) M. C. López, W. Clegg, I. Demirtas, M. R. J. Elsegood, H. Maskill, *J. Chem. Soc. Perkin Trans. 2* **2000**, 85-92.

[8] B. Löhr, S. Orlich, H. Kunz, *Synlett* **1999**, 1136-1138.
[9] C. G. Swain, T. E. C. Knee, A. MacLachlan, *J. Am. Chem. Soc.* **1960**, *82*, 6101-6104.
[10] C. A. Bunton, A. Konasiewicz, *J. Chem. Soc.* **1955**, 1354-1359.
[11] G. S. Hammond, J. T. Rudesill, *J. Am. Chem. Soc.* **1950**, *72*, 2769-2770.
[12] S. G. Smith, *Tetrahedron Lett.* **1970**, *11*, 4547-4549.
[13] C. G. Swain, G. Tsuchihashi, *J. Am. Chem. Soc.* **1962**, *84*, 2021-2022.
[14] S. Winstein, B. R. Appel, *J. Am. Chem. Soc.* **1964**, *86*, 2720-2721.
[15] R. A. McClelland, V. M. Kanagasabapathy, N. S. Banait, S. Steenken, *J. Am. Chem. Soc.* **1989**, *111*, 3966-3972.
[16] R. A. Diffenbach, K. Sano, R. W. Taft, *J. Am. Chem. Soc.* **1966**, *88*, 4747-4749.
[17] C. D. Ritchie, *Can. J. Chem.* **1986**, *64*, 2239-2249.
[18] E. Gelles, E. D. Hughes, C. K. Ingold, *J. Chem. Soc.* **1954**, 2918-2929.
[19] For common ion rate depression see: S. Winstein, E. Clippinger, A. H. Fainberg, R. Heck, G. C. Robinson, *J. Am. Chem. Soc.* **1956**, *78*, 328-335.
[20] a) P. Mendes, *Comput. Appl. Biosci.* **1993**, *9*, 563-571; b) P. Mendes, *Trends Biochem. Sci.* **1997**, *22*, 361–363; c) P. Mendes, D. Kell, *Bioinformatics* **1998**, *14*, 869-883. Further information about GEPASI: www.gepasi.org.
[21] N. Streidl, A. Antipova, H. Mayr, *J. Org. Chem.* **2009**, *74*, 7328-7334.
[22] For salt effects see: L. C. Manege, T. Ueda, M. Hojo, M. Fujio, *J. Chem. Soc., Perkin Trans. 2* **1998**, 1961-1965.
[23] M. Hojo, T. Ueda, E. Ueno, T. Hamasaki, D. Fujimura, *Bull. Chem. Soc. Jpn.* **2006**, *79*, 751-760.
[24] K.-T. Liu, M.-Y. Kuo, Y. Wang, *J. Phys. Org. Chem.* **1988**, *1*, 241-245.
[25] H. F. Schaller, A. A. Tishkov, X. Feng, H. Mayr, *J. Am. Chem. Soc.* **2008**, *130*, 3012-3022.
[26] Y. Okamoto, H. C. Brown, *J. Org. Chem.* **1957**, *22*, 485-494.
[27] a) M. K. Uddin, M. Fujio, H.-J. Kim, Z. Rappoport, Y. Tsuno, *Bull. Chem. Soc. Jpn.* **2002**, *75*, 1371-1379; b) Y. Tsuno, M. Fujio, *Adv. Phys. Org. Chem.* **1999**, *32*, 267-385.

[28] a) S. I. Miller, *J. Am. Chem. Soc.* **1959**, *81*, 101-106; b) M. O'Brien, R. A. More O'Ferrall, *J. Chem. Soc. Perkin Trans. 2* **1978**, 1045-1053.
[29] C. Hansch, A. Leo, R. W. Taft, *Chem. Rev.* **1991**, *91*, 165-195.
[30] E. Grunwald, S. Winstein, *J. Am. Chem. Soc.* **1948**, *70*, 846-854.
[31] T. W. Bentley, G. Llewellyn, *Prog. Phys. Org. Chem.* **1990**, *17*, 121-158.
[32] G. S. Hammond, *J. Am. Chem. Soc.* **1955**, *77*, 334-338.
[33] T. W. Bentley, J. P. Dau-Schmidt, G. Llewellyn, H. Mayr, *J. Org. Chem.* **1992**, *57*, 2387-2392.
[34] A. H. Fainberg, S. Winstein, *J. Am. Chem. Soc.* **1956**, *78*, 2770-2777.
[35] N. C. Deno, A. Schriesheim, *J. Am. Chem. Soc.* **1955**, *77*, 3051-3054.
[36] a) C. F. Bernasconi, *Acc. Chem. Res.* **1987**, *20*, 301-308; b) C. F. Bernasconi, *Acc. Chem. Res.* **1992**, *25*, 9-16.
[37] H. F. Schaller, H. Mayr, *Angew. Chem.* **2008**, *120*, 4022-4025; *Angew. Chem. Int. Ed.* **2008**, *47*, 3958-3961.

## 2. Electrophilicity versus Electrofugality of Tritylium Ions in Aqueous Acetonitrile[**]

### 2.1. Introduction

Stabilities (more precisely: Lewis acidities) of carbocations[1] are commonly associated with the rates of their formation in solvolysis reactions and the rates of their reactions with nucleophiles. It was reported that the tris(*p*-methoxy)tritylium ion and the *p*-(dimethylamino)tritylium ion are formed with almost equal rates from the corresponding trityl acetates and benzoates despite the $10^3$ fold higher stability ($pK_{R+}$) of the latter carbocation.[2]

In order to elucidate the origin of this unique breakdown of a rate-equilibrium relationship, we have now investigated the electrophilic reactivities of differently substituted tritylium ions in aqueous acetonitrile, i.e., the same solvent, which was used for the solvolysis studies. These investigations were accompanied by quantum chemical calculations.

Early studies on the electrophilic reactivities of tritylium ions focused on stabilized amino and methoxy substituted species. In particular malachite green, $(Me_2N)_2Tr^+$, and derivatives thereof were subjects of many investigations.[3] Rate constants for the reactions of $(Me_2N)_2Tr^+$, *p*-nitromalachite green, and crystal violet, $(Me_2N)_3Tr^+$, with water, hydroxide, and cyanide in aqueous solution[4] were the foundation of Ritchie's well-known constant selectivity relationship log $(k/k_0) = N_+$.[5]

Bunton and Hill studied the kinetics of the reactions of the tris(*p*-methoxy)tritylium cation with water and hydroxide in aqueous solution. While Bunton's work concentrated on the salt effects for these reactions,[6] Hill's studies of kinetic isotope effects showed that the first step, addition of water to the carbocation, rather than the subsequent proton transfer, is rate-determining.[7]

Taft reported that the rates of the reactions of methoxy and dimethylamino substituted tritylium ions with water are not closely correlated with their thermodynamic stabilities,[8] and suggested separation into families. Later investigations confirmed Taft's experimental results, but demonstrated that the deviations from the linear log $k_w/pK_{R+}$ correlation are marginal, when an extended series of compounds is considered.[9,5c]

Reactivities of less stabilized systems were studied by McClelland who used laser-flash techniques for the *in situ* generation of the carbocations.[9,10]

---

[**] This part has been published in: M. Horn, H. Mayr, *Chem. Eur. J.* **2010**, *16*, 7478-7487.

## 2.2. Results and Discussion

### 2.2.1. Kinetics

Rate constants $k_w$ for the reactions of water with the tritylium ions (Scheme 2.1) listed in Table 2.1 were determined at 20 °C in aqueous acetonitrile by photometric monitoring of the decays of the tritylium ions, which have absorption maxima between 420 and 504 nm. Different techniques have been utilized for this purpose.

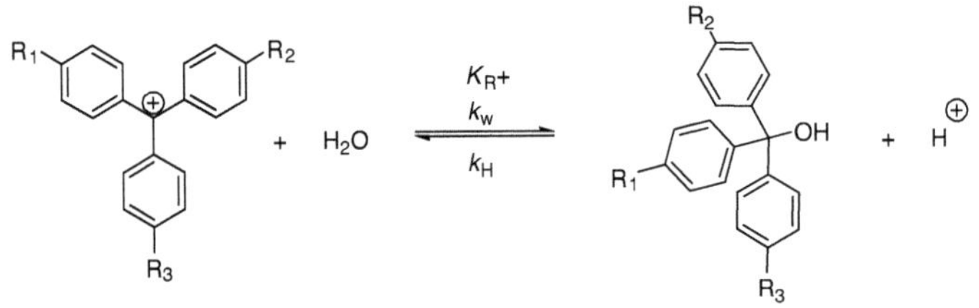

Scheme 2.1. Reactions of tritylium ions with water.

Table 2.1. Tritylium ions with corresponding p$K_{R+}$ values and electrophilicity parameters $E$.

| $R^1, R^2, R^3$ [a] | Abbreviation | p$K_{R+}$ [b] | $E$ [c] |
|---|---|---|---|
| H, H, H | Tr$^+$ | −6.63 | 0.51 |
| Me, H, H | MeTr$^+$ | −5.41 | −0.13 |
| Me, Me, H | Me$_2$Tr$^+$ | −4.71 | −0.70 |
| Me, Me, Me | Me$_3$Tr$^+$ | −3.56 | −1.21 |
| MeO, H, H | (MeO)Tr$^+$ | −3.40 | −1.87 |
| MeO, MeO, H | (MeO)$_2$Tr$^+$ | −1.24 | −3.04 |
| MeO, MeO, MeO | (MeO)$_3$Tr$^+$ | 0.82 | −4.35 |
| Me$_2$N, H, H | (Me$_2$N)Tr$^+$ | 3.88 [d] | −7.93 |
| Me$_2$N, MeO, H | (Me$_2$N)(MeO)Tr$^+$ | 4.86 [e] | −7.98 |
| Me$_2$N, Me$_2$N, H | (Me$_2$N)$_2$Tr$^+$ | 6.94 [e] | −10.29 |
| Me$_2$N, Me$_2$N, Me$_2$N | (Me$_2$N)$_3$Tr$^+$ | 9.39 [e] | −11.26 |

[a] For the location of the substituents see Scheme 2.1; [b] from ref. [9]; [c] empirical electrophilicity parameters from ref. [11]; [d] from ref. [8]; [e] from ref. [5c].

The more reactive ions, Tr$^+$, MeTr$^+$, Me$_2$Tr$^+$, Me$_3$Tr$^+$, and (MeO)Tr$^+$, were generated *in situ* by laser-flash photolysis of the corresponding trityl acetates in aqueous acetonitrile. The other tritylium ions were introduced as tetrafluoroborate salts, and their reactions were followed by stopped-flow or conventional photospectrometry. Because in all cases water was

## 2. Electrophilicity versus Electrofugality of Tritylium Ions

used in large excess, pseudo-first-order rate laws were obeyed, as illustrated in Figure 2.1. The obtained curves were fitted to the mono-exponential function $A_t = A_0 e^{-kt} + C$ by the method of least squares.

Aqueous solutions of the amino substituted systems $(Me_2N)Tr^+BF_4^-$ and $(Me_2N)(MeO)Tr^+ BF_4^-$ in acetonitrile did not decolorize completely. When small amounts of tetra-$n$-butylammonium acetate or benzoate were added to trap the generated protons, complete consumption of these carbocations was achieved. The tritylium ions $(Me_2N)_2Tr^+$ and $(Me_2N)_3Tr^+$ are so stable in aqueous solution, that even the addition of large amounts of carboxylates did not lead to noticeable changes of the carbocation concentrations. For that reason we constrained our studies to systems with p$K_{R+}$ < 5.

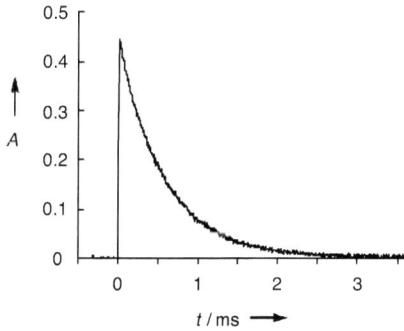

Figure 2.1. Absorbance decay of $(MeO)Tr^+$ at 472 nm. The carbocation was generated from the corresponding acetate ($c_0 = 1.00 \times 10^{-4}$ mol L$^{-1}$) via a laser pulse (7 ns, 266 nm, 60 mJ) in 50/50 (v/v) acetonitrile/water, 20 °C.

Are the rate constants for the reactions of tritylium ions with water ($k_w$) affected by the presence of carboxylate ions? It is well known, that tertiary amines like 1,4-diaza-bicyclo-[2.2.2]octane (DABCO), triethylamine, or quinuclidine act as general base catalysts for the addition of water to tritylium ions in pure water,[12] while general base catalysis by acetate has not been detected.[6,7]

In order to investigate the influence of carboxylate ions on the reaction kinetics in aqueous acetonitrile, the consumption rates of $(MeO)_3Tr^+BF_4^-$ in 90/10 (v/v) acetonitrile/water have been studied in the presence of variable amounts of ($n$-Bu)$_4$N$^+$AcO$^-$ and ($n$-Bu)$_4$N$^+$BzO$^-$ at 25 °C. Figure 2.2 shows that an increase of the concentration of AcO$^-$ or BzO$^-$ accelerated the mono-exponential decay of $(MeO)_3Tr^+$ linearly. Similar experiments have been performed with $(Me_2N)Tr^+BF_4^-$, and Table 2.2 gives an overview of all parameters which were obtained from regression lines as depicted in Figure 2.2.

## 2. Electrophilicity versus Electrofugality of Tritylium Ions

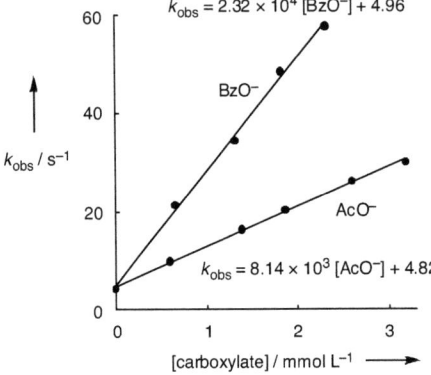

Figure 2.2. Plot of $k_{obs}$ for the decay of the absorbance of $(MeO)_3Tr^+BF_4^-$ ($c_0$ = (2.62 to 3.31) × $10^{-5}$ mol $L^{-1}$) in 90/10 (v/v) acetonitrile/water vs. the concentration of added tetra-$n$-butylammonium carboxylate, $\lambda$ = 484 nm, 25 °C.

The small slopes for the reactions of $(Me_2N)Tr^+$ with $AcO^-$ and $BzO^-$ (entries 7-9) indicate that only in the presence of high concentrations of carboxylate ions, significant effects of $[RCO_2^-]$ on the rates of consumption of this tritylium ion can be observed.

Table 2.2. Kinetics of the consumption of tritylium ions in aqueous acetonitrile with different additives in excess at 25 °C.

| entry | electrophile | additive | solvent [a] | slope / L $mol^{-1}$ $s^{-1}$ | intercept / $s^{-1}$ |
|---|---|---|---|---|---|
| 1 | $(MeO)_3Tr^+$ [b] | $AcO^-$ [c] | 90AN10W | 8.14 × $10^3$ | 4.82 |
| 2 | | | 50AN50W | 1.31 × $10^2$ | 6.85 |
| 3 | | $BzO^-$ [c] | 90AN10W | 2.32 × $10^4$ | 4.96 |
| 4 | | DABCO | 90AN10W | 4.78 × $10^2$ | 4.34 |
| 5 | | | 50AN50W | 9.82 × $10^1$ | 7.32 |
| 6 | | $OH^-$ [c] | 90AN10W | 2.62 × $10^5$ | _ [d] |
| 7 | $(Me_2N)Tr^+$ [b] | $AcO^-$ [c] | 90AN10W | 1.94 × $10^{-1}$ | 3.55 × $10^{-3}$ |
| 8 | | | 50AN50W | 1.94 × $10^{-2}$ [e] | 3.66 × $10^{-3}$ [e] |
| 9 | | $BzO^-$ [c] | 90AN10W | 3.52 × $10^{-1}$ | 3.48 × $10^{-3}$ |
| 10 | | DABCO | 90AN10W | 1.35 × $10^{-1}$ | 3.55 × $10^{-3}$ |

[a] The solvent is given in vol-%, AN = acetonitrile, W = water. 90AN10W = 90/10 (v/v) acetonitrile/water, etc; [b] the counterion was $BF_4^-$; [c] the counterion was $(n\text{-}Bu)_4N^+$; [d] not reliable, because intercept << $k_{obs}$; [e] at 20 °C.

As only 4.6 to 8.7 equivalents of carboxylate (0.15 to 4.9 mmol $L^{-1}$) were employed to achieve complete consumption of $(Me_2N)Tr^+$ and $(Me_2N)(MeO)Tr^+$ in the kinetic experiments at 20 °C, their influence on the rate constants $k_w$ were neglected.

## 2. Electrophilicity versus Electrofugality of Tritylium Ions

All rate constants are summarized in Table 2.3. An interpretation of the slopes of Figure 2.2 and Table 2.2 will be given in the section "Common Ion Return of Carboxylate Anions?".

Table 2.3. First-order rate constants $k_w$ for the reactions of tritylium ions with water in aqueous acetonitrile[a] at 20 °C.[b]

| | $k_w$ / s$^{-1}$ | | | | |
|---|---|---|---|---|---|
| | 90AN10W | 80AN20W | 60AN40W | 50AN50W | 33AN66W [c] |
| Tr$^+$ | 1.19 × 10$^5$ | 1.58 × 10$^5$ | 1.69 × 10$^5$ | 1.62 × 10$^5$ | 1.6 × 10$^5$ |
| MeTr$^+$ | 2.44 × 10$^4$ | 3.60 × 10$^4$ | 4.29 × 10$^4$ | 4.08 × 10$^4$ | 3.7 × 10$^4$ |
| Me$_2$Tr$^+$ | 7.85 × 10$^3$ | 9.35 × 10$^3$ | 9.84 × 10$^3$ | 9.89 × 10$^3$ | 1.1 × 10$^4$ |
| Me$_3$Tr$^+$ | 2.77 × 10$^3$ | 3.01 × 10$^3$ | 3.17 × 10$^3$ | 2.83 × 10$^3$ | 3.6 × 10$^3$ |
| (MeO)Tr$^+$ | 1.17 × 10$^3$ | 1.43 × 10$^3$ | 1.75 × 10$^3$ | 1.73 × 10$^3$ | 1.4 × 10$^3$ |
| (MeO)$_2$Tr$^+$ | 4.16 × 10$^1$ | 5.61 × 10$^1$ | 5.47 × 10$^1$ | 5.81 × 10$^1$ | 8.6 × 10$^1$ |
| (MeO)$_3$Tr$^+$ | 3.73 | 4.78 | 4.93 | 4.88 | 1.0 × 10$^1$ |
| (Me$_2$N)Tr$^+$ | 2.57 × 10$^{-3}$ | 3.43 × 10$^{-3}$ | 3.77 × 10$^{-3}$ | 3.77 × 10$^{-3}$ | - |
| (Me$_2$N)(MeO)Tr$^+$ | 1.53 × 10$^{-3}$ | 1.97 × 10$^{-3}$ | 2.16 × 10$^{-3}$ | 2.14 × 10$^{-3}$ | - |

[a] Solvents are given in vol%, AN = acetonitrile, W = water; [b] note that the rate constants in this table refer to a different temperature than those in Table 2.2; [c] from ref. [10a].

The presence of dications, i.e., N-protonated dimethylamino substituted tritylium ions, has been excluded in neutral aqueous solutions.[12c] Because in the present work, dimethylamino substituted systems have been studied in the presence of carboxylate ions, i.e., under slightly basic conditions, the contribution of dicationic species could be neglected.

### 2.2.2. Quantum Chemical Calculations

A calculated geometry for the parent tritylium cation has previously been reported.[13a] Aizman, Contreras, and Pérez have performed DFT calculations of substituted tritylium ions on the B3LYP/6-31G(d) level in order to determine their Parr electrophilicity parameters.[13b] Because neither geometries nor energies have been reported, we have now optimized geometries of tritylium ions, trityl alcohols and 1,1,1-triarylethanes on the B3LYP/6-31G(d,p) level. Thermochemical corrections for 298.15 K have been calculated for all minima from unscaled vibrational frequencies obtained at the same level, and combined with single-point energies on the MP2(FC)/6-31+G(2d,p) level to yield enthalpies $H_{298}$ and free energies $G_{298}$.

For tritylium ions carrying two or three *para*-methoxy groups and alcohols or ethanes carrying at least one *para*-methoxy group, different conformations have been considered. For example, Figure 2.3 shows the conformers of (MeO)$_2$Tr$^+$ and (MeO)$_2$Tr–Me along with their relative energies. The order of conformers with regard to their relative energies depends on the level of theory. Even within B3LYP/6-31G(d,p) it may change when going from the total energies to the thermally corrected ones. However, as can be seen in Figure 2.3, the differences are only marginal. The structural parameters of the energetically best conformers of the carbocations are summarized in Table 2.4.[††]

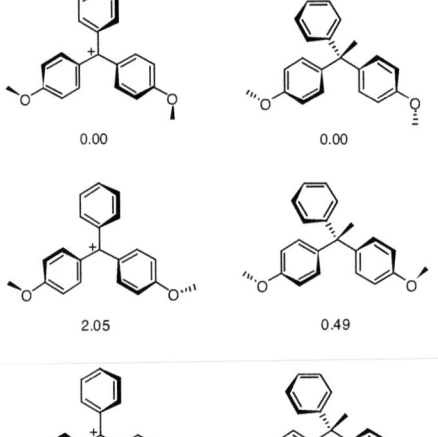

Figure 2.3. Conformations of (MeO)$_2$Tr$^+$ (left) and its methyl anion adduct (right). Relative energies ($G_{298}$ in kJ mol$^{-1}$) are given for the MP2(FC)/6-31+G(2d,p)//B3LYP/6-31G(d,p) level.

When an ion contains three equal *para*-substituents (H, Me, MeO, or Me$_2$N), the dihedral angle decreases slightly as the electron-donating ability of the substituent increases (33.6°, 33.1°, 32.9°, 32.4°, respectively) and the bond lengths remain constant (1.45 Å). For tritylium ions with differently substituted rings, the rings are distorted out of the plane to a different extent. The better the electron-donating ability of the *para*-substituent, the smaller the dihedral angle of the corresponding ring, and the shorter the distance between the ring and the central carbon.

---

[††] As no parameters were held fixed during geometry optimizations, in some cases the 4 central atoms did not form a perfect plane. Furthermore even the 6 atoms of a phenyl ring often weren't planar. As a consequence, the dihedral angles of Table 4 represent averaged values.

## 2. Electrophilicity versus Electrofugality of Tritylium Ions 33

Table 2.4. Structural parameters[a] of tritylium ions, B3LYP/6-31G(d,p).

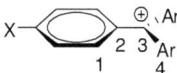

|  | dihedral angle $C_1C_2C_3C_4$ / ° |  |  | bond length $C_2$–$C_3$ / Å |  |  |
|---|---|---|---|---|---|---|
| $Tr^+$ | 33.6 (H)[b] |  |  | 1.45 (H) |  |  |
| $MeTr^+$ | 34.4 (H) | 31.2 (Me) |  | 1.45 (H) | 1.44 (Me) |  |
| $Me_2Tr^+$ | 35.3 (H) | 32.2 (Me) |  | 1.45 (H) | 1.45 (Me) |  |
| $Me_3Tr^+$ | 33.1 (Me) |  |  | 1.45 (Me) |  |  |
| $(MeO)Tr^+$ | 36.1 (H) | 28.0 (MeO) |  | 1.46 (H) | 1.43 (MeO) |  |
| $(MeO)_2Tr^+$ | 38.4 (H) | 30.5 (MeO) |  | 1.46 (H) | 1.44 (MeO) |  |
| $(MeO)_3Tr^+$ | 32.9 (MeO) |  |  | 1.45 (MeO) |  |  |
| $(Me_2N)Tr^+$ | 38.3 (H) | 24.0 ($Me_2N$) |  | 1.46 (H) | 1.42 ($Me_2N$) |  |
| $(Me_2N)(MeO)Tr^+$ | 40.0 (H) | 32.8 (MeO) | 26.5 ($Me_2N$) | 1.47 (H) | 1.45 (MeO) | 1.43 ($Me_2N$) |
| $(Me_2N)_2Tr^+$ | 41.9 (H) | 28.7 ($Me_2N$) |  | 1.47 (H) | 1.43 ($Me_2N$) |  |
| $(Me_2N)_3Tr^+$ | 32.4 ($Me_2N$) |  |  | 1.45 ($Me_2N$) |  |  |

[a] The values are assigned to the rings with the substituents in parentheses. [b] In ref. [14 a] value of 32.4° was determined for tritylium perchlorate by X-ray diffraction.

For $(Me_2N)(MeO)Tr^+$, the smallest dihedral angle is calculated for the ring carrying the dimethylamino group (26.5°), an intermediate angle for the methoxy substituted ring (32.8°), and the unsubstituted ring is twisted by 40.0°. The decreasing resonance contribution from the dimethylamino over the methoxy substituted to the unsubstituted ring is also indicated by the corresponding bond lengths $C_2$-$C_3$, which increase from 1.43 to 1.45 to 1.47 Å, respectively.

Though these calculations refer to the gas phase, spectroscopic investigations in solution confirm these structural assignments. The extraordinarily high resonance contribution of the dimethylamino group can be directly observed in the $^{13}$C-NMR spectrum of $(Me_2N)(MeO)Tr^+BF_4^-$ (Figure 2.4). The two *ortho* carbons as well as the two *meta* carbons of the dimethylamino substituted ring are not isochronous due to the high rotational barrier of this ring. As none of the two signal pairs coalesces in acetonitrile solution at 70 °C (100 MHz), a lower limit of 75 kJ mol$^{-1}$ can be estimated for the interconversion barrier of these carbon atoms. The equivalence of the corresponding carbon signals in the other two phenyl rings even at ambient temperature indicates a fast rotation of these rings on the NMR timescale. Details of the dynamic behavior of tritylium cations have previously been investigated by several groups.[15]

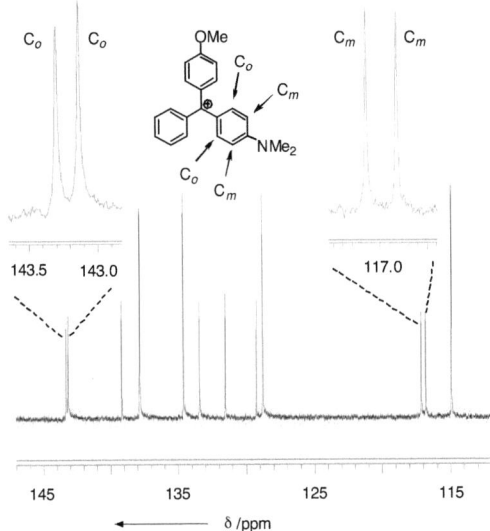

Figure 2.4. Part of the $^{13}$C-NMR spectrum of $(Me_2N)(MeO)Tr^+BF_4^-$ (CDCl$_3$, 150 MHz, 27 °C).

Gas phase hydroxide and methyl anion affinities (Table 2.5) have been calculated according to equations (2.1) and (2.2) and are shown to correlate linearly (Figure 2.5).

$$R_3C^+ + OH^- \rightarrow R_3C-OH \quad (2.1)$$

$$R_3C^+ + Me^- \rightarrow R_3C-Me \quad (2.2)$$

Table 2.5. Theoretical gas phase hydroxide and methyl anion affinities $\Delta H_{298}$, MP2(FC)/6-31+G(2d,p)//B3LYP/6-31G(d,p).

|  | $\Delta H_{298}$ (eq. 2.1) / kJ mol$^{-1}$ | $\Delta H_{298}$ (eq. 2.2) / kJ mol$^{-1}$ |
|---|---|---|
| Tr$^+$ | −715.54 | −883.72 |
| MeTr$^+$ | −704.42 | −871.61 |
| Me$_2$Tr$^+$ | −694.23 | −862.00 |
| Me$_3$Tr$^+$ | −684.90 | −852.30 |
| (MeO)Tr$^+$ | −688.99 | −857.22 |
| (MeO)$_2$Tr$^+$ | −667.15 | −835.36 |
| (MeO)$_3$Tr$^+$ | −649.64 | −817.64 |

## 2. Electrophilicity versus Electrofugality of Tritylium Ions

Table 2.5. Continued.

|  | $\Delta H_{298}$ (eq. 2.1) / kJ mol$^{-1}$ | $\Delta H_{298}$ (eq. 2.2) / kJ mol$^{-1}$ |
|---|---|---|
| (Me$_2$N)Tr$^+$ | −648.75 | −816.71 |
| (Me$_2$N)(MeO)Tr$^+$ | −632.90 | −799.91 |
| (Me$_2$N)$_2$Tr$^+$ | −605.42 | −774.20 |
| (Me$_2$N)$_3$Tr$^+$ | −574.12 | −740.66 |

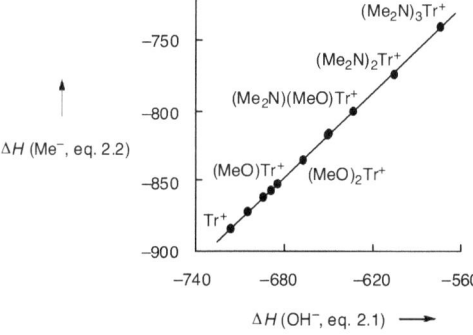

Figure 2.5. Correlation of theoretical Me$^-$ and OH$^-$ affinities ($\Delta H_{298}$ in kJ mol$^{-1}$), MP2(FC)/6-31+G(2d,p)/ /B3LYP/ 6-31G(d,p), slope 1.00, $R^2$ 1.00.

The slope of unity implies that structural variations of the tritylium ions affect their affinities toward Me$^-$ and OH$^-$ to the same extent. A similar behavior has previously been reported for benzhydrylium ions.[16] In each of the three subseries Me$_x$Tr, (MeO)$_x$Tr, and (Me$_2$N)$_x$Tr, the substituent effect on the anion affinity decreases as one goes from mono- to di- to trisubstituted systems (Figure 2.6).

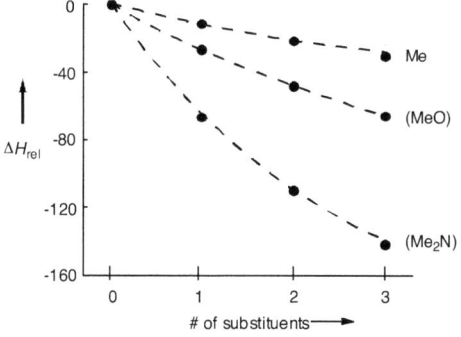

Figure 2.6. Plot of relative OH$^-$ affinities ($\Delta H_{298}$ in kJ mol$^{-1}$, Tr$^+$ = 0 kJ mol$^{-1}$) versus the number of identical substituents in Ar$_x$Ph$_{(3-x)}$C$^+$, MP2(FC)/6-31+G(2d,p)//B3LYP/6-31G(d,p).

Here, the common saturation effects in multi-substituted systems are enforced by the propeller-like conformations of the tritylium ions, which allow the first substituent to stabilize the carbocation more efficiently by planarizing the donor-substituted ring and squeezing the less electron-donating rings out of the plane.

### 2.2.3. Linear Free Energy Relationships

The comparison of hydroxide affinities in solution (p$K_{R+}$) with the corresponding gas phase data shows a good correlation for the methyl and methoxy substituted systems including the tris(dimethylamino) substituted tritylium ion. However, the unsymmetrical dimethylamino substituted systems (Me$_2$N)Tr$^+$, (Me$_2$N)$_2$Tr$^+$, and (Me$_2$N)(OMe)Tr$^+$ (open circles in Figure 2.7) are more stable in solution than expected on the basis of their gas phase hydroxide affinities. These deviations indicate exceptionally high solvation enthalpies of the mono- and diamino substituted tritylium ions which may account for the high intrinsic barriers for the formation and reactions of these tritylium ions (see below). The slope of the line drawn in Figure 2.7 (0.66) indicates that the stabilizing effects of the substituents in the gas phase are reduced to 66 % in aqueous solution. A value of 72 % has been reported for benzhydrylium ions.[16]

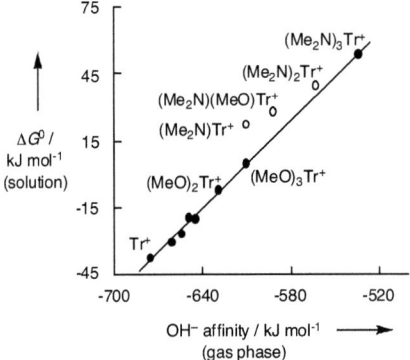

Figure 2.7. Correlation of $\Delta G^0$ (= 2.303$RT$p$K_{R+}$) of water attack at tritylium ions vs. calculated OH$^-$ affinities (gas phase, $\Delta G_{298}$, MP2(FC)/6-31+G(2d,p)//B3LYP/6-31G(d,p)); the open circles have been omitted from the linear regression, slope 0.66.

For constructing the rate-equilibrium relationship in Figure 2.8, the $k_w$ values of Table 2.1 were complemented by the rate constants for (Me$_2$N)$_2$Tr$^+$ and (Me$_2$N)$_3$Tr$^+$ which have previously been reported in the literature. Cigén has determined the rate constant of the reaction of (Me$_2$N)$_2$Tr$^+$ with water in pure water at 20 °C as $1.08 \times 10^{-4}$ s$^{-1}$.[3a] Ritchie reported a rate constant of $1.94 \times 10^{-5}$ s$^{-1}$ for (Me$_2$N)$_3$Tr$^+$ at 25 °C.[12c] With a value of 73.3 kJ mol$^{-1}$ for the free energy of activation (calculated with data taken from ref. [12c]), one can calculate a rate constant of $1.15 \times 10^{-5}$ s$^{-1}$ at 20 °C on the basis of the Eyring equation.

Table 2.3 shows that the nucleophilic reactivity of water in aqueous acetonitrile remains almost constant as the amount of water exceeds 20 vol%. This observation is in agreement with McClelland's[9] and our previous observations[17] that carbocations are trapped with almost equal rates in different acetonitrile/water mixtures containing 20 to 100 % water. We can therefore assume that the $k_w$ values determined by Cigén and Ritchie in pure water also hold for 50/50 (v/v) acetonitrile/water, and include them in the correlation of Figure 2.8.

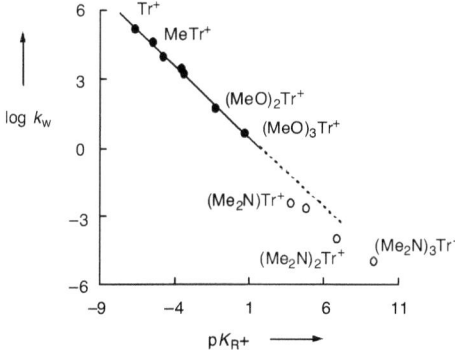

Figure 2.8. Plot of log $k_w$ (50NA50W, 20 °C) vs. p$K_{R+}$ (from Table 2.1); log $k_w$ of (Me$_2$N)$_2$Tr$^+$ and (Me$_2$N)$_3$Tr$^+$ = –3.97 and –4.94, respectively (from ref. [3a,12c], see text); the open circles have been omitted from the linear regression, slope -0.62, $R^2$ = 0.99.

Like the ionization rate constants of trityl acetates (see Chapter 1) only the electrophilicities of the methyl and methoxy substituted tritylium ions (log $k_w$) depend linearly on their Lewis acidities (p$K_{R+}$). The unsymmetrical dimethylamino substituted tritylium ions react more slowly than expected from their thermodynamic stabilities in aqueous solution, indicating higher intrinsic barriers for the reactions of these systems. In line with these findings, highly resonance stabilized carbocations have previously been reported to show low intrinsic reactivities.[18]

### 2.2.4. Hammett Analysis

Hammett-Brown parameters $\sigma_p^+$ were designed for reactions involving a positively charged center in conjugation to the substituents under consideration.[19] Figure 2.9 shows that the four symmetric systems Tr, Me₃Tr, (MeO)₃Tr, and (Me₂N)₃Tr correlate perfectly linearly with $\Sigma\sigma_p^+$. Unsymmetrically substituted systems deviate, and the magnitude of the deviations increases with increasing electron-donating ability of the *para*-substituents. A similar behavior has been found for the correlation between the ionization rates of trityl esters and $\Sigma\sigma_p^+$ (see Chapter 1).

Many examples have shown that in case of multiple ring substitution in di- and triarylcarbenium ions, $\sigma^+$ parameters are non-additive[9,20] for reasons which have previously been discussed.[20]

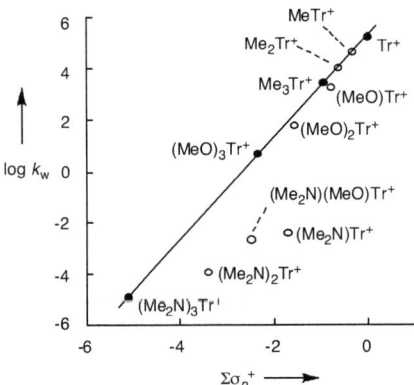

Figure 2.9. Plot of log $k_w$ (50AN50W, 20 °C) vs. $\Sigma\sigma_p^+$ ($\sigma_p^+$ = 0 (H), –0.31 (Me), –0.78 (OMe), and –1.70 (NMe₂) from ref. [21]; for $k_w$ of (Me₂N)₂Tr⁺ and (Me₂N)₃Tr⁺: see text; the regression line is drawn through the 4 symmetrically substituted systems Ar₃Tr⁺, $\rho$ = 1.99, $R^2$ = 1.00.

### 2.2.5. Electrophilicity Parameters of Tritylium Ions

Equation (2.3), where $E$ is an electrophile-specific reactivity parameter, while $s_N$ and $N$ are nucleophile-specific parameters, was designed to correlate bimolecular rate constants $k_2$ for electrophile-nucleophile combinations.[22] A set of benzhydrylium ions and quinone methides were used as reference electrophiles for determining $N$ and $s$ parameters of a large number of σ-, n- and π-nucleophiles.[23]

$$\log k_2 \,(20\,°C) = s_N(N + E) \qquad (2.3)$$

Because steric effects are not specifically included, equation (2.3) only provides reliable predictions of rate constants for reactions of nucleophiles with carbocations and Michael

## 2. Electrophilicity versus Electrofugality of Tritylium Ions 39

acceptors, if bulky systems are excluded. It was discussed that reactions of tritylium ions can only be treated by equation (2.3), if nucleophiles with negligible steric requirements, e.g., primary amines and alcohols or hydride donors, are considered.[11]

Figure 2.10 shows a good linear correlation of log $k_w$ with the electrophilicity parameters $E$ of tritylium ions (Table 2.1), which have previously been derived[11] from rate constants for their reactions with water and $n$-PrNH$_2$ taken from the literature.

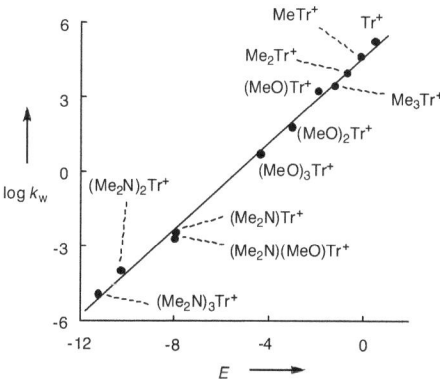

Figure 2.10. Plot of log $k_w$ vs. $E$-parameters of tritylium ions; for $k_w$ of (Me$_2$N)$_2$Tr$^+$ and (Me$_2$N)$_3$Tr$^+$: see text; 50/50 (v/v) acetonitrile/water, 20 °C, $R^2 = 0.99$.

The quality of the correlation shown in Figure 2.10 corroborates the reliability of the previously reported $E$-parameters. Even more, if Figure 2.10 had been employed to determine the nucleophilicity parameters of 50AN50W, values of $N = 5.32$ and $s_N = 0.86$ would have been obtained, close to the parameters derived from reactions with benzhydrylium ions ($N = 5.05$ and $s_N = 0.89$).[17] The internal consistency of our reactivity parameters is thus demonstrated.

### 2.2.6. Comparison of Electrofugality and Electrophilicity

It is generally assumed that highly stabilized carbocations are generated rapidly in solvolysis reactions, and react slowly with nucleophiles. Recently, we have reported that the inverse relationship between electrofugality (rate of formation of R$^+$ in a heterolytic process) and electrophilicity (rate of reactions of R$^+$ with nucleophiles), which holds for non-stabilized and slightly stabilized benzhydrylium ions, breaks down for amino substituted benzhydrylium ions.[24] A comparably poor correlation between electrophilicity and electrofugality for tritylium ions is shown in Figure 2.11.

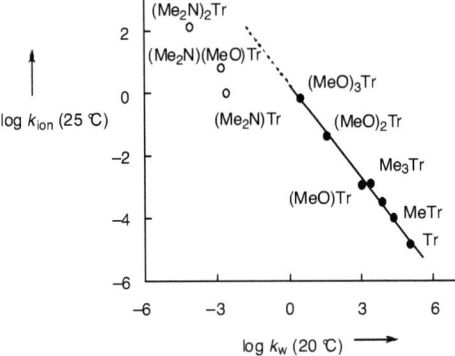

Figure 2.11. Plot of log $k_w$ vs. log $k_{ion}$ (acetates, data from Chapter 1), 90AN10W; log $k_w$ for $(Me_2N)_2Tr^+$ for 90AN10W has been calculated by dividing $1.08 \times 10^{-4}$ ($k_w$ in 50AN50W) by 1.36 (average ratio $k_w$(50AN50W)/$k_w$(90AN10W) for all other tritylium ions in Table 2.3).

Like in the series of benzhydrylium ions, electrophilicity is inversely correlated with electrofugality for methyl and methoxy substituted systems, while amino substituted systems ionize much more slowly than expected from the rates of their reactions with nucleophiles.

The same argument that has been used to rationalize the deviation of the amino substituted tritylium ions from the log $k_{ion}$/p$K_{R^+}$ correlation (Figure 1.12), i.e., late development of the resonance stabilization by the amino group on the reaction-coordinate of the ionization process, can be used to explain the correlation in Figure 2.11.

Because these deviations turn up in each of the correlations log $k_{ion}$/p$K_{R^+}$, log $k_w$/p$K_{R^+}$ (Figure 2.8) and log $k_{ion}$/log $k_w$ (Figure 2.11), they cannot be due to errors in one of the data sets.

### 2.2.7. Common Ion Return of Carboxylate Anions?

In order to determine rate constants $k_w$ for the reactions of dimethylamino substituted tritylium ions in aqueous acetonitrile we had to add AcO⁻ and BzO⁻ as proton traps to suppress reionization of the generated triarylmethanols. We will now analyze the origin of the rate-enhancement by these carboxylate ions, which was described in Figure 2.2 and Table 2.2.

Four different processes may account for the consumption of tritylium ions in aqueous solvents in the presence of a base B (carboxylate or amine). Apart from the reaction of the carbocation with water ($k_w$ in Scheme 2.2 and equation 2.4), the alcohol can be produced by attack of hydroxide ($k_{OH}$). Furthermore, the base can either act as a nucleophile and directly attack the carbocationic center ($k_{-ion}$), or catalyze the addition of water by abstracting a proton in a concerted manner ($k_{cat}$).

$$-d[R^+]/dt = [R^+] (k_w + k_{OH}[OH^-] + k_{-ion}[B] + k_{cat}[B]) \qquad (2.4)$$

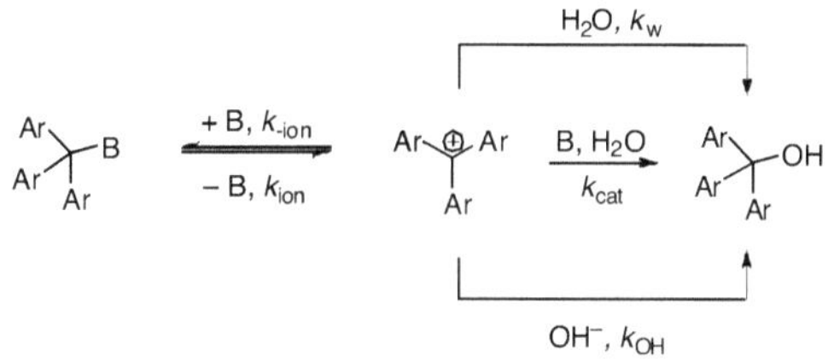

Scheme 2.2.

If the reaction with the base is reversible, a fifth term has to be considered, which refers to the ionization of the adduct ($k_{ion}$). In this case, equation (2.4) transforms into equation (2.5).

$$-d[R^+]/dt = [R^+](k_w + k_{OH}[OH^-] + k_{-ion}[B] + k_{cat}[B]) - k_{ion}[R-B] \qquad (2.5)$$

While the intercepts of the lines shown in Figure 2.2 can unambiguously be assigned to $k_w$ (4.71 s$^{-1}$ in the case of (MeO)$_3$Tr$^+$ in 90AN10W, average of entries 1, 3, 4 of Table 2.2), the interpretation of the slopes is less straightforward.

For the reaction of (MeO)$_3$Tr$^+$ with hydroxide in pure water, a second-order rate constant of $k_{OH}$ = 8200 M$^{-1}$ s$^{-1}$ has been reported,[6] which is considerably smaller than the value of $k_{OH}$ = 2.62 × 10$^5$ M$^{-1}$ s$^{-1}$ in 90AN10W (Table 2.2, entry 6). This difference can be explained by the better solvation of hydroxide in water than in acetonitrile. However, at the low concentrations of hydroxide present under the conditions of these experiments, $k_{OH}$[OH$^-$] appears to be negligible, because comparison of entries 1, 3, and 4 of Table 2.2 shows that the acceleration by DABCO is smaller than that by AcO$^-$ and BzO$^-$, in spite of the much higher basicity of DABCO, which must lead to higher OH$^-$ concentrations.

A similar argument allows for excluding that carboxylates act as general base catalysts ($k_{cat}$). DABCO and several quinuclidines have been reported to catalyze the addition of water to tritylium ions in pure water, and it has been shown that log $k_{cat}$ correlate linearly with the corresponding p$K_{aH}$ values of the amines.[12b] Because AcO$^-$ is much less basic than DABCO, it should catalyze the addition of water less efficiently. The converse observation that AcO$^-$

and BzO⁻ accelerate the consumption of $(MeO)_3Tr^+$ more than DABCO (Table 2.2, cf. slopes in entries 1, 3, 4 or 2, 5) leads to the conclusion that carboxylate ions attack the carbenium ion directly. These reactions are reversible and do not occur in pure water, where the consumption rate of $(MeO)_3Tr^+$ has been found to be independent of the concentration of AcO⁻.[6,7] Since the concentrations of AcO⁻ in the experiments of Figure 2.2 are considerably larger than they had been during the ionization studies of $(MeO)_3Tr\text{-}OAc$,[2] the equilibria of Scheme 2.2 lie almost completely on the side of the covalent ester. For this reason, in the experiments of Figure 2.2, the absorbances of $(MeO)_3Tr^+$ decreased mono-exponentially to zero with $k_{obs} = k_w + k_{-ion}[B]$, i.e., the slopes given in entries 1 and 3 of Table 2.2 reflect $k_{-ion}$.

This analysis of Figure 2.2 is supported by the following considerations. Nucleophilicity parameters of benzoate in 90AN10W have been determined as $N_{25} = 11.3$ and $s_{N,25} = 0.72$ at 25 °C.[24] By employing equation (2.3), the rate constant for the bimolecular reaction of $(MeO)_3Tr^+$ with BzO⁻ can be calculated as $1.01 \times 10^5$ M⁻¹ s⁻¹. Comparison with the experimental value of $2.32 \times 10^4$ M⁻¹ s⁻¹ (Table 2.2, entry 3) reveals agreement within a factor of 4.4, showing that the observed rate constant is in the correct order of magnitude. The deviation is within the range of tolerance of equation (2.3), especially when sterically demanding electrophiles like tritylium ions are involved.

An analogous calculation for the reaction of $(Me_2N)Tr^+$ with BzO⁻ yields a rate constant of 267 M⁻¹s⁻¹ (from equation (2.3)) which is much bigger than the slope of 0.352 M⁻¹s⁻¹ given in entry 9 of Table 2.2, suggesting that this slope cannot reflect the rate constant for the attack of BzO⁻ at $(Me_2N)Tr^+$. Division of the larger of these two numbers ($k_{-ion} = 267$ M⁻¹s⁻¹) by $k_{ion}$ (5.37 s⁻¹) yields the upper limit for the equilibrium constant of the ester formation $K = k_{-ion}/k_{ion} = 50$ M⁻¹. Hence, for [BzO⁻] = $10^{-3}$ M ($\approx$ 20 fold excess in our experiment), the ratio $[(Me_2N)Tr\text{-}OBz]/[(Me_2N)Tr^+]$ must be smaller than 0.05, which implies that the reaction of $(Me_2N)Tr^+$ with BzO⁻ in 90AN10W is thermodynamically unfavorable and can, therefore, not be observed. The small accelerations of the consumption of $(Me_2N)Tr^+$ in the presence of carboxylates are concluded to be due to the sum of $k_{OH}$ and $k_{cat}$.

On the other hand, a ratio $[(MeO)_3Tr\text{-}OBz]/[(MeO)_3Tr^+] = 6.1$ (with $k_{-ion} = 2.32 \times 10^4$ M⁻¹ s⁻¹, $k_{ion} = 3.79$ s⁻¹, and [BzO⁻] = $10^{-3}$ M) is calculated for the equilibrium mixture obtained from $(MeO)_3Tr^+$ and BzO⁻ in 90AN10W, showing that in this system the predominant species is the ester.

## 2.2.8. Complete Free Energy Diagrams for the Hydrolyses of Trityl Carboxylates

The ionization rate constants of trityl carboxylates and the rate constants for the consumption of tritylium ions shall now be combined to construct free energy diagrams for the hydrolyses of tritylium carboxylates in aqueous acetonitrile. The good correlation between log $k_w$ and the previously derived electrophilicity parameters $E$ (Figure 2.10) shows that the $E$ values for tritylium ions can be used to calculate rate constants for their combinations with $n$-nucleophiles by equation (2.3). Therefore all information for the construction of free energy diagrams is now available.

In Figure 2.12, the covalent trityl benzoates are set on the same level. The Eyring equation was used to calculate $\Delta G^\ddagger$ values for the first step of the reaction cascade from directly measured ionization rate constants $k_{ion}$ (Table 1.8).[25]

The rate constants for the combinations of the tritylium ions with benzoate $k_{-ion}$ in 90AN10W were calculated by equation (2.3) from the $E$-parameters given in Table 2.1 and the known nucleophilicity parameters of BzO⁻ (see above). One thus arrives at the positions of the tritylium ions in Figure 2.12. As discussed above, the directly measured rate constant for the reaction of (MeO)$_3$Tr⁺ with BzO⁻ is 4.4 times smaller than that calculated by equation (2.3). Taking this factor into account would lower the positions of all tritylium ions in Figure 2.12 by 3-4 kJ mol⁻¹, a negligible correction in view of the total spread of the reactivities. The thermodynamic stability order of the tritylium ions, which thus has been obtained on an entirely kinetic basis ($k_{ion}$ and $k_{-ion}$) is in line with the hydroxide stability scale p$K_{R+}$, which is based on equilibrium measurements in water and aqueous sulfuric acid.

The activation free energies for the last step in the reaction cascade of Figure 2.12 can again be derived from the Eyring equation. Because the relative stabilities of trityl benzoates and triarylmethanols are almost independent of the nature of the substituents on the aryl rings, all energy profiles will converge on the right of Figure 2.12. Readers should not be confused by the fact that Ar$_3$C–OH is located at approximately the same level as Ar$_3$C–OBz. The driving force of the hydrolysis reactions is not the lower value of $\Delta G^0$ of the alcohols, but the high concentration of water used for these reactions.

Figure 2.13 shows that the $\Delta G^0$ values derived from kinetic ($k_{ion}$ and $k_{-ion}$) and thermodynamic (p$K_{R+}$) data correlate with a slope of 1.07 with no obvious deviations. The consistency of our kinetic data is thus confirmed.

## 2. Electrophilicity versus Electrofugality of Tritylium Ions

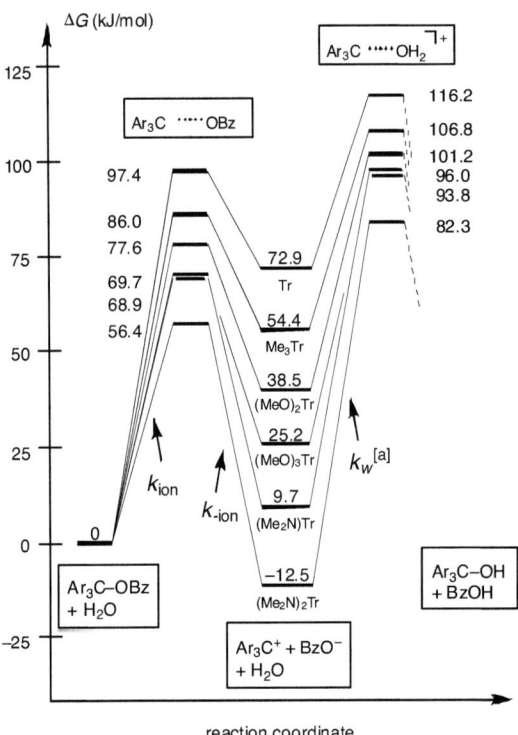

Figure 2.12. Free energy profiles for the hydrolyses of substituted trityl benzoates in 90AN10W, 25 °C; constructed for [BzO$^-$] = 1 mol L$^{-1}$; [a] $k_w$ taken from Table 2.3, determined at 20 °C.

Figure 2.12 can now be used to rationalize the kinetic phenomena reported in this and the preceding chapter. First of all, one can recognize that the transition state of the ionization step changes significantly as one goes from trityl benzoate to the bis(dimethylamino) substituted trityl derivative. The more electron-donors are attached, the less carbocation-like the transition state. The origin of the observed irregularities between methyl and methoxy substituted tritylium systems on one side and dimethylamino substituted ones on the other side is well visualized by Figure 2.12. As discussed earlier, the intrinsic barriers for the reactions of the highly resonance stabilized amino substituted tritylium ions are particularly high. As a consequence, the transition states for the ionizations of (MeO)$_3$Tr-OBz and (Me$_2$N)Tr-OBz (68.9 kJ mol$^{-1}$ ≈ 69.7 kJ mol$^{-1}$) as well as for the reactions of the corresponding cations with water (93.8 kJ mol$^{-1}$ ≈ 96.0 kJ mol$^{-1}$) almost coincide, although (Me$_2$N)Tr$^+$ is the much better stabilized cation. Hence, (MeO)$_3$Tr-OBz and (Me$_2$N)Tr-OBz ionize with similar rates, while (Me$_2$N)Tr$^+$ reacts much more slowly with nucleophiles than (MeO)$_3$Tr$^+$.

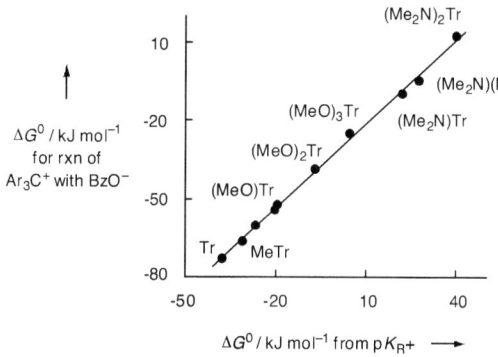

Figure 2.13. Correlation of free energies $\Delta G^0$ (combination of tritylium ions with BzO$^-$, data taken from Figure 2.12) with $\Delta G^0$ from p$K_{R+}$; slope 1.07, $R^2 = 1.00$.

## 2.3. Conclusion

Although the linear free energy relationship log $k_2$ (20 °C) = $s_N(N + E)$ cannot generally be applied to sterically shielded systems (e.g., reactions of tritylium ions with alkenes), it was now found that it works perfectly for the decays of tritylium ions in aqueous acetonitrile. The breakdown of the inverse correlation between the electrofugalities of carbocations (rates of ionization of R–X) and their electrophilicities (rates of reactions of R$^+$ with nucleophiles) for highly stabilized carbocations appears to be a general phenomenon. As previously reported for hydrolyses of benzhydrylium carboxylates,[24] an excellent inverse correlation between the electrophilic reactivities of methyl and methoxy substituted tritylium ions and the ionization rates of the corresponding trityl carboxylates was observed. However, amino substituted tritylium ions strongly deviate from this correlation. Because unsymmetrical amino substituted tritylium ions, in particular (Me$_2$N)Tr$^+$, are much better stabilized in solution than expected from their hydroxide affinities in the gas phase (Figure 2.7), they must experience special solvation effects. Reorganization of these strongly associated solvent molecules may account for the high intrinsic barriers encountered in reactions forming ($k_{ion}$) and quenching ($k_w$) these tritylium ions.

Apart from their use as protecting groups, tritylium ions also have many practical applications as hydride abstracting agents. This work has shown that there are excellent linear correlations between electrophilic reactivities (log $k_w$, $E$), $pK_{R^+}$ values and calculated hydroxide affinities in the gas phase if the amino substituted tritylium ions are excluded. The parent, methyl, and methoxy substituted systems can therefore be used as reference compounds for converting the manifold of published hydride abstraction rates by differently substituted tritylium ions into a common activity scale for hydride donors.

## 2.4. References

[1]  a) J. Hine, *Structural Effects on Equilibria in Organic Chemistry*, Wiley, New York, **1981**; b) H. Mayr, A. Ofial, in *Carbocation Chemistry* (Eds: G. A. Olah, G. K. S. Prakash), Wiley-Interscience, Hoboken, **2004**, pp. 331–358.

[2]  M. Horn, H. Mayr, *Chem. Eur. J.* **2010**, *16*, 7469-7477.

[3]  a) R. Cigén, *Acta Chem. Scand.* **1959**, *13*, 1113–1123; b) C. G. Ekström, *Acta Chem. Scand.* **1966**, *20*, 444–458.

[4]  C. D. Ritchie, G. A. Skinner, V. G. Badding, *J. Am. Chem. Soc.* **1967**, *89*, 2063–2071.

[5]  a) C. D. Ritchie, *Acc. Chem. Res.* **1972**, *5*, 348–354; b) C. D. Ritchie, P. O. I. Virtanen, *J. Am. Chem. Soc.* **1972**, *94*, 4966–4971; c) C. D. Ritchie, *Can. J. Chem.* **1986**, *64*, 2239–2249.

[6]  C. A. Bunton, S. K. Huang, *J. Am. Chem. Soc.* **1972**, *94*, 3536–3544.

[7]  E. A. Hill, W. J. Miller, *Tetrahedron Lett.* **1968**, *21*, 2565–2569.

[8]  R. A. Diffenbach, K. Sano, R. W. Taft, *J. Am. Chem. Soc.* **1966**, *88*, 4747–4749.

[9]  R. A. McClelland, V. M. Kanagasabapathy, N. S. Banait, S. Steenken, *J. Am. Chem. Soc.* **1989**, *111*, 3966–3972.

[10] a) R. A. McClelland, V. M. Kanagasabapathy, N. S. Banait, S. Steenken, *J. Am. Chem. Soc.* **1991**, *113*, 1009–1014; b) R. A. McClelland, N. S. Banait, S. Steenken, *J. Am. Chem. Soc.* **1986**, *108*, 7023–7027.

[11] S. Minegishi, H. Mayr, *J. Am. Chem. Soc.* **2003**, *125*, 286–295.

[12] a) J. N. Ride, P. A. H. Wyatt, Z. M. Zochowski, *J. Chem. Soc., Perkin Trans. 2* **1974**, 1188–1189; b) J. R. Gandler, *J. Am. Chem. Soc.* **1985**, *107*, 8218–8223; c) C. D. Ritchie, D. J. Wright, D.-S. Huang, A. A. Kamego, *J. Am. Chem. Soc.* **1975**, *97*, 1163–1170.

[13]  a) B. Reindl, T. Clark, P. v. R. Schleyer, *J. Phys. Chem. A* **1998**, *102*, 8953–8963; b) A. Aizman, R. Contreras, P. Pérez, *Tetrahedron* **2005**, *61*, 889–895.

[14]  A. H. Gomes de Mesquita, C. H. MacGillavry, K. Eriks, *Acta Cryst.* **1965**, *18*, 437–443.

[15]  a) R. Breslow, L. Kaplan, D. LaFollette, *J. Am. Chem. Soc.* **1968**, *90*, 4056–4064; b) J. W. Rakshys, Jr., S. V. McKinley, H. H. Freedman, *J. Am. Chem. Soc.* **1970**, *92*, 3518–3520; c) J. W. Rakshys, Jr., S. V. McKinley, H. H. Freedman, *J. Am. Chem. Soc.* **1971**, *93*, 6522–6529; d) I. I. Schuster, A. K. Colter, R. Kurland, *J. Am. Chem. Soc.* **1968**, *90*, 4679–4687.

[16]  C. Schindele, K. N. Houk, H. Mayr, *J. Am. Chem. Soc.* **2002**, *124*, 11208–11214.

[17]  S. Minegishi, S. Kobayashi, H. Mayr, *J. Am. Chem. Soc.* **2004**, *126*, 5174–5181.

[18]  J. P. Richard, T. L. Amyes, M. M. Toteva, *Acc. Chem. Res.* **2001**, *34*, 981–988.

[19]  Y. Okamoto, H. C. Brown, *J. Org. Chem.* **1957**, *22*, 485–495.

[20]  a) S. Nishida *J. Org. Chem.* **1967**, *32*, 2697–2701; b) M. K. Uddin, M. Fujio, H.-J. Kim, Z. Rappoport, Y. Tsuno, *Bull. Chem. Soc. Jpn.* **2002**, *75*, 1371–1379; c) Y. Tsuno, M. Fujio, *Adv. Phys. Org. Chem.* **1999**, *32*, 267–385.

[21]  C. Hansch, A. Leo, R. W. Taft, *Chem. Rev.* **1991**, *91*, 165–195.

[22]  H. Mayr, M. Patz, *Angew. Chem. Int. Ed.* **1994**, *33*, 938–957.

[23]  a) H. Mayr, T. Bug, M. F. Gotta, N. Hering, B. Irrgang, B. Janker, B. Kempf, R. Loos, A. R. Ofial, G. Remennikov, H. Schimmel, *J. Am. Chem. Soc.* **2001**, *123*, 9500–9512; b) H. Mayr, B. Kempf, A. R. Ofial, *Acc. Chem. Res.* **2003**, *36*, 66–77; c) H. Mayr, A. R. Ofial, *Pure Appl. Chem.* **2005**, *77*, 1807–1821; d) H. Mayr, A. R. Ofial, *J. Phys. Org. Chem.* **2008**, *21*, 584–595.

[24]  H. F. Schaller, A. A. Tishkov, X. Feng, H. Mayr, *J. Am. Chem. Soc.* **2008**, *130*, 3012–3022.

[25]  The ionization rate constant for $(Me_2N)_2Tr-OBz$ was derived from the correlation shown in Figure 1.6.

# 3. Electrophilicities of Acceptor-Substituted Tritylium Ions

## 3.1. Introduction

The linear free energy relationship (3.1), where $E$ is an electrophilicity, $N$ is a nucleophilicity, and $s_N$ is a nucleophile-specific sensitivity parameter, is based on the reactions of a set of carbon nucleophiles with a set of benzhydrylium ions and structurally related quinone methides, which were employed as reference electrophiles.[1,2]

$$\log k = s_N(N + E) \qquad (3.1)$$

As demonstrated in the past, this equation allows to predict the bimolecular rate constants $k$ for numerous electrophile-nucleophile combinations, and at present, it covers a reactivity range of 40 orders of magnitude. Hereby, the reactivities of the reference diaryl carbenium ions can be gradually adjusted by suitable substitution patterns in the *para*- and/or *meta*-positions of the rings, without changing the steric shielding at the reactive site.

As steric effects are not explicitly treated by equation (3.1), it was previously assumed that this correlation cannot be employed for reactions of tritylium ions.[3] However, it was found that the reactions of donor-substituted tritylium ions with water, primary amines, and hydride donors follow this correlation. Thus it was possible to calculate the rate constants for such reactions from the published $E$ parameters (Table 2.1) and the $N$ and $s_N$ parameters of the corresponding nucleophiles. Since only very few $E$ parameters of acceptor-substituted tritylium ions have previously been derived,[3] we have now determined the electrophilic reactivities of fluoro-substituted trityl systems,[4,5] which may be used as efficient hydride abstractors.[6,7]

## 3.2. Results

### 3.2.1. Rates of Hydride Transfers

The rate constants of hydride transfers from triphenylsilane to the tritylium ions listed in Table 3.1 have been determined in dichloromethane solution using conventional or stopped-flow spectrophotometrical techniques (Scheme 3.1).

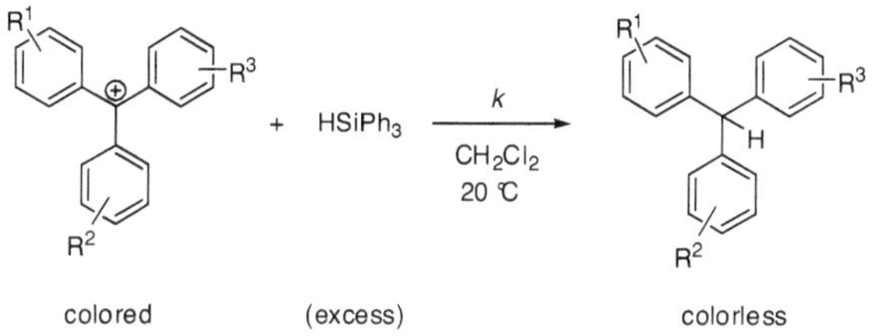

## 3. Electrophilicities of Acceptor-Substituted Tritylium Ions 49

Scheme 3.1. Reduction of tritylium ions by triphenylsilane.

Table 3.1. Tritylium ions studied in this work.

| $R^1, R^2, R^3$ [a] | abbreviation | $pK_{R+}$ | $E$ [b] |
|---|---|---|---|
| 2 × m-F, 2 × m-F, 2 × m-F | $(mF)_6Tr^+$ | −14.2 [c] | - |
| 2 × m-F, 2 × m-F, m-F | $(mF)_5Tr^+$ | - | - |
| 2 × m-F, m-F, m-F | $(mF)_2(mF)'(mF)''Tr^+$ | - | - |
| m-F, m-F, m-F | $(mF)(mF)'(mF)''Tr^+$ | −10.72 [c] | - |
| m-F, m-F, H | $(mF)(mF)'Tr^+$ | −9.17 [d] | - |
| m-F, H, H | $(mF)Tr^+$ | −7.81 [d] | - |
| p-F, H, H | $(pF)Tr^+$ | −6.41 [d] | - |
| p-F, p-F, H | $(pF)_2Tr^+$ | −6.22 [d] | - |
| p-F, p-F, p-F | $(pF)_3Tr^+$ | −6.05 [c] | - |
| H, H, H | $Tr^+$ | −6.63 [e] | 0.51 |
| p-Me, H, H | $MeTr^+$ | −5.41 [e] | −0.13 |
| p-Me, p-Me, H | $Me_2Tr^+$ | −4.71 [e] | −0.70 |
| p-Me, p-Me, p-Me, | $Me_3Tr^+$ | −3.56 [e] | −1.21 |
| p-MeO, H, H | $(MeO)Tr^+$ | −3.40 [e] | −1.59 [f] |
| p-MeO, p-MeO, H | $(MeO)_2Tr^+$ | −1.24 [e] | −3.04 |

[a] For the substitution pattern see Scheme 3.1; [b] empirical electrophilicity parameters from ref. [3]; [c] from ref. [8]; [d] from ref. [4]; [e] from ref. [9]; [f] the previously reported value of $E = -1.87$ (in ref. [3]) was based on only two reactions and should be revised.

The unsubstituted as well as donor-substituted tritylium ions have been employed as stable salts (usually $BF_4^-$ as counterion). Tetrafluoroborate salts of *meta*-fluoro substituted tritylium ions that were isolated in substance decomposed within hours, even in a protecting gas atmosphere. Therefore, destabilized tritylium ions have been generated in solution by combining the corresponding trityl chlorides or bromides with an excess of Lewis acid, usually $GaCl_3$.

All reactions were found to be of first-order in tritylium ion and triphenylsilane, according to the rate law (3.2). After addition of the silane to a tritylium ion, the decrease of the absorbance of $Ar_3C^+$ was followed photometrically (Figure 3.1). As the nucleophile was used in large excess, pseudo-first-order rate constants $k_{obs}$ (equation 3.3) were obtained by fitting the time-dependent absorbances to the mono-exponential equation (3.4), assuming the validity of Lambert-Beer's law.

$$-d[Ar_3C^+]/dt = k[Ar_3C^+][HSiPh_3] \qquad (3.2)$$

$$k_{obs} = k[HSiPh_3] \quad \text{for } [HSiPh_3]_0 \gg [Ar_3C^+]_0 \qquad (3.3)$$

$$[Ar_3C^+] = [Ar_3C^+]_0 \exp(-k_{obs}t) + \text{const.} \qquad (3.4)$$

For each electrophile a series of runs with different concentrations of triphenylsilane were performed, and plots of $k_{obs}$ versus the silane concentrations were linear (insert of Figure 3.1) with the slopes of the lines representing the second-order rate constants $k$ (Table 3.2).

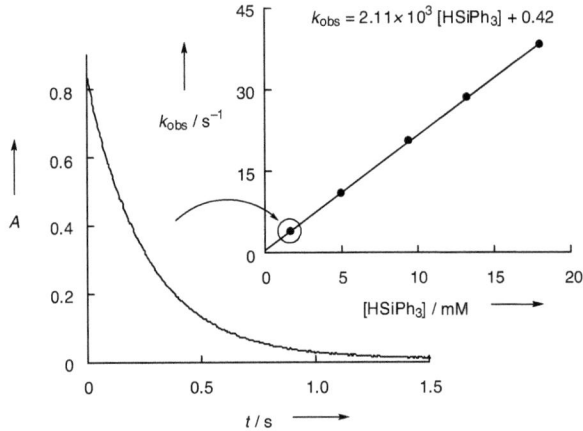

Figure 3.1. Time-dependent absorbance at λ = 420 nm for the reaction of $(mF)(mF)'(mF)''Tr^+$ with HSiPh$_3$ ($c_0$ = 1.63 × 10$^{-3}$ mol L$^{-1}$), CH$_2$Cl$_2$, 20 °C; $k_{obs}$ = 3.80 s$^{-1}$; the carbocation was generated from $(mF)(mF)'(mF)''$TrBr (5.57 × 10$^{-5}$ mol L$^{-1}$) and 53 equiv. of GaCl$_3$. Insert: Plot of $k_{obs}$ vs. [HSiPh$_3$]$_0$; $k$ = 2.11 × 10$^3$ L mol$^{-1}$ s$^{-1}$.

## 3. Electrophilicities of Acceptor-Substituted Tritylium Ions 51

Table 3.2. Second-order rate constants $k$ for hydride transfers from HSiPh$_3$ to tritylium ions (CH$_2$Cl$_2$, 20 °C).

| electrophile | source | $k$ [L mol$^{-1}$ s$^{-1}$] |
|---|---|---|
| ($m$F)$_2$($m$F)'($m$F)''Tr$^+$ | Ar$_3$CBr + GaCl$_3$ | $6.11 \times 10^3$ |
| ($m$F)($m$F)'($m$F)''Tr$^+$ | Ar$_3$CBr + GaCl$_3$ | $2.11 \times 10^3$ |
| ($m$F)($m$F)'Tr$^+$ | Ar$_3$CBr + GaCl$_3$ | $6.57 \times 10^2$ |
| ($m$F)Tr$^+$ | Ar$_3$CBr + GaCl$_3$ | $1.99 \times 10^2$ |
| ($p$F)Tr$^+$ | Ar$_3$C$^+$BF$_4^-$ | $4.58 \times 10^1$ |
| ($p$F)$_2$Tr$^+$ | Ar$_3$C$^+$BF$_4^-$ | $3.08 \times 10^1$ |
| ($p$F)$_3$Tr$^+$ | Ar$_3$CCl + GaCl$_3$ | $2.36 \times 10^1$ |
| Tr$^+$ | Ar$_3$C$^+$BF$_4^-$ | $7.55 \times 10^1$ |
|  | Ar$_3$C$^+$SbF$_6^-$ | $7.16 \times 10^1$ |
| MeTr$^+$ | Ar$_3$C$^+$BF$_4^-$ | $1.56 \times 10^1$ |
| Me$_2$Tr$^+$ | Ar$_3$C$^+$BF$_4^-$ | 4.02 |
| Me$_3$Tr$^+$ | Ar$_3$C$^+$BF$_4^-$ | 1.21 |
| (MeO)Tr$^+$ | Ar$_3$C$^+$BF$_4^-$ | $6.44 \times 10^{-1}$ |
| (MeO)$_2$Tr$^+$ | Ar$_3$C$^+$BF$_4^-$ | $2.61 \times 10^{-2}$ |

Hydride transfers from organic silanes to benzhydrylium ions were found to be prone to catalysis by BCl$_3$.[10] By contrast, variation of the concentration of GaCl$_3$ did not lead to significant changes of the rate constants for the reactions of tritylium ions with HSiPh$_3$ (Table 3.3). In agreement with previous results of Chojnowski,[6b] Table 3.2 furthermore demonstrates the independence of the rate constants of the counterions, as use of BF$_4^-$ gave the same result as SbF$_6^-$. From these observations one cannot only derive that the counterion is not involved in the rate-determining step, but also that the hydride donor is stable in the presence of the Lewis acid GaCl$_3$ in CH$_2$Cl$_2$ solution.

Table 3.3. Influence of the concentration of GaCl$_3$ on the rate constants of the reactions between fluoro-substituted tritylium ions and HSiPh$_3$ (CH$_2$Cl$_2$, 20 °C).

| precursor | [Ar$_3$C$^+$]$_0$/M | [GaCl$_3$]$_0$/M | [GaCl$_3$]$_0$/[Ar$_3$C$^+$]$_0$ | $k$ [L mol$^{-1}$ s$^{-1}$] |
|---|---|---|---|---|
| ($m$F)TrBr | $3.52 \times 10^{-5}$ | $1.43 \times 10^{-4}$ | 4 | $1.99 \times 10^{2}$ [a] |
|  | $8.85 \times 10^{-5}$ | $8.75 \times 10^{-3}$ | 99 | $1.89 \times 10^2$ |
| ($m$F)($m$F)'TrBr | $4.76 \times 10^{-5}$ | $1.64 \times 10^{-4}$ | 3 | $6.57 \times 10^{2}$ [a] |
|  | $7.24 \times 10^{-5}$ | $7.95 \times 10^{-3}$ | 110 | $6.37 \times 10^2$ |

[a] From Table 3.2.

Complete conversion of the fluoro-substituted trityl halides to the corresponding carbocations was achieved by successive addition of small parts of dissolved Lewis acids to the solutions of the trityl halides until the monitored absorbances reached a plateau. The reactions of $(mF)_6Tr^+$ and $(mF)_5Tr^+$ with HSiPh$_3$ were too fast to be followed even by the stopped-flow technique. Hence, we refrained from kinetic investigations of these two tritylium ions.

### 3.2.2. Rates of Reactions with Water

While the rates for the reactions of donor-substituted tritylium ions with water have been reported in Chapter 2, the rates of water-attack at $(mF)Tr^+$ and $(mF)(mF)'Tr^+$ have now been determined in aqueous acetonitrile at 20 °C. The carbocations have been generated by laser-flash photolysis of the corresponding acetates, and UV-vis spectrometry was employed to follow the progress of the reactions (Figure 3.2). The observation of mono-exponential curves indicated the trapping of the carbocations by water, while acetate anions, which varied in concentration, were not involved. The pseudo-first-order rate constants $k_w$ listed in Table 3.4 represent the mean values of 2-4 independent experiments.

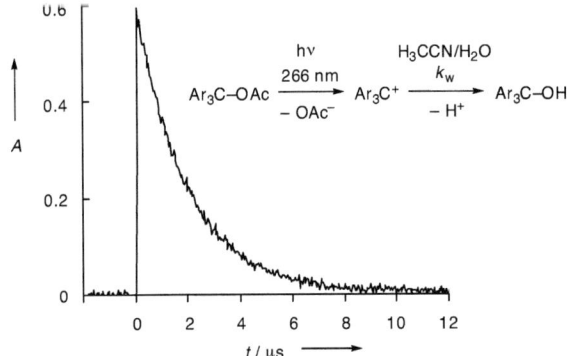

Figure 3.2. Decay of the absorbance of $(mF)Tr^+$ at 415 nm. The carbocation was generated from $(mF)$TrOAc ($c_0$ = 6.04 × 10$^{-4}$ mol L$^{-1}$) by a laser pulse (7 ns, 266 nm, 60 mJ) in 50/50 (v/v) acetonitrile/water, 20 °C.

3. Electrophilicities of Acceptor-Substituted Tritylium Ions                53

Table 3.4. First-order rate constants $k_w$ for the attack of water at tritylium ions in aqueous acetonitrile (20 °C).

| solvent[a] | $k_w$ [s$^{-1}$] | |
|---|---|---|
| | $(m\mathrm{F})\mathrm{Tr}^+$ | $(m\mathrm{F})(m\mathrm{F})'\mathrm{Tr}^+$ |
| 90AN10W | $3.88 \times 10^5$ | $1.18 \times 10^6$ |
| 80AN20W | $5.03 \times 10^5$ | $1.60 \times 10^6$ |
| 60AN40W | $5.02 \times 10^5$ | $1.65 \times 10^6$ |
| 50AN50W | $5.09 \times 10^5$ | $1.56 \times 10^6$ |

[a] Solvents are given in vol%, AN = acetonitrile, W = water.

### 3.2.3. Theoretical Calculations

Geometries of the fluoro-substituted tritylium ions, the corresponding alcohols and the 1,1,1-triarylethanes have been optimized by hybrid DFT methods (B3LYP/6-31G(d,p)). In some cases, different conformations had to be considered. Thermo-chemical corrections for 298.15 K have been calculated for all minima from unscaled vibrational frequencies obtained at the same level.

$$\mathrm{Ar_3C^+ + OH^- \rightarrow R_3C-OH} \qquad (3.5)$$

$$\mathrm{Ar_3C^+ + Me^- \rightarrow R_3C-Me} \qquad (3.6)$$

Table 3.5. Calculated gas phase hydroxide and methyl anion affinities $\Delta H_{298}$ of fluoro-substituted tritylium ions; MP2(FC)/6-31+G(2d,p)//B3LYP/6-31G(d,p).[a]

| system | $\Delta H_{\mathrm{OH}^-}$ (eq. 3.5) [kJ mol$^{-1}$] | $\Delta H_{\mathrm{Me}^-}$ (eq. 3.6) [kJ mol$^{-1}$] |
|---|---|---|
| $(m\mathrm{F})_6\mathrm{Tr}^+$ | –801.1 | –973.5 |
| $(m\mathrm{F})_5\mathrm{Tr}^+$ | –785.6 | –958.6 |
| $(m\mathrm{F})_2(m\mathrm{F})'(m\mathrm{F})''\mathrm{Tr}^+$ | –772.1 | –944.0 |
| $(m\mathrm{F})(m\mathrm{F})'(m\mathrm{F})''\mathrm{Tr}^+$ | –758.7 | –929.2 |
| $(m\mathrm{F})(m\mathrm{F})'\mathrm{Tr}^+$ | –743.2 | –914.1 |
| $(m\mathrm{F})\mathrm{Tr}^+$ | –728.2 | –898.4 |
| $(p\mathrm{F})\mathrm{Tr}^+$ | –720.9 | –890.3 |
| $(p\mathrm{F})_2\mathrm{Tr}^+$ | –727.4 | –897.0 |
| $(p\mathrm{F})_3\mathrm{Tr}^+$ | –734.2 | –903.4 |
| $\mathrm{Tr}^+$ [b] | –715.5 | –883.7 |

[a] For total energies and thermochemical corrections needed for the calculations see the Experimental Section; [b] from Table 2.5.

Apart from the triphenyl carbenium ion, which has been studied theoretically and experimentally with respect to its hydride affinity,[11] Zhu has reported theoretical hydride affinities of some mono-*para*-substituted tritylium ions in acetonitrile solution.[12] As we were interested in the extension of the existing scale of anion affinities of donor-substituted tritylium ions, we calculated the hydroxide and methyl anion affinities [equations (3.5) and (3.6), Table 3.5] in the gas-phase by combining single-point energies on the MP2(FC)/6-31+G(2d,p) level with the thermochemical corrections.

### 3.2.4. Product Study

When HSiPh$_3$ was added to a solution of MeTr$^+$BF$_4^-$ in CH$_2$Cl$_2$, the green color of the tritylium ion faded with concomitant gas evolution (BF$_3$). After workup, GC/MS analysis of the product mixture revealed the presence of FSiPh$_3$, HSiPh$_3$ and MeTrH, which could not be separated. For the detailed procedure see the Experimental Part (p. 202).

### 3.3. Discussion

Figure 3.3, which includes previously reported results for donor-substituted tritylium ions (Chapter 2), shows a good correlation between calculated hydroxide and methyl anion affinities. The slope of almost unity demonstrates a negligible interaction between the OH or CH$_3$ group at the C$_{sp3}$ center and the substituents in the aromatic rings.

As seen in Table 3.5, each *para*-fluoro substituent destabilizes the carbocation by about 7 kJ mol$^{-1}$ in the gas phase, and two *para*-fluoro substituents have approximately the same effect as one *meta*-fluoro substituent. The consecutive introduction of *meta*-fluoro leads to an almost constant destabilization of the carbocation by approximately 15 kJ mol$^{-1}$ per fluorine. In contrast, the effects of donor-substituents were found to be non-additive in Chapter 2.

A striking difference between the stabilization of tritylium ions in the gas phase and in solution is shown in Figure 3.4, where $\Delta G^0$ of water-attack in water is plotted against the hydroxide affinities in the gas phase. Whereas *para*-fluoro substitution destabilizes trityl cations in the gas phase, it has a stabilizing effect in aqueous solution.

## 3. Electrophilicities of Acceptor-Substituted Tritylium Ions 55

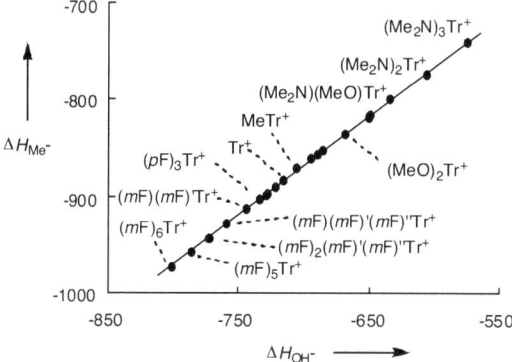

Figure 3.3: Correlation of calculated Me⁻ and OH⁻ affinities ($\Delta H_{298}$ in kJ mol⁻¹), MP2(FC)/6-31+G(2d,p)//B3LYP/6-31G(d,p), $\Delta H_{Me^-} = 1.03\Delta H_{OH^-} - 151$ kJ mol⁻¹, $n = 20$, $R^2 = 0.9997$; values for donor-substituted systems and Tr⁺ from Chapter 2.

If the systems designated by open circles in Figure 3.4 are neglected, the slope of the correlation line indicates that the stability differences between the systems in the gas phase are diminished to 60 % in aqueous solution. We earlier reported a value of 66 % when only methyl and methoxy-substituted systems were considered (see Figure 2.7).

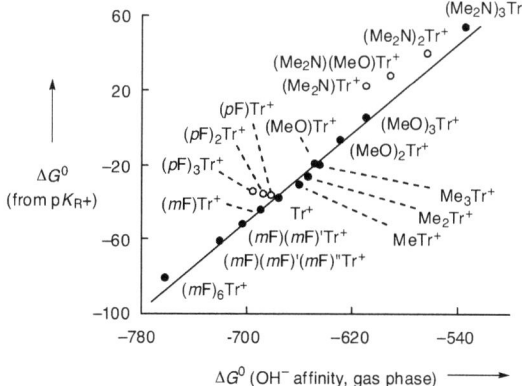

Figure 3.4. Correlation of $\Delta G^0$ (= 2.303$RT$ p$K_{R^+}$) of water-attack (in water) with calculated OH⁻ affinities $\Delta G_{298}$ (gas phase, MP2(FC)/6-31+G(2d,p)//B3LYP/6-31G(d,p)), in kJ mol⁻¹; data for donor-substituted systems from Chapter 2; the open circles were not included in the correlation; slope: 0.60.

# 3. Electrophilicities of Acceptor-Substituted Tritylium Ions

As can be seen in Figure 3.5, log $k$ for hydride transfers from HSiPh$_3$ to tritylium ions in CH$_2$Cl$_2$ solution correlate with log $k_w$ for the water-attack in 50AN50W with a slope of unity. Despite the very different solvents and the very different nature of the nucleophiles, the slope of unity suggests comparable carbocationic characters of the transition states in these two reaction series. The unity slope in Figure 3.5 furthermore implies that the nucleophile-specific sensitivity parameter $s_N$ for the reactions of HSiPh$_3$ with tritylium ions must be identical to that for the solvent mixture 50AN50W (0.89).[13] Further investigations have to show whether the $N$ and $s_N$ parameters of silanes can be employed for their reactions with tritylium ions and other types of carbocations.

Because of the good linear correlation in Figure 3.5 and the fact that the $E$ parameters in Table 3.1 are mainly based on reactions with water, the kinetic data collected in Table 3.2 can be used to derive electrophilicity parameters $E$ for fluoro-substituted tritylium ions. Figure 3.6 shows that the rates of the reactions of HSiPh$_3$ with donor-substituted tritylium ions (log $k$) correlate linearly with the known $E$ parameters of these electrophiles. The resulting correlation equation is given in the caption of Figure 3.6. Extrapolation of the correlation line to the rate constants for the fluoro-substituted systems delivers the corresponding $E$-parameters, as illustrated for $(m\text{F})\text{Tr}^+$ and $(m\text{F})(m\text{F})'\text{Tr}$.

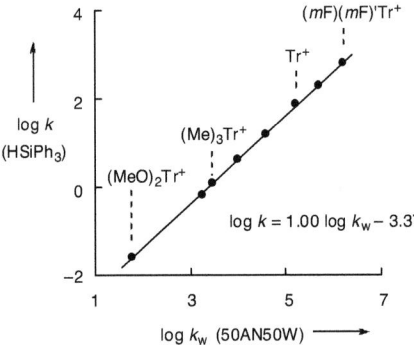

Figure 3.5. Plot of log $k$ for the reactions of triarylmethyl cations with HSiPh$_3$ (CH$_2$Cl$_2$, 20 °C) versus log $k_w$ for the reactions with water (50AN50W, 20 °C); data for log $k_w$ of Tr$^+$ and donor-substituted systems from Chapter 2; $n = 8$, $R^2 = 0.9995$.

Table 3.6 lists the individual $E$-parameters for fluoro-substituted tritylium ions obtained by substitution of the rate constants in Table 3.2 into the correlation equation of Figure 3.6. The electrophilicity of 2.54 for $(m\text{F})_2(m\text{F})'(m\text{F})''\text{Tr}^+$ is similar to that reported for $(p\text{CF}_3)_2\text{Tr}^+$ ($E$ = 2.28),[3] indicating that four $m$-F substituents exert a comparable effect as two $p$-CF$_3$ groups.

## 3. Electrophilicities of Acceptor-Substituted Tritylium Ions 57

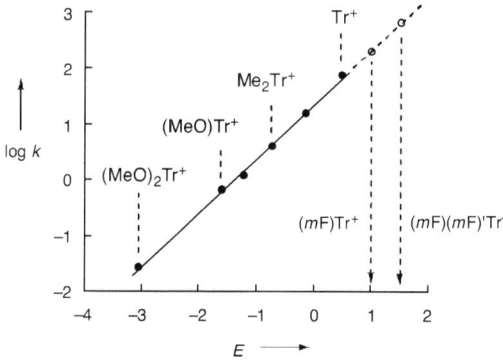

Figure 3.6. Plot of log $k$ vs. $E$ for the reactions of HSiPh$_3$ with tritylium ions (CH$_2$Cl$_2$, 20 °C); correlation equation (only the filled circles): log $k = 0.97E + 1.32$; $n = 6$, $R^2 = 0.9983$.

Table 3.6. Derived electrophilicity parameters $E$ for fluoro-substituted tritylium ions.

| electrophile | $E$ |
|---|---|
| $(mF)_2(mF)'(mF)''Tr^+$ | 2.54 |
| $(mF)(mF)'(mF)''Tr^+$ | 2.07 |
| $(mF)(mF)'Tr^+$ | 1.54 |
| $(mF)Tr^+$ | 1.01 |
| $Tr^+$ | 0.51 [a] |
| $(pF)Tr^+$ | 0.35 |
| $(pF)_2Tr^+$ | 0.17 |
| $(pF)_3Tr^+$ | 0.05 |

[a] From Table 3.1.

The influence of multiple substitution on the $E$-parameter of the triphenyl-methyl cation is shown in Figure 3.7, where $\Delta E$ for consecutive substitution is plotted against the number of substituents. While the first *para*-methyl group decreases the electrophilicity of Tr$^+$ by 0.64, the second and third *para*-methyl groups lead to a reduction of $E$ by only 0.57 and 0.51, respectively. This phenomenon can be explained by the propeller-like structure of tritylium ions and saturation effects. By contrast, each *meta*-fluoro substituent increases the $E$-value of the tritylium ion by approximately the same amount. Obviously, no saturation effects are operating in this series, in agreement with the finding that the anion affinities of triarylmethyl carbenium ions increase almost linearly with the number of *m*-F substituents (see above).

The LUMO energy of a carbocation can be interpreted as an indicator for the ease with which it reacts with an electron donating species.

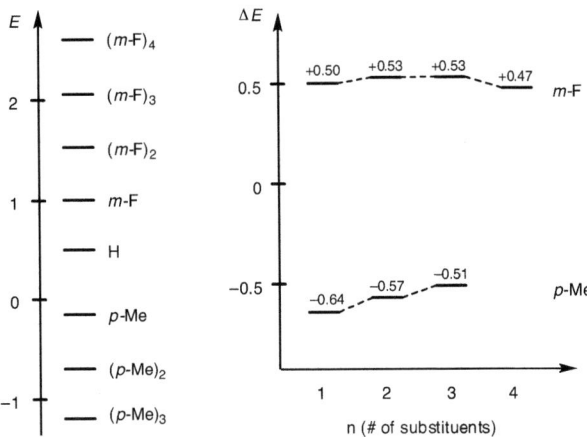

Figure 3.7. Electrophilicity parameters $E$ of substituted tritylium ions (left) and variation of $E$ by consecutive introduction of $m$-F and $p$-Me substituents (right): $\Delta E = E$[tritylium ion with n substituents] – $E$[tritylium ion with (n–1) substituents].

Figure 3.8 plots the $E$ parameters of tritylium and benzhydrylium ions against the corresponding LUMO energies, which have been calculated on the B3LYP/6-31G(d,p) level of theory.

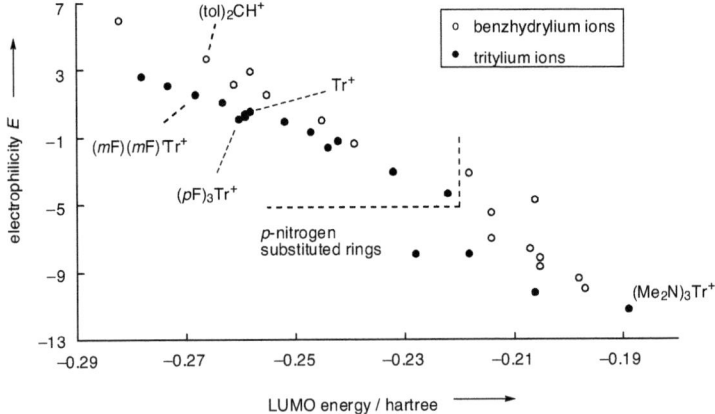

Figure 3.8. Plot of empirical electrophilicity parameters $E$ of tritylium and benzhydrylium ions against their LUMO energies [B3LYP/6-31G(d,p), gas phase]; data for benzhydrylium ions from ref. [14].

While the $E$ parameters refer to solution, the LUMO energies refer to the gas phase. Therefore, the considerable scattering observed for carbocations, which are stabilized by amino groups in the *para*-positions of the rings, is presumably due to solvation effects.

Two clearly diverging trends exist for the two series of electrophiles when the reactivities are increased. Although the LUMO energy of $(mF)(mF)'Tr^+$ is more negative than that of $(tol)_2CH^+$, meaning that electrons – either free or as part of a nucleophile – perceive a higher thermodynamic force to react with the tritylium ion, the observable electrophilic reactivity of the diarylmethyl ion exceeds that of the triarylmethyl ion by a factor of 100. Steric hindrance caused by the additional aryl ring in tritylium ions is the reason for this phenomenon.

The *para*-fluoro substituted tritylium ions follow the same opposing trend that was already observed in Figure 3.4.

## 3.4. Conclusion

Although the nucleophilicity parameters $N/s_N$, which have been derived from the reactions of the corresponding nucleophiles with benzhydrylium ions, are not generally applicable to reactions with tritylium ions, the correlation equation $\log k = s_N(N + E)$ (eq. 3.1) has been reported to hold for reactions of tritylium ions with aqueous solvents, primary amines, and hydride donors (ref. [3,15], Chapter 2). Electrophilicity parameters $E$ of fluoro-substituted tritylium ions could be evaluated from the rate constants of hydride abstractions from $HSiPh_3$. These electrophilicities increase by approximately 0.5 units of $E$ per *meta*-fluoro substituent, which is about the same amount as they decrease per *para*-methyl substituent. Fluoro-substitution in *para*-position has opposite effects in the gas phase and in solution. According to quantum chemical calculations, *para*-fluoro increases the $OH^-$ and $Me^-$ affinities in the gas phase, whereas kinetic investigations in $CH_2Cl_2$ show that *para*-fluoro has a stabilizing effect on tritylium ions in solution. The next chapter deals with the effects of *meta*- and *para*-fluoro-substitution on the heterolysis rates of trityl derivatives.

## 3.5. References

[1] H. Mayr, T. Bug, M. F. Gotta, N. Hering, B. Irrgang, B. Janker, B. Kempf, R. Loos, A. R. Ofial, G. Remennikov, H. Schimmel, *J. Am. Chem. Soc.* **2001**, *123*, 9500-9512.

[2] H. Mayr, A. R. Ofial, *J. Phys. Org. Chem.* **2008**, *21*, 584-595.

[3] S. Minegishi, H. Mayr, *J. Am. Chem. Soc.* **2003**, *125*, 286-295.
[4] I. I. Schuster, A. K. Colter, R. J. Kurland, *J. Am. Chem. Soc.* **1968**, *90*, 4679-4687.
[5] J. W. Rakshys, Jr., S. V. McKinley, H. H. Freedman, *J. Am. Chem. Soc.* **1971**, *93*, 6522-6529.
[6] a) J. Chojnowski, L. Wilczek, W. Fortuniak, *J. Organomet. Chem.* **1977**, *135*, 13-22; b) J. Chojnowski, W. Fortuniak, W. Stanczyk, *J. Am. Chem. Soc.* **1987**, *109*, 7776-7781.
[7] a) P. Huszthy, K. Lempert, *J. Chem. Soc., Perkin Trans. 2* **1982**, 1671-1674; b) M. Green, S. Greenfield, M. Kersting, *J. Chem. Soc., Chem. Commun.* **1985**, 18-20; c) D. Mandon, L. Toupet, D. Astruc, *J. Am. Chem. Soc.* **1986**, *108*, 1320-1322; d) G. Karabatsos, M. Tornaritis, *Tetrahedron Lett.* **1989**, *30*, 5733-5736; e) S. Słomkowski, S. Penczek, *Chem. Commun.* **1970**, 1347-1348; f) M. P. Dreyfuss, J. C. Westphal, P. Dreyfuss, *Macromolecules* **1968**, *1*, 437-441; g) K.-ud-Din, P. H. Plesch, *J. Chem. Soc., Perkin Trans. 2* **1978**, 937-938; h) S. Penczek, *Makromol. Chem.* **1974**, *175*, 1217-1252; i) R. Damico, C. D. Broaddus, *J. Org. Chem.* **1966**, *31*, 1607-1612; j) F. A. Carey, H. S. Tremper, *J. Am. Chem. Soc.* **1968**, *90*, 2578-2583; k) F. A. Carey, C.-L. Wang Hsu, *J. Organomet. Chem.* **1969**, *19*, 29-41; l) T.-Y. Cheng, R. M. Bullock, *Organometallics*, **1995**, *14*, 4031-4033; m) T.-Y. Cheng, R. M. Bullock, *Organometallics* **2002**, *21*, 2325-2331; n) K.-T. Smith, M. Tilset, *J. Organomet. Chem.* **1992**, *431*, 55-64; o) C. A. Mullen, M. R. Gagné, *J. Am. Chem. Soc.* **2007**, *129*, 11880-11881.
[8] S. V. Kulkarni, R. Schure, R. Filler, *J. Am. Chem. Soc.* **1973**, *95*, 1859-1864.
[9] R. A. McClelland, V. M. Kanagasabapathy, N. S. Banait, S. Steenken, *J. Am. Chem. Soc.* **1989**, *111*, 3966–3972.
[10] H. Mayr, N. Basso, G. Hagen, *J. Am. Chem. Soc.* **1992**, *114*, 3060-3066.
[11] a) D. G. Gusev, O. V. Ozerov, *Chem. Eur. J.* **2011**, *17*, 634-640; b) X.-M. Zhang, J. W. Bruno, E. Enyinnaya, *J. Org. Chem.* **1998**, *63*, 4671-4678; c) J.-P. Cheng, K. L. Handoo, V. D. Parker, *J. Am. Chem. Soc.* **1993**, *115*, 2655-2660.
[12] X.-Q. Zhu, C.-H. Wang, *J. Phys. Chem. A* **2010**, *114*, 13244-13256.
[13] S. Minegishi, S. Kobayashi, H. Mayr, *J. Am. Chem. Soc.* **2004**, *126*, 5174-5181.
[14] T. Singer, *PhD Thesis*, **2008**, Munich.
[15] D. Richter, H. Mayr, *Angew. Chem.* **2009**, *121*, 1992-1995; *Angew. Chem. Int. Ed.* **2009**, *48*, 1958-1961.

# 4. Electrofugalities of Acceptor-Substituted Tritylium Ions

## 4.1. Introduction

In analogy to the procedure used for the construction of comprehensive nucleophilicity and electrophilicity scales,[1] we suggested to employ the linear free energy relationship (4.1) for correlating rates of heterolyses.[2] The carbocation (electrofuge) is hereby characterized by a single, solvent-independent parameter, $E_f$, whereas the couple of leaving group and solvent is characterized by the two parameters $N_f$ and $s_f$. A series of differently substituted benzhydrylium ions were chosen as reference electrofuges, and by now, linear correlations in a reactivity range spanning 25 powers of ten have been employed to characterize more than 30 nucleofuges in a variety of solvents.[2,3]

$$\log k_{\text{ion}}(25\ °C) = s_f(E_f + N_f) \qquad (4.1)$$

The additional aryl group in triarylmethyl (trityl) derivatives can be expected to accelerate the ionization by two effects: the better stabilization of the generated carbenium ion, as well as the steric repulsion in the ground state (back strain effect).[4] While the solvolytic generation of the unsubstituted triphenylmethyl cation has been analyzed in several kinetic studies,[5] substituted tritylium derivatives have only rarely been investigated. As substituted trityl moieties are widely used as protecting groups,[6] knowledge of their leaving group abilities (electrofugalities) is important.

Chapter 1 of this thesis reported about solvolysis rates of donor-substituted trityl carboxylates in aqueous acetonitrile and acetone. These investigations are now extended to acceptor-substituted trityl derivatives.

Ionization rate constants of trityl halides (chlorides, bromides, fluorides) and carboxylates (acetates, benzoates, *para*-nitrobenzoates) have been determined in aqueous and neat acetonitrile and acetone (Scheme 4.1). As the nucleofuge-specific parameters $N_f$ and $s_f$ of the halide and carboxylate anions have previously been reported,[3] the kinetic data should enable us to test whether equation (4.1), which is based on solvolyses of benzhydryl derivatives, is also applicable to ionizations of trityl derivatives.

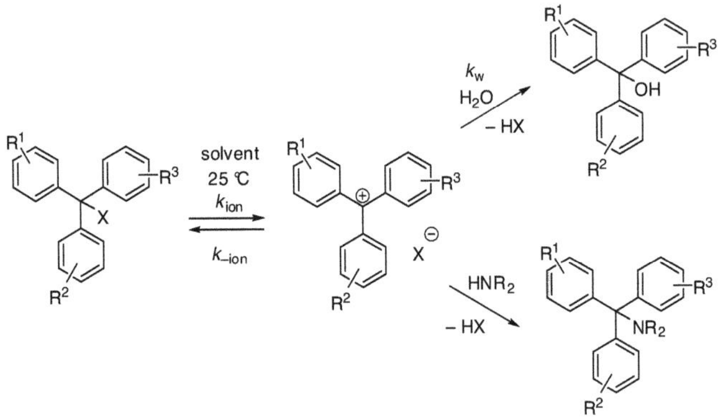

# 4. Electrofugalities of Acceptor-Substituted Tritylium Ions

Scheme 4.1. Solvolyses of substituted trityl derivatives.

Table 4.1. The tritylium ions studied in this work.

| $R^1, R^2, R^3$ [a] | abbreviation | $pK_{R+}$ |
|---|---|---|
| p-Me, p-Me, p-Me | $Me_3Tr^+$ | −3.56 [b] |
| p-Me, p-Me, H | $Me_2Tr^+$ | −4.71 [b] |
| p-Me, H, H | $MeTr^+$ | −5.41 [b] |
| H, H, H | $Tr^+$ | −6.63 [b] |
| p-Cl, H, H | $(pCl)Tr^+$ | - |
| p-F, H, H | $(pF)Tr^+$ | −6.41 [c] |
| p-F, p-F, H | $(pF)_2Tr^+$ | −6.22 [c] |
| p-F, p-F, p-F | $(pF)_3Tr^+$ | −6.05 [d] |
| m-F, H, H | $(mF)Tr^+$ | −7.81 [c] |
| m-F, m-F, H | $(mF)(mF)'Tr^+$ | −9.17 [c] |
| (m-F)$_2$, H, H | $(mF)_2Tr^+$ | - |
| m-F, m-F, m-F | $(mF)(mF)'(mF)''Tr^+$ | −10.72 [d] |
| (m-F)$_2$, (m-F)$_2$, H | $(mF)_2(mF)'_2Tr^+$ | - |
| (m-F)$_2$, m-F, m-F | $(mF)_2(mF)'(mF)''Tr^+$ | - |
| (m-F)$_2$, (m-F)$_2$, (m-F)$_2$ | $(mF)_6Tr^+$ | −14.2 [d] |

[a] For the substitution pattern, see Scheme 4.1; [b] from ref. [7]; [c] from ref. [8]; [d] from ref. [9].

## 4.2. Results

Unlike in previous kinetic investigations of donor-substituted trityl derivatives (see Chapter 1), carbenium ions did not accumulate during the solvolysis reactions, i.e., $k_w$ and/or $k_{-ion}$ are fast compared to $k_{ion}$ (typical $S_N1$ reaction, Scheme 4.1).

## 4. Electrofugalities of Acceptor-Substituted Tritylium Ions 63

The generation of the Brønsted acids HX suggested the use of conductimetry as method of choice for monitoring the progress of the reactions. As shown by the calibration curve in Figure 4.1, the solution's conductivity $\kappa$ is proportional to the amount of completely solvolyzed substrate in the concentration range employed. Only relative conductivities $\kappa_{rel}$ were needed for the evaluation of the kinetic data, and we have not calibrated the conductivity cell for determining absolute values of $\kappa$.

Reactions with half-lives > 10 s were studied by conventional methods, while faster reactions were followed by a stopped-flow apparatus. All recorded curves were fitted according to equation (4.2) by the method of least squares.

$$\kappa_{rel} = \kappa_{max}(1 - \exp(-k_{ion}t)) + \text{const.} \qquad (4.2)$$

In the reaction sequence of Scheme 4.1 two cases have to be distinguished. First, if the trapping of the carbocation by water ($k_w$) is fast compared to the ion return ($k_{-ion}[X^-]$) the generation of the alcohol and of HX will follow the exponential rate law in equation (4.2). If, however, the ion return proceeds with a similar rate or even faster than the attack of water, its importance will grow during the course of the reactions due to increasing $[X^-]$. As a consequence, the increase of concentrations does not follow an exponential function, and the evaluation of the kinetic parameters from the recorded conductivity data becomes more complicated.

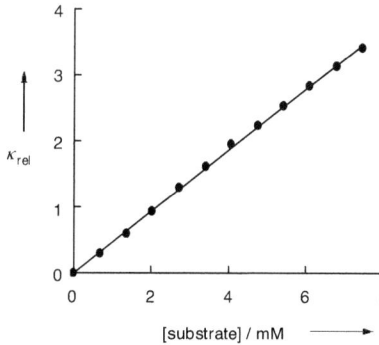

Figure 4.1. Calibration curve for the solvolysis of $(mF)(mF)'TrCl$ in 80/20 (vol-%) $CH_3CN/H_2O$ at 25 °C; plot of $\kappa_{rel}$ at $t_\infty$ against the amount of substrate, conventional conductimetry, [piperidine] = $1.02 \times 10^{-2}$ mol L$^{-1}$.

Common ion rate depression is a well-known phenomenon,[5h,5i,10] that has recently been studied systematically in solvolyses of benzhydryl chlorides.[11,12] It was shown that the determination of the ionization rate constants $k_{ion}$ can be simplified by adding large amounts of amines which rapidly trap the generated benzhydrylium ions, and suppress the return of chloride ions (Scheme 4.1).[11a] This procedure enforces first-order kinetics, thus circumventing the need for sophisticated mathematical evaluations. It resembles the methodology of Winstein and Appel, who used azide anions instead of amines,[5g] but has the advantage that the reactions can now be followed conductimetrically, whereas the number of ions remains unchanged when common ion return is suppressed by azide ions.

In the present study of solvolyses of trityl derivatives we also observed common ion rate depression in many cases, especially when trityl halides were hydrolyzed in solvent mixtures with low fractions of water. Therefore, we adopted the aforementioned technique and followed the kinetics in the presence of amines.

For each substrate, several measurements with increasing concentrations of amine were performed. Although small amounts of amine often did not lead to perfect first-order kinetics because common ion return was not fully suppressed, fits according to equation (4.2) were enforced in order to get $k_{obs}$ values as shown in Figure 4.2a. When the amine concentrations were increased, trapping of the carbocations by the amines became more and more effective, and the quality of the mono exponential fits improved. Figure 4.2b exemplifies an almost perfect agreement between the experimental and simulated increase of conductivity for the ionization of ($p$F)TrCl in 90/10 (vol-%) acetonitrile/water.

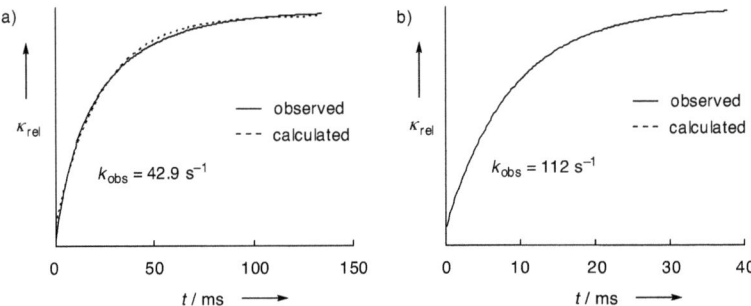

Figure 4.2. Increase of conductivity during the ionization of ($p$F)TrCl ($c_0$ = 1.00 × $10^{-3}$ mol $L^{-1}$) in 90/10 (vol-%) $CH_3CN/H_2O$ at 25 °C a) without piperidine; b) in the presence of 7.30 × $10^{-2}$ mol $L^{-1}$ piperidine.

# 4. Electrofugalities of Acceptor-Substituted Tritylium Ions 65

Plots of $k_{obs}$ against the amine concentrations showed an initial increase of the observed rate constants until plateaus were reached (Figure 4.3). The non-linearity of these graphs excludes the operation of an $S_N2$ mechanism, and the plateaus correspond to the ionization rate constants $k_{ion}$ at which the generated carbenium ions are quantitatively trapped by the amines before they can recombine with the leaving group.

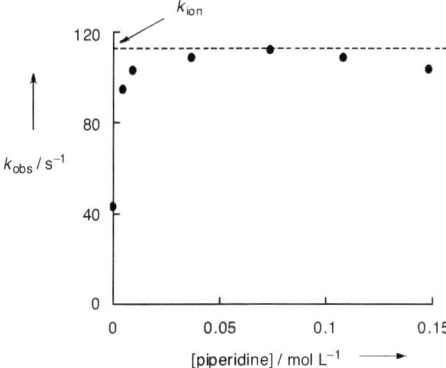

Figure 4.3. Observed rate constants $k_{obs}$ for the ionization of $(p\text{F})\text{TrCl}$ ($c_0 = 1.00 \times 10^{-3}$ mol $L^{-1}$) in 90/10 (vol-%) $CH_3CN/H_2O$ at variable concentrations of piperidine, 25 °C.

This method also allowed the determination of ionization rate constants in aprotic solvents, like pure acetonitrile and acetone.[13] Figure 4.4 shows plots of $k_{obs}$ for the ionization of $(m\text{F})(m\text{F})'\text{TrBr}$ versus the concentrations of different amines in pure acetonitrile. It illustrates a small effect of the nature of the trapping amine on the height of the plateaus, similar to observations previously made for ionizations of benzhydryl chlorides.[11a]

While use of piperidine and butylamine essentially results in the same ionization rate constant, diethylamine gives rise to smaller values of $k_{obs}$, and higher concentrations were needed to fully suppress the ion return, possibly because of the greater steric demand of diethylamine.

Although the difference of the plateaus (1.8 s$^{-1}$ vs. 1.4 s$^{-1}$) is noticeable, it is marginal in view of the 8 powers of 10 that span the reactivity range of the substrates investigated in this work. A possible explanation might be the change of the solvent polarity that is induced by high amine concentrations. This effect might also be responsible for the small decrease of $k_{obs}$ when very high concentrations of amine are used (Figure 4.3).

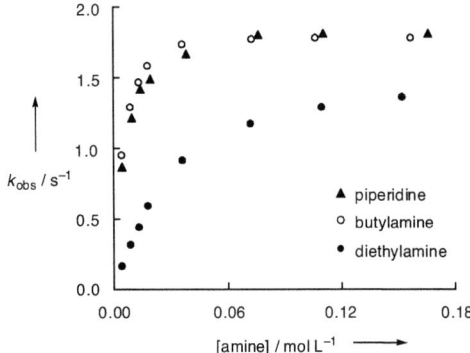

Figure 4.4. Observed rate constants $k_{obs}$ for the ionization of $(mF)(mF)'TrBr$ ($c_0 = 1.0 \times 10^{-3}$ mol L$^{-1}$) in pure acetonitrile at 25 °C. Butylamine, piperidine, and diethylamine were used as trapping agents.

As carboxylates are weak nucleophiles in aqueous solution, amines are not required to suppress common ion return in solvolyses of trityl carboxylates in aqueous acetone or acetonitrile. However, as weak acids like acetic acid are not fully dissociated into ions in the solvent mixtures used, amines were used here as auxiliary bases to ensure complete dissociation of the generated carboxylic acids, following the protocol in Chapter 1. In this way, the sensitivity of the conductivity measurements was increased. A base was also required during the hydrolyses of trityl fluoride, which is known to be catalyzed by protons.[5e] Here, addition of small amounts of triethylamine or piperidine (9-10 equivalents) led to almost identical ionization rate constants. Common ion return was not observed in this case, due to the lower nucleophilicity of fluoride compared to chloride or bromide. All obtained rate constants of substituted trityl derivatives are summarized in Table 4.2.

Table 4.2. Ionization rate constants of trityl halides and carboxylates, 25 °C.

| electrofuge | additive[a] | nucleofuge | solvent[b] | $k_{ion}$ [s$^{-1}$] |
|---|---|---|---|---|
| Me$_3$Tr$^+$ | pip | Cl$^-$ | 100AN | $2.50 \times 10^2$ |
| | pip | | 100A | $7.68 \times 10^{-1}$ |
| Me$_2$Tr$^+$ | pip | Cl$^-$ | 100AN | $5.19 \times 10^1$ |
| | pip | | 100A | $8.11 \times 10^{-2}$ |
| MeTr$^+$ | pip | Cl$^-$ | 100AN | 5.91 |
| | pip | | 90AN10W | $4.77 \times 10^2$ |
| | pip | Br$^-$ | 100A | $3.91 \times 10^1$ |

## 4. Electrofugalities of Acceptor-Substituted Tritylium Ions

Table 4.2. Continued.

| electrofuge | additive | nucleofuge | solvent | $k_{ion}$ [s$^{-1}$] |
|---|---|---|---|---|
| Tr$^+$ | pip | F$^-$ | 80AN20W | $4.02 \times 10^{-5}$ |
| | pip/TEA [c] | | 60AN40W | $4.37 \times 10^{-4}$ |
| | pip/TEA [c] | | 50AN50W | $1.30 \times 10^{-3}$ |
| | - | | 80A20W | $4.23 \times 10^{-6}$ [d] |
| | - | | 70A30W | $1.60 \times 10^{-5}$ [d] |
| | - | | 50A50W | $9.83 \times 10^{-4}$ [d] |
| | pip | Cl$^-$ | 100AN | $4.91 \times 10^{-1}$ |
| | pip | | 90AN10W | $8.09 \times 10^{1}$ |
| | pip | | 80AN20W | $2.52 \times 10^{2}$ |
| | - | | 90A10W | $1.93$ [e] |
| | - | | 80A20W | $2.19 \times 10^{1}$ [e] |
| | pip | Br$^-$ | 100AN | $6.04 \times 10^{2}$ |
| ($p$Cl)Tr$^+$ | pip | Cl$^-$ | 80AN20W | $1.30 \times 10^{2}$ |
| | - | | 60AN40W | $4.89 \times 10^{2}$ |
| | - | | 50AN50W | $9.52 \times 10^{2}$ |
| | pip | | 90A10W | $1.59$ |
| | pip | | 80A20W | $1.28 \times 10^{1}$ |
| | - | | 60A40W | $1.66 \times 10^{2}$ |
| | - | | 50A50W | $5.90 \times 10^{2}$ |
| | TEA | PNB$^-$ [f] | 90AN10W | $7.24 \times 10^{-4}$ |
| | TEA | | 80AN20W | $2.22 \times 10^{-3}$ |
| | TEA | | 60AN40W | $6.10 \times 10^{-3}$ |
| | TEA | | 50AN50W | $1.04 \times 10^{-2}$ |
| | pip | BzO$^-$ | 60A40W | $1.27 \times 10^{-4}$ |
| | pip | | 50A50W | $3.58 \times 10^{-4}$ |
| ($p$F)Tr$^+$ | pip | Cl$^-$ | 100AN | $6.47 \times 10^{-1}$ |
| | pip | | 90AN10W | $1.12 \times 10^{2}$ |
| | pip | | 80AN20W | $3.11 \times 10^{2}$ |
| | - | | 60AN40W | $1.11 \times 10^{3}$ |
| | pip | | 90A10W | $5.35$ |
| | pip | | 80A20W | $3.06 \times 10^{1}$ |
| | - | | 60A40W | $4.46 \times 10^{2}$ |
| (pF)$_2$Tr$^+$ | pip | Cl$^-$ | 90AN10W | $1.55 \times 10^{2}$ |
| (pF)$_3$Tr$^+$ | pip | Cl$^-$ | 100AN | $1.02$ |
| | pip | | 90AN10W | $1.70 \times 10^{2}$ |
| ($m$F)Tr$^+$ | pip | Cl$^-$ | 100AN | $2.95 \times 10^{-2}$ |
| | pip | | 90AN10W | $9.51$ |
| | pip | | 80AN20W | $3.27 \times 10^{1}$ |

## 4. Electrofugalities of Acceptor-Substituted Tritylium Ions

Table 4.2. Continued.

| electrofuge | additive | nucleofuge | solvent | $k_{ion}$ [s$^{-1}$] |
|---|---|---|---|---|
| (mF)Tr$^+$ | - | Cl$^-$ | 60AN40W | 1.32 × 10$^2$ |
|  | - |  | 50AN50W | 2.54 × 10$^2$ |
|  | pip | Br$^-$ | 100AN | 3.45 × 10$^1$ |
|  | TEA | AcO$^-$ | 60AN40W | 3.45 × 10$^{-5}$ |
| (mF)(mF)'Tr$^+$ | pip | Cl$^-$ | 100AN | 1.46 × 10$^{-3}$ |
|  | pip |  | 90AN10W | 8.26 × 10$^{-1}$ |
|  | pip |  | 80AN20W | 3.60 |
|  | - |  | 60AN40W | 1.60 × 10$^1$ |
|  | - |  | 50AN50W | 3.22 × 10$^1$ |
|  | pip | Br$^-$ | 100AN | 1.81 |
|  | pip |  | 90AN10W | 1.44 × 10$^2$ |
|  | pip |  | 90A10W | 5.41 |
| (mF)$_2$Tr$^+$ | pip | Cl$^-$ | 100AN | 2.12 × 10$^{-3}$ |
|  | pip |  | 90AN10W | 1.25 |
|  | pip |  | 80AN20W | 4.80 |
| (mF)(mF)'(mF)"Tr$^+$ | pip | Cl$^-$ | 100AN | 7.10 × 10$^{-5}$ |
|  | pip |  | 90AN10W | 6.38 × 10$^{-2}$ |
|  | pip |  | 80AN20W | 2.74 × 10$^{-1}$ |
|  | pip |  | 90A10W | 1.53 × 10$^{-3}$ |
|  | pip |  | 80A20W | 1.65 × 10$^{-2}$ |
|  | pip | Br$^-$ | 100AN | 8.27 × 10$^{-2}$ |
|  | pip |  | 90AN10W | 9.91 |
|  | pip |  | 80AN20W | 3.90 × 10$^1$ |
|  | pip |  | 60AN40W | 1.50 × 10$^2$ |
| (mF)$_2$(mF)'(mF)"Tr$^+$ | pip | Br$^-$ | 100AN | 5.20 × 10$^{-3}$ |
|  | pip |  | 90AN10W | 7.85 × 10$^{-1}$ |
|  | pip |  | 80AN20W | 3.48 |
|  | pip |  | 60AN40W | 1.63 × 10$^1$ |
|  | pip |  | 90A10W | 1.51 × 10$^{-2}$ |
| (mF)$_2$(mF)'$_2$Tr$^+$ | pip | Cl$^-$ | 100AN | 6.3 × 10$^{-6}$ [g] |
|  | pip |  | 90AN10W | 1.01 × 10$^{-2}$ |
|  | pip |  | 80AN20W | 4.76 × 10$^{-2}$ |
| (mF)$_6$Tr$^+$ | pip | Br$^-$ | 100AN | 2.1 × 10$^{-5}$ [g] |
|  | pip |  | 90AN10W | 3.39 × 10$^{-3}$ |
|  | pip |  | 80AN20W | 1.68 × 10$^{-2}$ |

[a] pip = piperidine, TEA = triethylamine; [b] the solvent is given in vol-%, AN = acetonitrile, A = acetone, W = water; [c] measurements using either pip or TEA gave the same results; [d] from ref. [5c]; [e] from ref. [5f]; [f] PNB = *para*-nitrobenzoate; [f] very slow, approximate value.

## 4.3. Discussion

### 4.3.1. Leaving Groups

Similar to solvolyses of benzhydryl and phenethyl derivatives,[14] the ordering of leaving group abilities in trityl ionizations is Br$^-$ > Cl$^-$ > PNB$^-$ > BzO$^-$ > AcO$^-$ (Scheme 4.2).

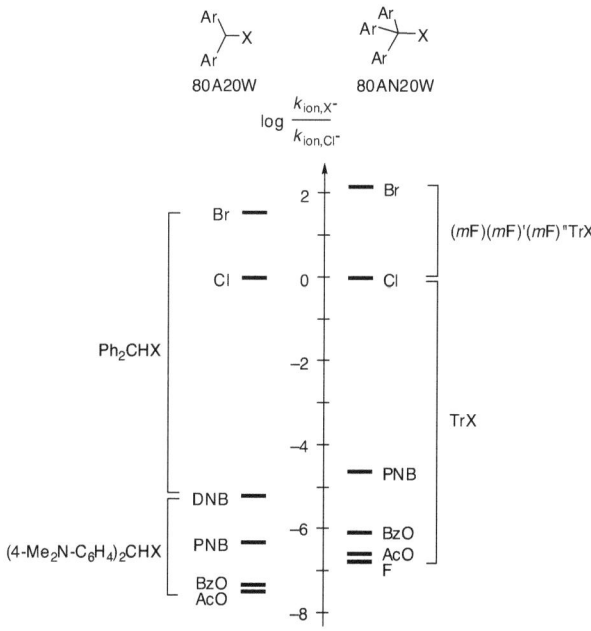

Scheme 4.2. Relative ionization rates of benzhydryl (from ref. [2a] and [15]) and trityl derivatives (from Table 4.2 and Chapter 1), DNB = 3,5-dinitrobenzoate, at 25 °C.

However, the ratio $k_{ion,Br}/k_{ion,Cl}$ increases when the electrofuge is changed from 30 for benzhydryl to 142 for trityl, which may be due to the higher ground state strain in trityl bromides compared to benzhydryl bromides (back strain).[4] The fact that the Cl/BzO ratio is only 1.5 × 10$^6$ in the trityl series, while it is 2.1 × 10$^7$ in the benzhydryl series (Scheme 4.2), can also be explained by the higher strain in trityl carboxylates.

## 4.3.2. Ion Recombination

The degree of ion return depends on several factors, which shall be analyzed in more detail. Common ion return is much more pronounced for ionizations of trityl bromides than for trityl chlorides. While piperidine concentrations of almost 0.1 mol L$^{-1}$ are needed to prevent ion return during the ionization of $(mF)(mF)'(mF)''$TrBr in pure acetonitrile, less than 0.007 mol L$^{-1}$ of piperidine sufficed in the case of $(mF)(mF)'(mF)''$TrCl (Figure 4.5).

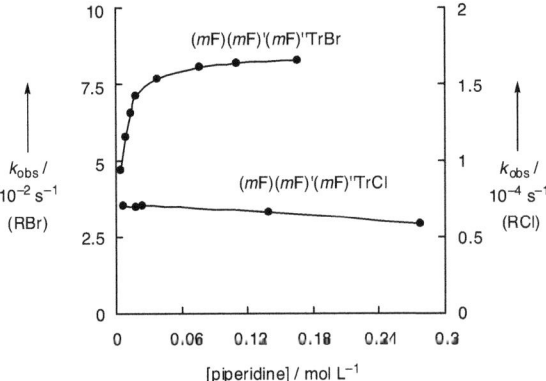

Figure 4.5. Influence of the leaving group on common ion return: Observed rate constants $k_{obs}$ for the ionizations of $(mF)(mF)'(mF)''$TrBr ($c_0$ = 1.0 × 10$^{-3}$ mol L$^{-1}$, left scale) and $(mF)(mF)'(mF)''$TrCl ($c_0$ = 1.0 × 10$^{-3}$ mol L$^{-1}$, right scale) in 100AN in dependence on the concentration of piperidine, 25 °C.

Textbooks often claim that the nucleophilicity order Br$^-$ > Cl$^-$ in protic solvents is reversed in aprotic solvents, where nucleophilicities are not predominantly controlled by solvation effects as in protic solvents, but by the different strengths of the developing C–X bonds.[16]

This general statement has to be revised. Indeed, nucleophilic substitutions of octyl mesylate in chlorobenzene[17a] and of hexyl tosylate in DMSO[17b] proceed 2-3 times faster with Cl$^-$ than with Br$^-$. The results in Figure 4.5 indicate, however, that Br$^-$ reacts much faster with carbocations than Cl$^-$ also in acetonitrile,[12] in line with direct rate measurements of laser-flash photolytically generated benzhydrylium ions with halide ions.[12] While we were able to determine rate constants for the reactions of several benzhydrylium ions with Cl$^-$ in acetonitrile, the corresponding reactions with Br$^-$ were so fast that they could not be followed with the equipment available at that time. Future work will be needed to analyze the relative

## 4. Electrofugalities of Acceptor-Substituted Tritylium Ions 71

nucleophilicities of halide ions not only as a function of the solvent but also of the reaction partner.

Figure 4.6 illustrates that ion return is less important for more electrophilic tritylium ions; thus recombination of the ions is not observed in the solvolysis of $(mF)_6TrBr$ in 90AN10W. In contrast, approximately 0.1 mol $L^{-1}$ piperidine is needed to reach the plateau with a mono-exponential increase of conductivity in the case of $(mF)(mF)'TrBr$.

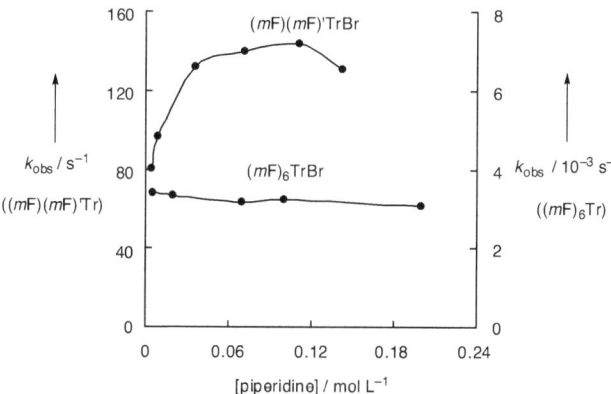

Figure 4.6. Influence of the electrofuge on common ion return: Observed rate constants $k_{obs}$ for the ionizations of $(mF)(mF)'TrBr$ ($c_0 = 1.0 \times 10^{-3}$ mol $L^{-1}$, left scale) and $(mF)_6TrBr$ ($c_0 = 8.0 \times 10^{-4}$ mol $L^{-1}$, right scale) in 90AN10W in dependence on the concentration of piperidine, 25 °C.

The analysis previously reported for benzhydryl halide solvolyses[12] shall now be employed to rationalize these observations. Table 3.4 in Chapter 3 shows that $(mF)(mF)'Tr^+$ reacts with 90AN10W with a rate constant of $1.2 \times 10^6$ s$^{-1}$. Equation (4.3),[1] which calculates rate constants (20 °C) of reactions between carbocations and nucleophiles from a carbocation-specific parameter $E$ and the nucleophile-specific parameters $N$ and $s_N$, can be used to calculate the rates for the reactions of tritylium ions with halide ions.

$$\log k = s_N(E + N) \qquad (4.3)$$

From the nucleophilicity parameters of Br$^-$ in various acetonitrile/water mixtures[12] one can extrapolate that its reaction with $(mF)(mF)'Tr^+$ proceeds with diffusion-control ($k \approx 10^{10}$ L mol$^{-1}$ s$^{-1}$) which means that already at bromide concentrations as low as $10^{-4}$ mol $L^{-1}$ this reaction becomes comparable to the rate of the reaction of the carbocation with the solvent. As shown in the upper curve of Figure 4.6, amine additives are needed to suppress common ion return.

On the other hand, Figure 3.7 in Chapter 3 allows one to extrapolate that $(mF)_6Tr^+$ reacts $10^2$ times faster with 90AN10W than $(mF)(mF)'Tr^+$, corresponding to a first-order rate constant of approximately $10^8$ s$^{-1}$ for its reaction with the solvent. In order to compete with the trapping by the solvent, bromide concentrations of more than $10^{-2}$ mol L$^{-1}$ would be needed, which cannot be reached at substrate concentrations of $10^{-3}$ mol L$^{-1}$. In line with this analysis, the lower graph in Figure 4.6 indicates the absence of common ion return.

The higher degree of common ion return in 90AN10W than in 80AN20W (Figure 4.7) can be explained analogously. Table 3.4 in Chapter 3 shows that trapping of $(mF)Tr^+$ by the solvent is only 1.3 times faster in 80AN20W than in 90AN10W. On the other hand, the nucleophilicity of Cl$^-$ in acetonitrile/water mixtures decreases dramatically when the content of water is increased,[12] resulting in a much bigger recombination rate constant in 90AN10W than in 80AN20W.

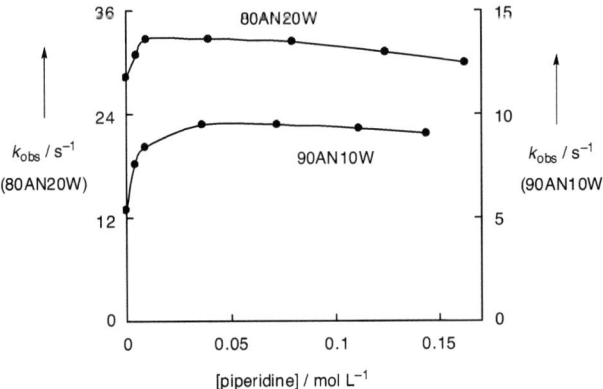

Figure 4.7. Influence of the solvent on common ion return: Observed rate constants $k_{obs}$ for the ionizations of $(mF)TrCl$ in 80AN20W ($c_0 = 9.8 \times 10^{-4}$ mol L$^{-1}$, left scale) and 90AN10W ($c_0 = 1.0 \times 10^{-3}$ mol L$^{-1}$, right scale) in dependence on the concentration of piperidine, 25 °C.

## 4.3.3. Linear Free Energy Relationships

The ionization rate constants of substituted trityl chlorides (Table 4.2) correlate linearly with the thermodynamic stabilities of the tritylium ions in aqueous solution (Figure 4.8).

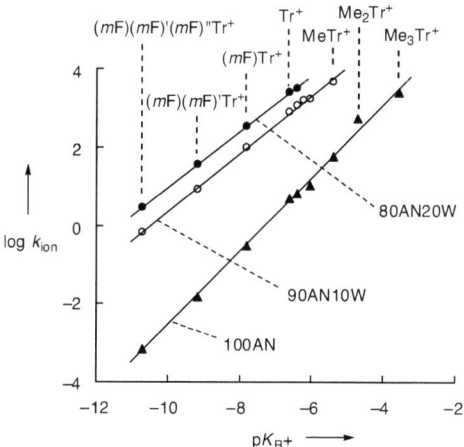

Figure 4.8. Correlation of ionization rates of triarylmethyl chlorides with p$K_{R+}$, slopes: 0.72 (80AN20W), 0.74 (90AN10W), 0.94 (100AN).

The slope of almost unity in the case of pure acetonitrile (0.94) implies that the energy differences of the transition states are similar to those of the free carbocations. From the nucleophile-specific parameters $N$ = 17.2 and $s_N$ = 0.6 for Cl$^-$ in pure acetonitrile,[12] one can calculate by equation (4.3) that carbocations with $E$ > –2.2 will undergo diffusion-controlled reactions with Cl$^-$ ($k_{calc}$ = $10^9$ L mol$^{-1}$ s$^{-1}$). As Me$_3$Tr$^+$, the least reactive system of Figure 4.8, has an $E$-value of –1.21,[18] it can be concluded that all carbocations of Figure 4.8 undergo barrier-free combinations with Cl$^-$ in pure acetonitrile. From the principle of microscopic reversibility one can, therefore, derive that the activation free energies of the ionizations in acetonitrile equal the reaction free energies (Figure 4.9, right). In line with this consideration, the slope of the line "100AN" in Figure 4.8 is close to unity.

In contrast, the reactions of Cl$^-$ with tritylium ions in solvents containing water proceed with a barrier (Figure 4.9, left), i.e., the carbocationic characters of the electrofuges are not fully developed in the transition states. Accordingly, the slope of the line "80AN20W" in Figure 4.8 is only 0.72.

# 4. Electrofugalities of Acceptor-Substituted Tritylium Ions

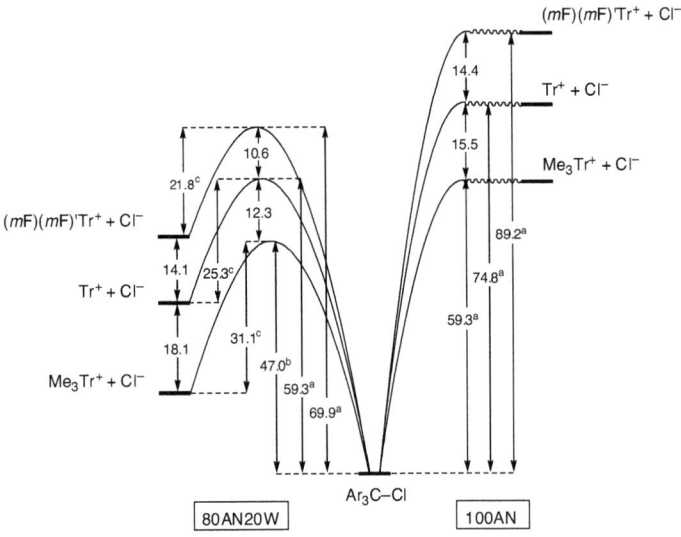

Figure 4.9. Free energy diagrams for the ionizations of substituted trityl chlorides in 80AN20W (left) and pure acetonitrile (right), 25 °C; a) from ionization rate constants (Table 4.2); b) the ionization rate constant of Me$_3$TrCl in 80AN20W (3.64 × 10$^4$ s$^{-1}$) has been extrapolated from the correlation line in Figure 4.8; c) calculated by equation (4.3) using $E$ from Chapter 3 and ref. [18], and $N/s_N$ from ref. [12].

The energy differences between the three tritylium ions in Figure 4.9 (14.1 and 18.1 kJ/mol for 80AN20W; 14.4 and 15.5 kJ/mol for 100AN), which have been evaluated by purely kinetic means, may be compared with the stability differences derived from p$K_{R+}$ values (Table 4.1). The latter give differences $\Delta\Delta G$ of 14.5 and 17.5 kJ mol$^{-1}$ for the couples (mF)(mF)'Tr$^+$/Tr$^+$ and Tr$^+$/Me$_3$Tr$^+$, respectively.

### 4.3.4. Winstein-Grunwald Analysis

In Figure 4.10 the ionization rates of trityl chlorides in aqueous acetonitrile are plotted against the corresponding ionizing powers $Y_{t\text{-BuCl}}$.[19] The trend of decreasing $m$-values with increasing stabilization of tritylium ions can be assigned to a Hammond shift of the transition states.

As shown in Figure 4.9 (left), the C–Cl bond of the trityl chlorides is not completely broken in the transition states of the ionization processes. In line with the Hammond

## 4. Electrofugalities of Acceptor-Substituted Tritylium Ions 75

postulate,[20] the transition state of the ionization step is the earlier on the reaction coordinate, the more stable the generated carbocation. The earlier, i.e., the less carbocation-like the transition states, the smaller the influence of the solvent ionizing power on the ionization rates.

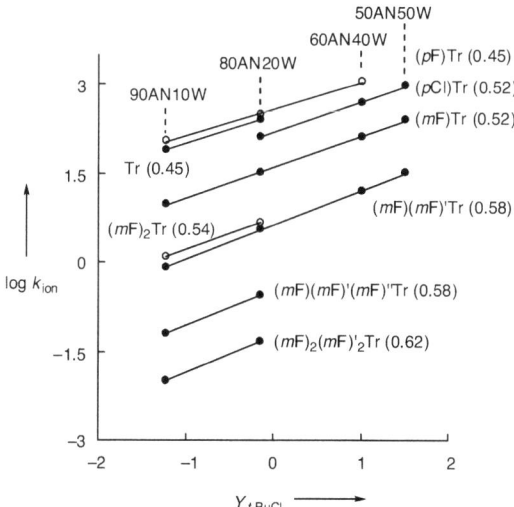

Figure 4.10. Winstein-Grunwald plot for the ionizations of triarylmethyl chlorides in aqueous acetonitrile, 25 °C. The numbers in parentheses represent the slopes $m$ of the correlation lines. Ionizing powers for solvent mixtures from ref. [19].

### 4.3.5. Hammett Analysis

The carbocationic character of the trityl residue in the transition state of the ionization does not only depend on the nature of the substitutents in the trityl moiety, but also on the solvent. This can be visualized by Hammett plots (Figures 4.11 and 4.12). In both series, trityl chlorides and bromides, the absolute values of $\rho$ increase with decreasing water fraction of the solvent, because the transition states become more carbocation-like (see also Figure 4.8).

It is interesting to note that the ionization rates of trityl bromides are generally more prone to substituent effects than those of trityl chlorides. Because of the higher nucleophilicities of bromide ions (see: 4.3.2. Ion Recombination), the transition states for trityl bromide ionizations are generally more carbocation-like, which is reflected by their more negative $\rho$ values, particularly in aqueous mixtures.

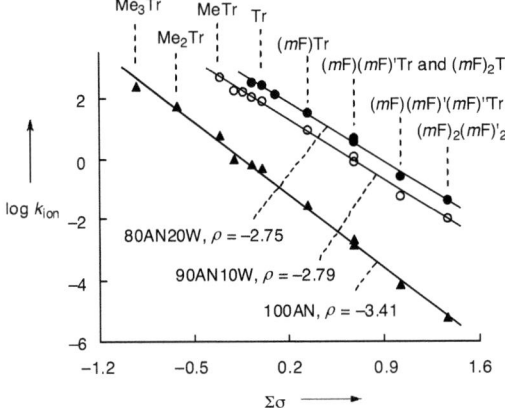

Figure 4.11. Hammett plot for the ionizations of triarylmethyl chlorides in aqueous acetonitrile, 25 °C; substituent constants (from ref. [21]), Cl: $\sigma_p^+ = 0.11$; F: $\sigma_m = 0.34$, $\sigma_p^+ = -0.07$; Me $\sigma_p^+ = -0.31$.

Because the ionization of a trityl bromide is generally faster than that of the corresponding chloride, the Bell-Evans-Polanyi principle[22] as well as the Hammond postulate predict earlier transition states for the ionizations of $Ar_3CBr$ than for $Ar_3CCl$.

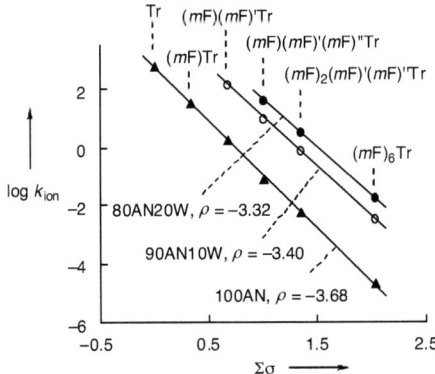

Figure 4.12. Hammett plot for the ionizations of triarylmethyl bromides in aqueous acetonitrile, 25 °C; substituent constants as in Figure 4.11.

However, the Hammett $\rho$ values in aqueous acetonitrile indicate a greater carbocationic character of the transition states in trityl bromide ionizations. This discrepancy is resolved by consideration of the reverse reactions: the faster ion recombinations with $Br^-$ have an earlier, i.e., more carbocation-like transition state than the combinations with $Cl^-$. It is the lower intrinsic barrier for the reactions with $Br^-$, which makes $Br^-$ both a better nucleofuge and a better nucleophile than $Cl^-$. As a consequence, the Bell-Evans-Polanyi principle and Hammond

## 4. Electrofugalities of Acceptor-Substituted Tritylium Ions 77

postulate must lead to contradicting predictions when the reactions are analyzed in the forward and backward sense.

### 4.3.6. Determination of Electrofugality Parameters $E_f$

Nucleofugality parameters $N_f$ and $s_f$ for a large number of leaving groups in various solvents have previously been derived from solvolyses of benzhydryl derivatives (Table 4.3).[3]

Table 4.3. Nucleofugality parameters of leaving groups $X^-$, derived from ionizations of benzhydryl derivatives.[a]

| nucleofuge | solvent | $N_f$ | $s_f$ |
|---|---|---|---|
| $F^-$ [b] | 80AN20W | −2.28 | 0.93 |
|  | 60AN40W | −1.43 | 0.84 |
|  | 80A20W | −2.73 | 1.07 |
| $Cl^-$ | 100AN [b] | 0.30 | 1.39 |
|  | 90AN10W | 2.23 | 1.08 |
|  | 80AN20W | 2.96 | 1.00 |
|  | 60AN40W | 3.84 | 0.96 |
|  | 100A [b] | −1.00 | 1.38 |
|  | 90A10W | 1.14 | 1.11 |
|  | 80A20W | 2.03 | 1.05 |
|  | 60A40W | 3.30 | 0.97 |
| $Br^-$ | 60AN40W | 5.23 | 0.99 |
|  | 90A10W | 2.29 | 1.01 |
| $AcO^-$ | 80AN20W | −4.52 | 1.11 |
|  | 60AN40W | −4.18 | 1.08 |
|  | 80A20W | −4.73 | 1.18 |
|  | 60A40W | −4.05 | 1.17 |
| $BzO^-$ | 80AN20W | −4.19 | 1.12 |
|  | 60AN40W | −3.92 | 1.02 |
|  | 80A20W | −4.46 | 1.17 |
|  | 60A40W | −3.89 | 1.15 |
| PNB [d] | 80AN20W | −3.41 | 0.98 |
|  | 60AN40W | −3.30 | 0.91 |
|  | 90A10W | −3.70 | 1.17 |
|  | 80A20W | −3.40 | 1.16 |
|  | 60A40W | −2.79 | 1.11 |

[a] From ref. [3] if not noted otherwise; [b] unpublished data; [c] from ref. [13]; [d] PNB = $p$-nitrobenzoate.

If equation (4.1) holds, plots of (log $k_{ion}$)/$s_f$ versus $N_f$ should yield straight lines with slopes of unity. Figure 4.13 shows that trityl chloride follows this correlation with a slope slightly bigger than unity (1.03). A good correlation is also found for the different carboxylates. However, this correlation has a larger slope (1.23) and does not coincide with the line for chloride. The data for trityl fluoride form a third line below the two other lines. It is the larger steric demand of the trityl group compared to the benzhydryl group, which makes carboxylates better leaving groups than chloride, and chloride a better leaving group than fluoride in trityl solvolyses than expected from the corresponding nucleofugalities based on benzhydryl ionizations (see Scheme 4.2).

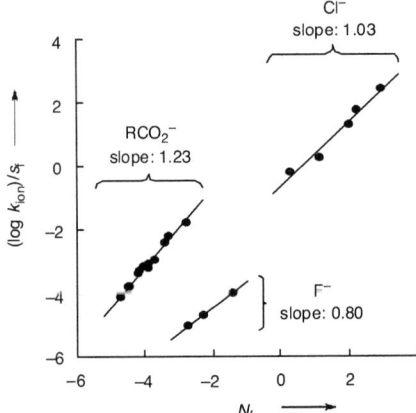

Figure 4.13. Plot of (log $k_{ion}$)/$s_f$ vs. $N_f$ for the ionizations of triphenylmethyl esters and halides; the kinetic data for the carboxylic ester ionizations are taken from Chapter 1.

Figure 4.14 shows that analogous plots for ring-substituted trityl chlorides and bromides also have slopes close to 1.0, which allows us to derive electrofugality parameters $E_f$ for tritylium ions (Table 4.4).[23] The last column of Table 4.4 demonstrates that $k_{exp}/k_{calc}$ varies between 0.3 and 4.8; hence, it is now possible to estimate absolute ionization rates of various trityl chlorides and bromides from the electrofugality parameters given in Table 4.4 and the previously reported nucleofugalitities $N_f$ and $s_f$ for Cl⁻ and Br⁻ in different solvents.

The power of this approach is highlighted by Table 4.5, where the nucleofugality parameters of Cl⁻ in methanol, ethanol, and 80 % aqueous ethanol (from ref. [3]), which have not been used for the evaluation of the tritylium electrofugalities, were employed to calculate ionization rate constants of trityl chloride in these solvents. As their deviations from Swain's experimental values[5f] are generally smaller than a factor of 3 (Table 4.5), one can conclude that the $E_f$ parameters in Table 4.4 combined with the previously reported $N_f$ and $s_f$

## 4. Electrofugalities of Acceptor-Substituted Tritylium Ions

parameters of $Cl^-$ and $Br^{-[3]}$ provide a simple method to estimate ionization rates of various trityl chlorides and bromides in a variety of different solvents.

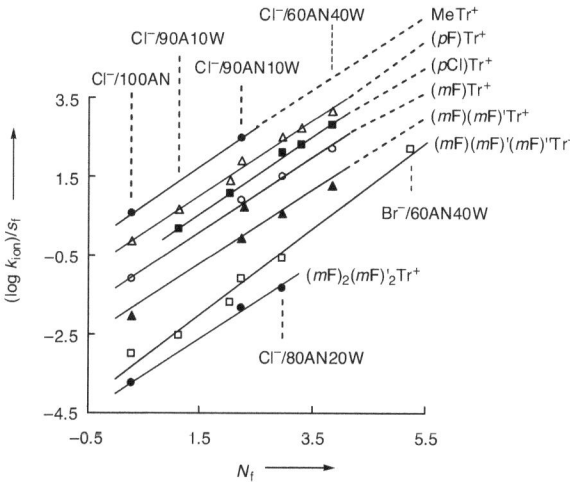

Figure 4.14. Plot of (log $k_{ion}$)/$s_f$ vs. $N_f$ for the ionizations of substituted trityl chlorides and bromides in aqueous acetonitrile and aqueous acetone, 25 °C. The slopes of the correlation lines vary from 0.93 to 1.08.

Table 4.4. Electrofugality parameters $E_f$ of substituted tritylium ions.

| electrofuge | $E_f$ | LG | solvent | $k_{calc}$ / $s^{-1}$ [a] | $k_{exp}/k_{calc}$ [b] |
|---|---|---|---|---|---|
| $Me_3Tr^+$ | 1.17 | $Cl^-$ | 100AN | $1.10 \times 10^2$ | 2.3 |
|  |  |  | 100A | 1.72 | 0.4 |
| $Me_2Tr^+$ | 0.57 | $Cl^-$ | 100AN | $1.62 \times 10^1$ | 3.2 |
|  |  |  | 100A | $2.55 \times 10^{-1}$ | 0.3 |
| $MeTr^+$ | 0.25 | $Cl^-$ | 100AN | 5.81 | 1.0 |
|  |  |  | 90AN10W | $4.77 \times 10^2$ | 1.0 |
| $Tr^+$ | −0.63 | $Cl^-$ | 100AN | $3.48 \times 10^{-1}$ | 1.4 |
|  |  |  | 90AN10W | $5.35 \times 10^1$ | 1.5 |
|  |  |  | 80AN20W | $2.14 \times 10^2$ | 1.2 |
|  |  |  | 90A10W | 3.68 | 0.5 |
|  |  |  | 80A20W | $2.95 \times 10^1$ | 0.7 |
| $(pCl)Tr^+$ | −0.96 | $Cl^-$ | 80AN20W | $1.00 \times 10^2$ | 1.3 |
|  |  |  | 60AN40W | $5.82 \times 10^2$ | 0.8 |

Table 4.4. Continued.

| electrofuge | $E_f$ | LG | solvent | $k_{calc} / s^{-1}$ [a] | $k_{exp}/k_{calc}$ [b] |
|---|---|---|---|---|---|
| (pCl)Tr$^+$ | –0.96 | Cl$^-$ | 90A10W | 1.58 | 1.0 |
| | | | 80A20W | $1.33 \times 10^1$ | 1.0 |
| | | | 60A40W | $1.86 \times 10^2$ | 0.9 |
| (pF)Tr$^+$ | –0.50 | Cl$^-$ | 100AN | $5.27 \times 10^{-1}$ | 1.2 |
| | | | 90AN10W | $7.39 \times 10^1$ | 1.5 |
| | | | 80AN20W | $2.88 \times 10^2$ | 1.1 |
| | | | 60AN40W | $1.61 \times 10^3$ | 0.7 |
| | | | 90A10W | 5.13 | 1.0 |
| | | | 80A20W | $4.04 \times 10^1$ | 0.8 |
| | | | 60A40W | $5.20 \times 10^2$ | 0.9 |
| (pF)$_2$Tr$^+$ | –0.20 | Cl$^-$ | 90AN10W | $1.55 \times 10^2$ | - |
| (pF)$_3$Tr$^+$ | –0.25 | Cl$^-$ | 100AN | 1.17 | 0.9 |
| | | | 90AN10W | $1.38 \times 10^2$ | 1.2 |
| (mF)Tr$^+$ | –1.43 | Cl$^-$ | 100AN | $2.69 \times 10^{-2}$ | 1.1 |
| | | | 90AN10W | 7.31 | 1.3 |
| | | | 80AN20W | $3.39 \times 10^1$ | 1.0 |
| | | | 60AN40W | $2.06 \times 10^2$ | 0.6 |
| (mF)(mF)'Tr$^+$ | –2.25 | Cl$^-$ | 100AN | $1.95 \times 10^{-3}$ | 0.7 |
| | | | 90AN10W | $9.51 \times 10^{-1}$ | 0.9 |
| | | | 80AN20W | 5.13 | 0.7 |
| | | | 60AN40W | $3.36 \times 10^1$ | 0.5 |
| | | Br$^-$ | 90A10W | 1.10 | 4.9 |
| (mF)$_2$Tr$^+$ | –2.21 | Cl$^-$ | 100AN | $2.21 \times 10^{-3}$ | 1.0 |
| | | | 90AN10W | 1.05 | 1.2 |
| | | | 80AN20W | 5.62 | 0.9 |
| (mF)(mF)'(mF)"Tr$^+$ | –3.42 | Cl$^-$ | 100AN | $4.40 \times 10^{-5}$ | 1.6 |
| | | | 90AN10W | $5.19 \times 10^{-2}$ | 1.2 |
| | | | 80AN20W | $3.47 \times 10^{-1}$ | 0.8 |
| | | | 90A10W | $2.95 \times 10^{-3}$ | 0.5 |
| | | | 80A20W | $3.47 \times 10^{-2}$ | 0.5 |
| | | Br$^-$ | 60AN40W | $6.19 \times 10^1$ | 2.4 |
| (mF)$_2$(mF)'(mF)"Tr$^+$ | –4.05 | Br$^-$ | 60AN40W | $1.47 \times 10^1$ | 1.1 |
| | | | 90A10W | $1.67 \times 10^{-2}$ | 0.9 |
| (mF)$_2$(mF)'$_2$Tr$^+$ | –4.11 | Cl$^-$ | 100AN | $5.06 \times 10^{-6}$ | 1.2 |
| | | | 90AN10W | $9.32 \times 10^{-3}$ | 1.1 |
| | | | 80AN20W | $7.08 \times 10^{-2}$ | 0.7 |

[a] From equation (4.1) using $E_f$ from this Table and $N_f/s_f$ from Table 4.3; [b] $k_{exp}$ from Table 4.2.

## 4. Electrofugalities of Acceptor-Substituted Tritylium Ions 81

While the ionization rates of trityl chlorides and bromides correlate with the nucleofugality paramaters $N_f/s_f$ derived from benzhydryl solvolyses (Figure 4.14), trityl carboxylates deviate. However, the synthetic chemist who might only be interested in the question whether a certain trityl carboxylate ionizes in a period of seconds (i.e., cannot be used), hours, or months (i.e., infinitely stable), can also employ the electrofugalities $E_f$ listed in Table 4.4 to derive approximate ionization rates of trityl carboxylates.

Table 4.5. Experimental ($k_{exp}$) and calculated ($k_{calc}$) ionization rate constants of triphenylmethyl chloride in different solvents, 25 °C.

| solvent | $k_{exp}$ / s$^{-1}$ [a] | $k_{calc}$ / s$^{-1}$ [b] | $k_{exp}/k_{calc}$ |
|---|---|---|---|
| MeOH | $1.59 \times 10^2$ | $1.81 \times 10^2$ | 0.9 |
| EtOH | 4.89 | $1.55 \times 10^1$ | 0.3 |
| 80EtOH20W | $5.36 \times 10^2$ | $3.84 \times 10^2$ | 1.4 |

[a] From ref. [5f]; [b] calculated with equation (4.1), $N_f/s_f$ for Cl$^-$ from ref. [3]: 2.91/0.99 (MeOH); 1.82/1.00 (EtOH); 3.24/0.99 (80EtOH20W).

Table 4.6 shows that the $E_f$ parameters for tritylium ions and $N_f/s_f$ parameters for carboxylate ions, which have been derived from benzhydryl solvolyses (Table 2 in ref. [3]), reproduce experimental ionization rate constants of trityl carboxylic esters with deviations of factors 15-71.

Table 4.6. Comparison between experimental and calculated ionization rate constants of substituted trityl carboxylates in different solvents, 25 °C.

| electrofuge | nucleofuge | solvent [a] | $k_{exp}$ / s$^{-1}$ | ref. | $k_{calc}$ / s$^{-1}$ [b] | $k_{exp}/k_{calc}$ |
|---|---|---|---|---|---|---|
| Me$_3$Tr$^+$ | AcO$^-$ | 80AN20W | $4.98 \times 10^{-3}$ | Table 1.7 | $1.91 \times 10^{-4}$ | 26 |
| | | 60AN40W | $1.77 \times 10^{-2}$ | Table 1.7 | $5.61 \times 10^{-4}$ | 32 |
| | BzO$^-$ | 80AN20W | $1.51 \times 10^{-2}$ | Table 1.8 | $4.15 \times 10^{-4}$ | 36 |
| | | 60AN40W | $4.55 \times 10^{-2}$ | Table 1.8 | $1.57 \times 10^{-3}$ | 29 |
| Me$_2$Tr$^+$ | AcO$^-$ | 80AN20W | $1.21 \times 10^{-3}$ | Table 1.7 | $4.13 \times 10^{-5}$ | 29 |
| | | 60AN40W | $5.62 \times 10^{-3}$ | Table 1.7 | $1.26 \times 10^{-4}$ | 45 |
| | BzO$^-$ | 80AN20W | $3.55 \times 10^{-3}$ | Table 1.8 | $8.82 \times 10^{-5}$ | 40 |
| | | 60AN40W | $1.05 \times 10^{-2}$ | Table 1.8 | $3.83 \times 10^{-4}$ | 27 |
| MeTr$^+$ | AcO$^-$ | 80AN20W | $3.59 \times 10^{-4}$ | Table 1.7 | $1.82 \times 10^{-5}$ | 20 |
| | | 60AN40W | $1.46 \times 10^{-3}$ | Table 1.7 | $5.70 \times 10^{-5}$ | 26 |
| | BzO$^-$ | 80AN20W | $8.08 \times 10^{-4}$ | Table 1.8 | $3.87 \times 10^{-5}$ | 21 |
| | | 60AN40W | $2.78 \times 10^{-3}$ | Table 1.8 | $1.81 \times 10^{-4}$ | 15 |

Table 4.6. Continued.

| electrofuge | nucleofuge | solvent [a] | $k_{exp}$ / s$^{-1}$ | ref. | $k_{calc}$ / s$^{-1}$ [b] | $k_{exp}/k_{calc}$ |
|---|---|---|---|---|---|---|
| Tr$^+$ | AcO$^-$ | 80AN20W | $5.88 \times 10^{-5}$ | Table 1.7 | $1.92 \times 10^{-6}$ | 31 |
| | | 60AN40W | $2.70 \times 10^{-4}$ | Table 1.7 | $6.39 \times 10^{-6}$ | 42 |
| | | 80A20W | $1.38 \times 10^{-5}$ | Table 1.10 | $4.73 \times 10^{-7}$ | 29 |
| | | 60A40W | $1.99 \times 10^{-4}$ | Table 1.10 | $3.35 \times 10^{-6}$ | 59 |
| | BzO$^-$ | 80AN20W | $1.67 \times 10^{-4}$ | Table 1.8 | $4.00 \times 10^{-6}$ | 42 |
| | | 60AN40W | $5.14 \times 10^{-4}$ | Table 1.8 | $2.29 \times 10^{-5}$ | 22 |
| | | 80A20W | $3.50 \times 10^{-5}$ | Table 1.10 | $1.08 \times 10^{-6}$ | 32 |
| | | 60A40W | $2.87 \times 10^{-4}$ | Table 1.10 | $6.34 \times 10^{-6}$ | 45 |
| | PNB$^-$ | 80AN20W | $4.19 \times 10^{-3}$ | Table 1.9 | $1.10 \times 10^{-4}$ | 38 |
| | | 60AN40W | $9.68 \times 10^{-3}$ | Table 1.9 | $2.65 \times 10^{-4}$ | 37 |
| | | 90A10W | $3.63 \times 10^{-4}$ | Table 1.10 | $8.59 \times 10^{-6}$ | 42 |
| | | 80A20W | $1.49 \times 10^{-3}$ | Table 1.10 | $2.11 \times 10^{-5}$ | 71 |
| | | 60A40W | $1.08 \times 10^{-2}$ | Table 1.10 | $1.60 \times 10^{-4}$ | 68 |
| ($p$Cl)Tr$^+$ | PNB$^-$ | 80AN20W | $2.22 \times 10^{-3}$ | Table 4.2 | $5.22 \times 10^{-5}$ | 43 |
| | | 60AN40W | $6.10 \times 10^{-3}$ | Table 4.2 | $1.33 \times 10^{-4}$ | 46 |
| ($m$F)Tr$^+$ | AcO$^-$ | 60AN40W | $3.45 \times 10^{-5}$ | Table 4.2 | $8.73 \times 10^{-7}$ | 40 |

[a] Solvents are given in v/v, AN = acetonitrile, A = acetone, W = water; [b] calculated with log $k_{ion} = s_f(N_f + E_f)$, $E_f$ parameters from Table 4.4, nucleofugality parameters $s_f/N_f$ of carboxylates from Table 4.3.

## 4.4. Conclusion

Ionization rate constants of substituted trityl chlorides, bromides, and carboxylates in various solvents could be determined by suppressing the recombination of separated ions by amine additives. Faster recombination of tritylium ions with Br$^-$ than with Cl$^-$ was not only observed in protic solvents but also in neat acetonitrile and acetone, indicating that chloride is not generally a stronger nucleophile than bromide in aprotic solvents, as generalized in common text books. Substituent variation in trityl halides affects the faster ionizations of trityl bromides more than the slower ionizations of trityl chlorides in contrast to the expectations based on the Bell-Evans-Polanyi principle and the Hammond postulate. The failure of these treatments is rationalized by the lower intrinsic barriers for the reactions involving bromide anions. Combination of the electrofugality parameters $E_f$ of tritylium ions in Table 4 with the benzhydrylium-derived nucleofuge-specific parameters $N_f$ and $s_f$ for Br$^-$ and Cl$^-$ have been found to yield reliable predictions of ionization rates of substituted trityl bromides and chlorides in different solvents (standard deviation of a factor of 1.4) on the basis

## 4. Electrofugalities of Acceptor-Substituted Tritylium Ions 83

of the correlation log $k_{ion} = s_f(E_f + N_f)$ [equation (4.1)]. Though the predictions for trityl carboxylates are less precise, a first orientation about ionization half-lives can be obtained from Figure 4.16. By arranging tritylium ions with increasing electrofugality from bottom to top, and leaving group/solvent combinations with increasing nucleofugality from right to left, one arrives at a presentation where trityl derivatives, which can be studied kinetically, are located in the green corridor, while highly labile or persistent systems are in the red and the blue section, respectively.

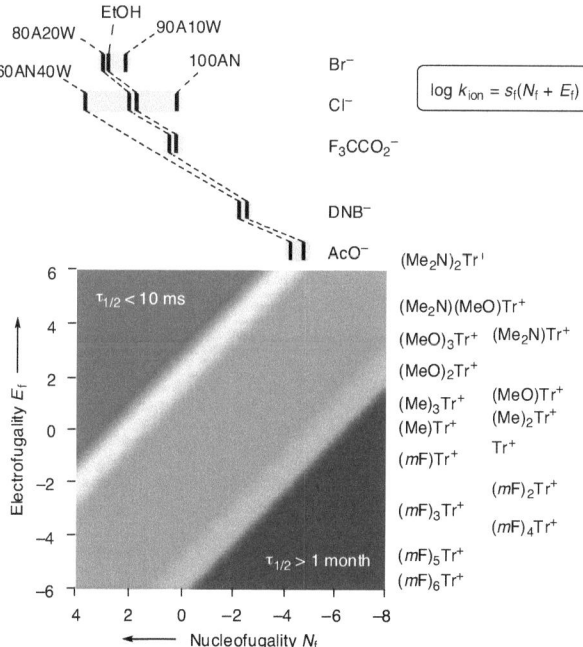

Figure 4.16. Graphical presentation of approximate ionization rates of trityl derivatives. The $E_f$ parameters for methoxy and dimethylamino-substituted tritylium ions have been estimated on the basis of their relative carboxylate hydrolyses (data in Chapter 1).

## 4.5. References

[1] a) H. Mayr, T. Bug, M. F. Gotta, N. Hering, B. Irrgang, B. Janker, B. Kempf, R. Loos, A. R. Ofial, G. Remennikov, H. Schimmel, *J. Am. Chem. Soc.* **2001**, *123*, 9500-9512; b) H. Mayr, A. R. Ofial, *J. Phys. Org. Chem.* **2008**, *21*, 584-595; c) H. Mayr, A. R. Ofial, *Nachr. Chem.* **2008**, *56*, 871-877; d) H. Mayr, *Angew. Chem.* **2011**, *123*, 3692-3698; *Angew. Chem. Int. Ed.* **2011**, *50*, 3612-3618.

[2] a) B. Denegri, S. Minegishi, O. Kronja, H. Mayr, *Angew. Chem.* **2004**, *116*, 2353-2356; *Angew. Chem. Int. Ed.* **2004**, *43*, 2302-2305; b) B. Denegri, A. Streiter, S. Juric, A. R. Ofial, O. Kronja, H. Mayr, *Chem. Eur. J.* **2006**, *12*, 1648-1656; *Chem. Eur. J.* **2006**, *12*, 5415; c) B. Denegri, A. Streiter, S. Juric, A. R. Ofial, O. Kronja, H. Mayr, *Chem. Eur. J.* **2006**, *12*, 1657-1666; d) C. Nolte, H. Mayr, *Eur. J. Org. Chem.* **2010**, 1435-1439.

[3] N. Streidl, B. Denegri, O. Kronja, H. Mayr, *Acc. Chem. Res.* **2010**, *43*, 1537-1549.

[4] a) H. C. Brown, *Science* **1946**, *103*, 385-387; b) C. J. M. Stirling, *Tetrahedron* **1985**, *41*, 1613-1666.

[5] a) A. C. Nixon, G. E. K. Branch, *J. Am. Chem. Soc.* **1936**, *58*, 492-498; b) G. S. Hammond, J. T. Rudesill, *J. Am. Chem. Soc.* **1950**, *72*, 2769-2770; c) C. G. Swain, R. B. Mosely, *J. Am. Chem. Soc.* **1955**, *77*, 3727-3731; d) C. G. Swain, R. B. Mosely, D. E. Bown, *J. Am. Chem. Soc.* **1955**, *77*, 3231-3737; e) C. G. Swain, T. E. C. Knee, A. MacLachlan, *J. Am. Chem. Soc.* **1960**, *82*, 6101-6104; f) C. G. Swain, A. MacLachlan, *J. Am. Chem. Soc.* **1960**, *82*, 6095-6101; g) S. Winstein, B. R. Appel, *J. Am. Chem. Soc.* **1964**, *86*, 2718-2720; h) S. Winstein, B. R. Appel, *J. Am. Chem. Soc.* **1964**, *86*, 2720-2721; i) K. T. Leffek, *Can. J. Chem.* **1970**, *48*, 1-6.

[6] a) M. Smith, D. H. Rammler, I. H. Goldberg, H. G. Khorana, *J. Am. Chem. Soc.* **1962**, *84*, 430-440; b) H. Schaller, G. Weimann, B. Lerch, H. G. Khorana, *J. Am. Chem. Soc.* **1963**, *85*, 3821-3827; c) C. Bleasdale, S. B. Ellwood, B. T. Golding, *J. Chem. Soc. Perkin Trans. 1* **1990**, 803-805; d) M. Sekine, T. Mori, T. Wada, *Tetrahedron Lett.* **1993**, *34*, 8289-8292; e) M. Sekine, T. Hata, *J. Org. Chem.* **1987**, *52*, 946-948; f) A. P. Henderson, J. Riseborough, C. Bleasdale, W. Clegg, M. R. J. Elsegood, B. T. Golding, *J. Chem. Soc. Perkin Trans. 1* **1997**, 3407-3414; g) P. G. M. Wuts, T. W. Greene, *Greene's Protective Groups in Organic Synthesis*, 4th Ed., Wiley-Interscience, Hoboken, **2007**.

[7] R. A. McClelland, V. M. Kanagasabapathy, N. S. Banait, S. Steenken, *J. Am. Chem. Soc.* **1989**, *111*, 3966-3972.

[8] I. I. Schuster, A. K. Colter, R. J. Kurland, *J. Am. Chem. Soc.* **1968**, *90*, 4679-4687.

[9] S. V. Kulkarni, R. Schure, R. Filler, *J. Am. Chem. Soc.* **1973**, *95*, 1859-1864.

[10] a) D. J. Raber, J. M. Harris, P. v. R. Schleyer, in *Ions and Ion Pairs in Organic Reactions* (Ed.: M. Swarc) Wiley, **1974**, Vol. 2, pp. 247-374; b) S. Winstein, E. Clippinger, A. H. Fainberg, R. Heck, G. C. Robinson, *J. Am. Chem. Soc.* **1956**, *78*, 328-335.

[11] a) N. Streidl, A. Antipova, H. Mayr, *J. Org. Chem.* **2009**, *74*, 7328-7334; b) H. Mayr, A. R. Ofial, *Pure Appl. Chem.* **2009**, *81*, 667-683.

[12] S. Minegishi, R. Loos, S. Kobayashi, H. Mayr, *J. Am. Chem. Soc.* **2005**, *127*, 2641-2649.

[13] N. Streidl, H. Mayr, *Eur. J. Org. Chem.* **2011**, 2498-2506.

[14] D. S. Noyce, J. A. Virgilio, *J. Org. Chem.* **1972**, *37*, 2643-2647.

[15] H. F. Schaller, A. A. Tishkov, X. Feng, H. Mayr, *J. Am. Chem. Soc.* **2008**, *130*, 3012-3022.

[16] a) F. A. Carey, R. J. Sundberg, in *Advanced Organic Chemistry, Part A: Structure and Mechanisms*, 5$^{th}$ Ed., Springer, **2007**, p. 412; b) M. B. Smith, J. March, in *March's Advanced Organic Chemistry*, 6$^{th}$ Ed., Wiley-Interscience, Hoboken, NJ, **2007**, p. 491.

[17] a) D. Landini, A. Maia, F. Montanari, *J. Am. Chem. Soc.* **1978**, *100*, 2796-2801; b) R. Fuchs, L. L. Cole, *J. Am. Chem. Soc.* **1973**, *95*, 3194-3197.

[18] S. Minegishi, H. Mayr, *J. Am. Chem. Soc.* **2003**, *125*, 286-295.

[19] T. W. Bentley, J.-P. Dau-Schmidt, G. Llewellyn, H. Mayr, *J. Org. Chem.* **1992**, *57*, 2387-2392.

[20] G. S. Hammond, *J. Am. Chem. Soc.* **1955**, *77*, 334-338.

[21] C. Hansch, A. Leo, R. W. Taft, *Chem. Rev.* **1991**, *91*, 165-195.

[22] M. J. S. Dewar, R. C. Dougherty, *The PMO Theory of Organic Chemistry*, Plenum Press, New York, **1975**.

[23] The non-linear solver What's*Best*! 4.0 (Lindo Systems Inc.) has been used for this purpose.

## 5. Towards a General Hydride Donor Ability Scale

### 5.1. Introduction

Hydride transfers play an important role in organic chemistry. They occur in biochemical redox processes like in the NAD(P)H/NAD(P)$^+$ couple,[1] and are found in initiation, chain transfer, and termination reactions of cationic polymerizations.[2] Key steps in the Cannizzaro, Meerwein-Ponndorf-Verley, Oppenauer, Tishchenko, Leuckart-Wallach, and the Sommelet reaction consist of hydride transfers, and the conversion of ketones or aldehydes into alcohols by use of metal hydrides such as LiAlH$_4$ or NaBH$_4$ is one of the most important methods in the chemist's toolbox.[3] An ordering principle for these reactions is now reported.

The linear free energy relationship (5.1), where $E$ is an electrophilicity, $N$ a nucleophilicity, and $s_N$ a nucleophile-dependent slope parameter, was introduced in 1994, and has been shown to be useful in predicting bimolecular rate constants in many electrophile-nucleophile combination reactions.[4,5]

Nucleophiles as different as allyl metal compounds, alkenes, alkynes, diazo compounds, amines, carbanions, phosphanes, halide anions, alcohols, and sulfur ylides have meanwhile been characterized by nucleophilicity parameters $N$ and $s_N$. These parameters are based on reactions with benzhydrylium ions and quinone methides as reference electrophiles, which are characterized by $E$, and the nucleophilicity scale thus established covers a reactivity range from $N \approx -5$ for non-activated arenes to $N \approx 30$ for cyano-substituted benzyl anions.

$$\log k \ (20 \ °C) = s_N(N + E) \quad\quad\quad\quad (5.1)$$

Several hydride donors have previously been studied in reactions with benzhydrylium ions, and nucleophilicity parameters $N/s_N$ have been evaluated for various dihydropyridines, hydrocarbons, silanes, germanes, stannanes and borohydrides.[6]

However, most comparisons of hydride donor abilities in the literature were based on rate constants for hydride transfers to the triphenylmethyl cation,[7] and the question arose whether it is possible to include these hydride donors in our comprehensive nucleophilicity scale. For this purpose it was necessary to analyze the validity of equation (5.1) for the reactions of tritylium ions with hydride donors.

## 5. Towards a General Hydride Donor Ability Scale

Whereas equation (5.1) does not hold for combinations of π-nucleophiles (e.g. alkenes) with the sterically shielded tritylium ions, it has successfully been employed for reactions of tritylium ions with n-nucleophiles, like water and amines.[8]

However, while the addition of the alkene allyltrimethylsilane proceeds much faster to the secondary *p*-anisylphenyl carbenium ion than to the tertiary diphenylneopentyl carbenium ion, both carbocations exhibit the same electrophilic reactivity toward the hydride-donor dimethyl-phenylsilane (Table 5.1).

Table 5.1. Second-order rate constants for the reactions of carbocations with $HSiMe_2Ph$ and allyltrimethylsilane (L mol$^{-1}$ s$^{-1}$, $CH_2Cl_2$, –70 °C, from ref. [9]).

| | $HSiMe_2Ph$ | 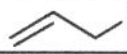 $SiMe_3$ (allyl) |
|---|---|---|
| 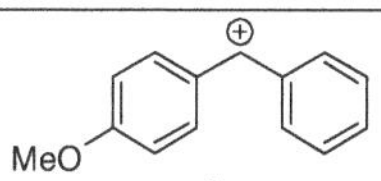 MeO-C₆H₄-CH⁺-Ph | 149 | 187 |
| 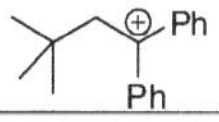 (neopentyl)C⁺(Ph)(Ph) | 148 | 0.11 |

This indicates that sterical arguments seem to be less relevant for hydride transfers from silanes to carbocations than for alkylations of carbocations by alkenes.

In previous work it has been demonstrated that the $N$ and $s_N$ parameters of N-benzyl-1,4-dihydronicotinamide, derived from its reactions with benzhydrylium ions, provide reliable predictions of the rates of its reactions with substituted tritylium ions.[6]

The rate constants for the reactions of a series of substituted tritylium ions with triphenylsilane as hydride donor have been used to evaluate $E$-parameters for fluoro-substituted tritylium ions (Chapter 3). These data will be implemented here to test the applicability of equation (5.1) with respect to hydride transfers to tritylium ions in a more general way (Scheme 5.1).

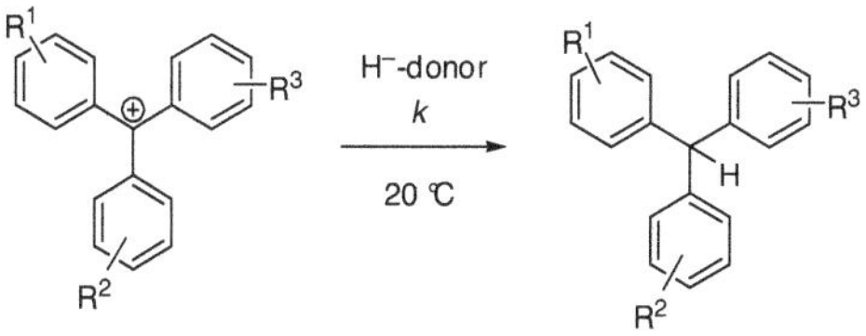

Scheme 5.1. Reduction of tritylium ions by hydride donors.

## 5.2. Methodology and Results

In order to determine reactivity parameters $N$ and $s_N$ of hydride donors, the kinetics of their reactions with substituted benzhydrylium ions (Table 5.2) were analyzed. The latter were either used as stable tetrafluoroborate salts, or, when such salts could not be isolated, were generated in solution by mixing the corresponding benzhydryl chloride with an excess of Lewis acid (GaCl$_3$). The amount of GaCl$_3$ was shown not to affect the rate constants of hydride transfers from HSiPh$_3$ to tritylium ions (see Chapter 3).

Table 5.2. Benzhydrylium ions and their electrophilicity parameters $E$ (from ref. [4]).

| system | abbreviation | $E$ |
|---|---|---|
|  | (ani)PhCH$^+$ | 2.11 |
|  | (ani)(tol)CH$^+$ | 1.48 |
|  | (ani)$_2$CH$^+$ | 0.00 |
|  | (fur)$_2$CH$^+$ | −1.36 |
|  | (pfa)$_2$CH$^+$ | −3.14 |
|  | (mfa)$_2$CH$^+$ | −3.85 |

All reactions were found to be of first-order in each reactant, according to the second-order rate law (5.2). As the concentrations of the hydride donors exceeded those of the benzhydrylium ions by factors > 10, mono-exponential decays of the carbocation absorbances were observed, from which the first-order rate constants $k_{obs}$ were determined (Figure 5.1a). Plots of $k_{obs}$ against the nucleophile concentrations (Figure 5.1b) were linear with the second-order rate constants $k$ (listed in Table 5.3) being the slopes of the correlation lines.

$$-d[\text{El}]/dt = k[\text{El}][\text{Nu}] \qquad (5.2)$$

# 5. Towards a General Hydride Donor Ability Scale

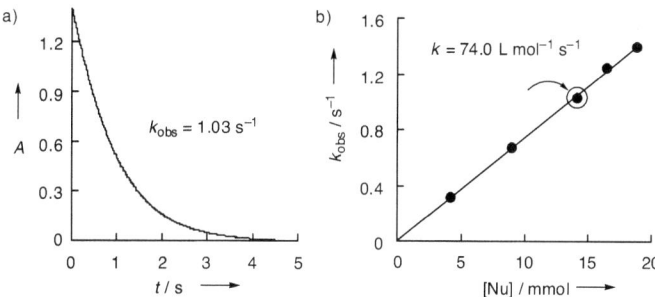

Figure 5.1. a) Decay of the absorbance at $\lambda = 512$ nm during the reaction of 4,4'-dimethoxybenzhydryl cations ($c_0 = 2.59 \times 10^{-5}$ M, generated from the corresponding chlorides with $GaCl_3$), with $HSiPh_3$ ($c_0 = 1.42 \times 10^{-2}$ M), $CH_2Cl_2$, 20 °C; b) plot of first-order rate constants $k_{obs}$ vs. $[HSiPh_3]_0$.

Table 5.3. Second-order rate constants of hydride transfers to substituted benzhydrylium ions ($CH_2Cl_2$, 20 °C), and nucleophilicity parameters $N / s_N$.

| hydride donor | acceptor | solvent | $k$ [L mol$^{-1}$ s$^{-1}$] | $N / s_N$ |
|---|---|---|---|---|
| $HSiBu_3$ | $(ani)(tol)CH^+$ | $CH_2Cl_2$ | $1.12 \times 10^4$ | 3.99 / 0.74 [a] |
|  | $(ani)_2CH^+$ | $CH_2Cl_2$ | $8.86 \times 10^2$ |  |
|  | $(fur)_2CH^+$ | $CH_2Cl_2$ | $8.85 \times 10^1$ |  |
| $HSiMe_2Ph$ | $(ani)(tol)CH^+$ | $CH_2Cl_2$ | $6.02 \times 10^3$ | 3.55 / 0.75 [a] |
|  | $(ani)_2CH^+$ | $CH_2Cl_2$ | $4.34 \times 10^2$ |  |
|  | $(fur)_2CH^+$ | $CH_2Cl_2$ | $4.49 \times 10^1$ |  |
| $HSiEt_3$ | $(ani)(tol)CH^+$ | $CH_2Cl_2$ | $4.87 \times 10^3$ | 3.48 / 0.74 [a] |
|  | $(ani)_2CH^+$ | $CH_2Cl_2$ | $3.98 \times 10^2$ |  |
|  | $(fur)_2CH^+$ | $CH_2Cl_2$ | $3.76 \times 10^1$ |  |
| $HSiPh_3$ | $(ani)(tol)CH^+$ | $CH_2Cl_2$ | $1.02 \times 10^3$ | 2.65 / 0.72 [a] |
|  | $(ani)_2CH^+$ | $CH_2Cl_2$ | $7.40 \times 10^1$ |  |
|  | $(fur)_2CH^+$ | $CH_2Cl_2$ | 9.01 |  |
| $HSi(SiMe_3)_3$ | $(ani)(tol)CH^+$ | $CH_2Cl_2$ | $8.72 \times 10^3$ | 3.59 / 0.81 |
|  | $(ani)_2CH^+$ | $CH_2Cl_2$ | $9.79 \times 10^2$ |  |
|  | $(fur)_2CH^+$ | $CH_2Cl_2$ | $8.89 \times 10^1$ |  |
|  | $(pfa)_2CH^+$ | $CH_2Cl_2$ | 1.99 |  |
|  | $(mfa)_2CH^+$ | $CH_2Cl_2$ | $5.33 \times 10^{-1}$ |  |
| $HSiH_2Ph$ | – | $CH_2Cl_2$ | – | 0.25 / 0.67 [b] |
| $HSnBu_3$ | – | $CH_2Cl_2$ | – | 9.96 / 0.55 [b] |
| $^nBu_4Sn$ | $(ani)PhCH^+$ | $CH_2Cl_2$ | $8.99 \times 10^1$ | –0.30 / 1.07 |
|  | $(ani)(tol)CH^+$ | $CH_2Cl_2$ | $1.74 \times 10^1$ |  |

Table 5.3. Continued.

| hydride donor | | acceptor | solvent | $k$ [L mol$^{-1}$ s$^{-1}$] | $N$ / $s_N$ |
|---|---|---|---|---|---|
| $^n$Bu$_4$Sn | | (ani)$_2$CH$^+$ | CH$_2$Cl$_2$ | 4.86 × 10$^{-1}$ | |
| Bu$_4$N$^+$BH$_4^-$ | | - | DMSO | - | 14.94 / 0.79 [c] |
| | 1a | - | CH$_2$Cl$_2$ | - | 0.52 / 0.97 [d] |
| | 1b | - | CH$_2$Cl$_2$ | - | 0.09 / 0.98 [d] |
| | 1c | - | CH$_2$Cl$_2$ | - | –0.07 / 1.03 [d] |
| | 1d | - | CH$_2$Cl$_2$ | - | –0.86 / 0.92 [d] |
| | 1e | (ani)PhCH$^+$ | CH$_2$Cl$_2$ | 7.36 | –1.06 / 0.81 |
| | | (ani)(tol)CH$^+$ | CH$_2$Cl$_2$ | 2.15 | |
| | | (ani)$_2$CH$^+$ | CH$_2$Cl$_2$ | 1.39 × 10$^{-1}$ | |
| | 1f | - | CH$_2$Cl$_2$ | - | –0.88 / 0.94 [d] |
| | 1g | - | CH$_2$Cl$_2$ | - | –0.74 / 0.99 [d] |
| | 1h | - | CH$_2$Cl$_2$ | - | 0.64 / 0.97 [d] |
| | 1i | - | CH$_2$Cl$_2$ | - | 8.67 / 0.82 [c] |
| | | - | 90W10AN | - | 11.35 / 0.66 [c] |

[a] Slightly deviating nucleophilicity parameters have been reported in ref. [4]; these parameters were based on measurements at –70 °C, and approximated activation parameters were used to derive rate constants for 20 °C. The values determined in the present work can be regarded as being more precise; [b] from ref. [4]; [c] from ref. [6]; [d] from ref. [10].

Plots of log $k$ vs. the electrophilicity parameters $E$ of the benzhydrylium ions (Figure 5.2) gave rise to linear correlations, and delivered the $N$ parameters of the hydride donors as negative intercepts on the abscissae, as well as the $s_N$ parameters as slopes of the correlation lines (last column of Table 5.3).

## 5. Towards a General Hydride Donor Ability Scale

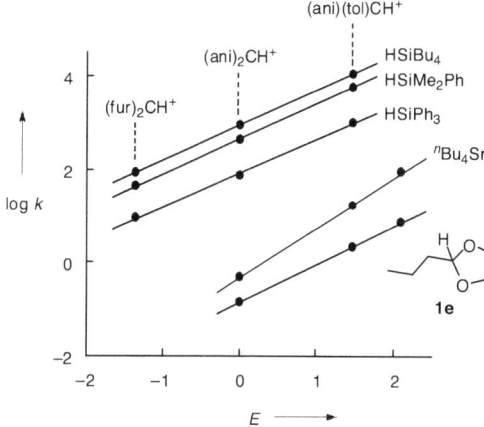

Figure 5.2. Plot of log $k$ for hydride transfers to substituted benzhydrylium ions vs. the corresponding electrophilicities $E$.

Table 5.4. Tritylium ions and their empirical electrophilicity parameters $E$.

| $R^1, R^2, R^{3\,[a]}$ | abbreviation | $E$ |
|---|---|---|
| $(m\text{-F})_2, m\text{-F}, m\text{-F}$ | $(m\text{F})_2(m\text{F})'(m\text{F})''\text{Tr}^+$ | 2.54 [b] |
| $(m\text{-F})_2, (m\text{-F})_2, \text{H}$ | $(m\text{F})_2(m\text{F})'_2\text{Tr}^+$ | 2.54 [c] |
| $m\text{-F}, m\text{-F}, m\text{-F}$ | $(m\text{F})(m\text{F})'(m\text{F})''\text{Tr}^+$ | 2.07 [b] |
| $m\text{-F}, m\text{-F}, \text{H}$ | $(m\text{F})(m\text{F})'\text{Tr}^+$ | 1.54 [b] |
| $(m\text{-F})_2, \text{H}, \text{H}$ | $(m\text{F})_2\text{Tr}^+$ | 1.54 [d] |
| $m\text{-F}, \text{H}, \text{H}$ | $(m\text{F})\text{Tr}^+$ | 1.01 [b] |
| $p\text{-F}, \text{H}, \text{H}$ | $(p\text{F})\text{Tr}^+$ | 0.35 [b] |
| $p\text{-F}, p\text{-F}, \text{H}$ | $(p\text{F})_2\text{Tr}^+$ | 0.17 [b] |
| $p\text{-F}, p\text{-F}, p\text{-F}$ | $(p\text{F})_3\text{Tr}^+$ | 0.05 [b] |
| $m\text{-Cl}, m\text{-Cl}, m\text{-Cl}$ | $(m\text{Cl})(m\text{Cl})'(m\text{Cl})''\text{Tr}^+$ | 1.99 [b] |
| $m\text{-Cl}, \text{H}, \text{H}$ | $(m\text{Cl})\text{Tr}^+$ | 1.06 [b] |
| $\text{H}, \text{H}, \text{H}$ | $\text{Tr}^+$ | 0.51 [e] |
| $p\text{-Me}, \text{H}, \text{H}$ | $\text{MeTr}^+$ | −0.13 [e] |
| $p\text{-Me}, p\text{-Me}, \text{H}$ | $\text{Me}_2\text{Tr}^+$ | −0.70 [e] |
| $p\text{-Me}, p\text{-Me}, p\text{-Me}$ | $\text{Me}_3\text{Tr}^+$ | −1.21 [e] |
| $p\text{-MeO}, \text{H}, \text{H}$ | $(\text{MeO})\text{Tr}^+$ | −1.59 [b] |
| $p\text{-MeO}, p\text{-MeO}, \text{H}$ | $(\text{MeO})_2\text{Tr}^+$ | −3.04 [e] |
| $p\text{-MeO}, p\text{-MeO}, p\text{-MeO}$ | $(\text{MeO})_3\text{Tr}^+$ | −4.35 [e] |
| $p\text{-Me}_2\text{N}, \text{H}, \text{H}$ | $(\text{Me}_2\text{N})\text{Tr}^+$ | −7.93 [e] |
| $p\text{-Me}_2\text{N}, p\text{-Me}_2\text{N}, \text{H}$ | $(\text{Me}_2\text{N})_2\text{Tr}^+$ | −10.29 [e] |
| $p\text{-Me}_2\text{N}, p\text{-Me}_2\text{N}, p\text{-Me}_2\text{N}$ | $(\text{Me}_2\text{N})_3\text{Tr}^+$ | −11.26 [e] |

[a] For the substitution pattern see Scheme 5.1; [b] from Chapter 3; [c] assumed to be the same as that for $(m\text{F})_2(m\text{F})'(m\text{F})''\text{Tr}^+$; [d] assumed to be the same as that for $(m\text{F})(m\text{F})'\text{Tr}^+$; [e] from ref. [8].

The rates of hydride transfers from the donors in Table 5.3 to the tritylium ions in Table 5.4 have been determined at 20 °C (Scheme 5.1), following the procedure described for benzhydrylium ions in Figure 5.1. Hereby, the *meta*-fluoro substituted systems have been generated in the same way as the benzhydrylium ions, i.e., the corresponding trityl halides (chlorides or bromides) were mixed in solution with an excess of Lewis acid ($GaCl_3$). The donor-substituted systems have been introduced as isolated salts, mainly tetrafluoroborates.

Table 5.5 compares the second-order rate constants thus obtained ($k_{exp}$) with those calculated ($k_{calc}$) by equation (5.1) from the electrophilicity parameters $E$ in Table 5.4 and the nucleophilicity parameters $N/s_N$ of the hydride donors in Table 5.3.

Table 5.5. Experimental and calculated second-order rate constants for hydride transfers to tritylium ions ($CH_2Cl_2$, 20 °C).

| acceptor | source | donor | $k_{exp}$ [L mol$^{-1}$ s$^{-1}$] | $k_{calc}$ [L mol$^{-1}$ s$^{-1}$] | $k_{calc}/k_{exp}$ |
|---|---|---|---|---|---|
| $(mF)_2(mF)'_2Tr^+$ | R-Cl + $GaCl_3$ | **1f** | 5.24 | 3.63 × 10$^1$ | 6.9 |
| $(mF)_2(mF)'(mF)''Tr^+$ | R-Br + $GaCl_3$ | HSiEt$_3$ | 4.62 × 10$^3$ | 2.85 × 10$^4$ | 6.2 |
| | R-Br + $GaCl_3$ | HSiPh$_3$ | 6.11 × 10$^{3\ [a]}$ | 5.46 × 10$^3$ | 0.9 |
| | R-Br + $GaCl_3$ | **1a** | 1.40 × 10$^3$ | 9.29 × 10$^2$ | 0.7 |
| $(mF)(mF)'(mF)''Tr^+$ | R-Br + $GaCl_3$ | HSiEt$_3$ | 2.05 × 10$^3$ | 1.28 × 10$^4$ | 6.2 |
| | R-Br + $GaCl_3$ | HSiPh$_3$ | 2.11 × 10$^{3\ [a]}$ | 2.50 × 10$^3$ | 1.2 |
| | R-Br + $GaCl_3$ | **1a** | 3.19 × 10$^2$ | 3.25 × 10$^2$ | 1.0 |
| $(mF)(mF)'Tr^+$ | R-Br + $GaCl_3$ | HSiEt$_3$ | 7.90 × 10$^2$ | 5.19 × 10$^3$ | 6.6 |
| | R-Br + $GaCl_3$ | HSiBu$_3$ | 1.47 × 10$^3$ | 1.24 × 10$^4$ | 8.4 |
| | R-Br + $GaCl_3$ | HSiPh$_3$ | 6.57 × 10$^{2\ [a]}$ | 1.04 × 10$^3$ | 1.6 |
| | R-Br + $GaCl_3$ | Bu$_4$Sn | 3.80 | 2.12 × 10$^1$ | 5.6 |
| | R-Br + $SnCl_4$ | **1a** | 4.02 × 10$^1$ | 9.96 × 10$^1$ | 2.5 |
| $(mF)_2Tr^+$ | R-Cl + $GaCl_3$ | Bu$_4$Sn | 3.09 | 2.12 × 10$^1$ | 6.9 |
| $(mF)Tr^+$ | R-Br + $GaCl_3$ | HSiEt$_3$ | 3.29 × 10$^2$ | 2.10 × 10$^3$ | 6.4 |
| | R-Br + $GaCl_3$ | HSiBu$_3$ | 6.41 × 10$^2$ | 5.01 × 10$^3$ | 7.8 |
| | R-Br + $GaCl_3$ | HSiPh$_3$ | 1.99 × 10$^{2\ [a]}$ | 4.32 × 10$^2$ | 2.2 |
| | R-Br + $GaCl_3$ | Bu$_4$Sn | 5.44 × 10$^{-1}$ | 5.75 | 11 |
| | R-Cl + $GaCl_3$ | **1a** | 9.51 | 3.05 × 10$^1$ | 3.2 |
| | R-Cl + $SnCl_4$ | **1a** | 9.07 | 3.05 × 10$^1$ | 3.4 |
| $(pF)Tr^+$ | BF$_4^-$ | HSiEt$_3$ | 7.75 × 10$^1$ | 6.83 × 10$^2$ | 8.8 |
| | BF$_4^-$ | HSiPh$_3$ | 4.58 × 10$^{1\ [a]}$ | 1.45 × 10$^2$ | 3.2 |
| | BF$_4^-$ | **1a** | 1.27 | 6.98 | 5.5 |
| $(pF)_2Tr^+$ | BF$_4^-$ | HSiEt$_3$ | 4.68 × 10$^1$ | 5.02 × 10$^2$ | 11 |
| | BF$_4^-$ | HSiBu$_3$ | 8.06 × 10$^1$ | 1.20 × 10$^3$ | 15 |

## 5. Towards a General Hydride Donor Ability Scale

Table 5.5. Continued.

| acceptor | source | donor | $k_{exp}$ [L mol$^{-1}$ s$^{-1}$] | $k_{calc}$ [L mol$^{-1}$ s$^{-1}$] | $k_{calc}/k_{exp}$ |
|---|---|---|---|---|---|
| (pF)$_2$Tr$^+$ | BF$_4^-$ | HSiPh$_3$ | $3.08 \times 10^1$ [a] | $1.07 \times 10^2$ | 3.5 |
| | BF$_4^-$ | 1a | $8.73 \times 10^{-1}$ | 4.67 | 5.3 |
| (pF)$_3$Tr$^+$ | R-Cl + GaCl$_3$ | HSiEt$_3$ | $2.65 \times 10^1$ | $4.09 \times 10^2$ | 15 |
| | R-Cl + GaCl$_3$ | HSiPh$_3$ | $2.36 \times 10^1$ [a] | $8.79 \times 10^1$ | 3.7 |
| (mCl)(mCl)'(mCl)''Tr$^+$ | R-Cl + TiCl$_4$ | 1b | $5.38 \times 10^1$ [b] | $1.09 \times 10^2$ | 2.0 |
| | R-Cl + TiCl$_4$ | 1g | $4.16$ [b] | $1.73 \times 10^1$ | 4.2 |
| | R-Cl + TiCl$_4$ | HSiH$_2$Ph | $4.59$ [b] | $3.17 \times 10^1$ | 6.9 |
| (mCl)Tr$^+$ | R-Cl + TiCl$_4$ | 1b | $2.33$ [b] | $1.34 \times 10^1$ | 5.8 |
| Tr$^+$ | BF$_4^-$ | HSiEt$_3$ | $1.36 \times 10^2$ [c] | $8.97 \times 10^2$ | 6.6 |
| | R-Cl + TMSOTf | HSiEt$_3$ | $1.29 \times 10^2$ | $8.97 \times 10^2$ | 7.0 |
| | BF$_4^-$ | HSiBu$_3$ | $2.75 \times 10^2$ | $2.14 \times 10^3$ | 7.8 |
| | BF$_4^-$ | HSiPh$_3$ | $7.55 \times 10^1$ [a] | $1.88 \times 10^2$ | 2.5 |
| | BF$_4^-$ | HSiMe$_2$Ph | $1.82 \times 10^2$ [d] | $1.11 \times 10^3$ | 6.1 |
| | BF$_4^-$ | HSi(SiMe$_3$)$_3$ | $9.78 \times 10^2$ | $2.09 \times 10^3$ | 2.1 |
| | BF$_4^-$ | Bu$_4$Sn | $6.7 \times 10^{-2}$ [e] | 1.68 | 25 |
| | BF$_4^-$ | 1a | $1.77$ [f] | 9.98 | 5.6 |
| | R-Cl + BCl$_3$ | 1b | $3.17 \times 10^{-1}$ [b] | 3.87 | 12 |
| | R-Cl + BCl$_3$/TiCl$_4$ | 1c | $8.15 \times 10^{-2}$ [b] | 2.84 | 35 |
| | R-Cl + BCl$_3$ | 1d | $1.43 \times 10^{-2}$ [b] | $4.76 \times 10^{-1}$ | 33 |
| | SbCl$_6^-$ | 1h | $1.2$ [g] | $1.30 \times 10^1$ | 11 |
| | AsF$_6^-$ | 1h | $1.9$ [g] | $1.30 \times 10^1$ | 6.8 |
| MeTr$^+$ | BF$_4^-$ | HSiEt$_3$ | $3.69 \times 10^1$ | $3.01 \times 10^2$ | 8.2 |
| | BF$_4^-$ | HSiBu$_3$ | $7.19 \times 10^1$ | $7.18 \times 10^2$ | 10 |
| | BF$_4^-$ | HSiPh$_3$ | $1.56 \times 10^1$ [a] | $6.52 \times 10^1$ | 4.2 |
| | BF$_4^-$ | HSiMe$_2$Ph | $5.36 \times 10^1$ | $3.67 \times 10^2$ | 6.8 |
| | BF$_4^-$ | HSi(SiMe$_3$)$_3$ | $3.19 \times 10^2$ | $6.35 \times 10^2$ | 2.0 |
| Me$_2$Tr$^+$ | BF$_4^-$ | HSiEt$_3$ | $1.08 \times 10^1$ | $1.14 \times 10^2$ | 11 |
| | BF$_4^-$ | HSiBu$_3$ | $2.31 \times 10^1$ [h] | $2.72 \times 10^2$ | 12 |
| | BF$_4^-$ | HSiPh$_3$ | $4.02$ [a] | $2.54 \times 10^1$ | 6.3 |
| | BF$_4^-$ | HSiMe$_2$Ph | $1.52 \times 10^1$ | $1.37 \times 10^2$ | 9.0 |
| | BF$_4^-$ | HSi(SiMe$_3$)$_3$ | $6.5 \times 10^1$ [e] | $2.19 \times 10^2$ | 3.4 |
| | BF$_4^-$ | HSnBu$_3$ | $6.19 \times 10^4$ | $1.24 \times 10^5$ | 2.0 |
| Me$_3$Tr$^+$ | BF$_4^-$ | HSiEt$_3$ | 3.43 | $4.78 \times 10^1$ | 14 |
| | BF$_4^-$ | HSiBu$_3$ | 6.08 | $1.14 \times 10^2$ | 19 |
| | BF$_4^-$ | HSiPh$_3$ | $1.21$ [a] | $1.09 \times 10^1$ | 9.0 |
| | BF$_4^-$ | HSiMe$_2$Ph | 4.71 | $5.69 \times 10^1$ | 12 |
| | BF$_4^-$ | HSnBu$_3$ | $2.96 \times 10^4$ | $6.49 \times 10^4$ | 2.2 |
| | BF$_4^-$ | 1i | $6.06 \times 10^5$ [i] | $1.31 \times 10^6$ | 2.2 |

Table 5.5. Continued.

| acceptor | source | donor | $k_{exp}$ [L mol$^{-1}$ s$^{-1}$] | $k_{calc}$ [L mol$^{-1}$ s$^{-1}$] | $k_{calc}/k_{exp}$ |
|---|---|---|---|---|---|
| (MeO)Tr$^+$ | BF$_4^-$ | HSiEt$_3$ | 1.57 | $2.50 \times 10^1$ | 16 |
| | BF$_4^-$ | HSiBu$_3$ | 3.44 | $5.97 \times 10^1$ | 17 |
| | BF$_4^-$ | HSiPh$_3$ | $6.44 \times 10^{-1}$ [a] | 5.80 | 9.0 |
| | BF$_4^-$ | HSiMe$_2$Ph | 2.44 | $2.95 \times 10^1$ | 12 |
| | BF$_4^-$ | HSnBu$_3$ | $1.32 \times 10^4$ | $4.01 \times 10^4$ | 3.0 |
| (MeO)$_2$Tr$^+$ | BF$_4^-$ | HSiEt$_3$ | $5.59 \times 10^{-2}$ | 2.12 | 38 |
| | BF$_4^-$ | HSiBu$_3$ | $1.29 \times 10^{-1}$ | 5.05 | 39 |
| | BF$_4^-$ | HSiPh$_3$ | $2.61 \times 10^{-2}$ [a] | $5.24 \times 10^{-1}$ | 20 |
| | BF$_4^-$ | HSiMe$_2$Ph | $1.07 \times 10^{-1}$ | 2.41 | 23 |
| | BF$_4^-$ | HSnBu$_3$ | $1.42 \times 10^3$ | $6.40 \times 10^3$ | 4.5 |
| (MeO)$_3$Tr$^+$ | BF$_4^-$ | HSnBu$_3$ | $2.27 \times 10^2$ | $1.22 \times 10^3$ | 5.4 |
| | BF$_4^-$ | 1i | $2.40 \times 10^3$ [i] | $3.49 \times 10^3$ | 1.5 |
| | not reported | 1i | $2.1 \times 10^5$ [j] | $4.17 \times 10^4$ | 0.2 |
| (Me$_2$N)$_2$Tr$^+$ | BF$_4^-$ | HSnBu$_3$ | $3.06 \times 10^{-1}$ | $6.58 \times 10^{-1}$ | 2.2 |
| | BF$_4^-$ | Bu$_4$N$^+$BH$_4^-$ | $3.74 \times 10^4$ [l] | $4.72 \times 10^3$ | 0.1 |
| | BF$_4^-$ | 1i | $1.46 \times 10^1$ [i,k] | 5.01 | 0.3 |
| | not reported | 1i | $2.7 \times 10^1$ [j] | 5.01 | 0.2 |
| (Me$_2$N)$_3$Tr$^+$ | Cl$^-$ | Bu$_4$N$^+$BH$_4^-$ | $1.61 \times 10^3$ [l] | $8.08 \times 10^2$ | 0.5 |

[a] From Chapter 3; [b] from ref. [11]; [c] a value of $1.10 \times 10^2$ L mol$^{-1}$ s$^{-1}$ (25 °C) was reported in ref. [7e]; [d] a value of $2.10 \times 10^2$ L mol$^{-1}$ s$^{-1}$ (25 °C) was reported in ref. [7d]; [e] kinetics of low quality; [f] from ref. [12]; [g] at 28 °C, from ref. [13]; [h] primary kinetic isotope effect $k_H/k_D$ = 1.6; [i] from ref. [6]; [j] in water at 25 °C, from ref. [14]; [l] in DMSO; [k] in 90W10AN.

*Reaction products.* Product studies have not been performed with all hydride donors used in this work. However, all analyzed reactions revealed the formation of the triarylmethane. When HSiPh$_3$ and HSiBu$_3$ were used as hydride donors, the corresponding fluorosilanes FSiR$_3$ could be detected. In the case of HSi(SiMe$_3$)$_3$ the products stemming from the silane could not be identified. For details see the Experimental Part (p. 202).

## 5.3. Discussion

Considering group 14 elements, Sn–H hydride donors are more reactive than Si–H donors, which in turn are more reactive than C–H donors, which is in accord with the order of electronegativities of these elements. This order is not only reflected by the *N*-parameters in Table 5.3, but also by the rate constants of hydride transfers to tritylium ions in Table 5.5.

## 5. Towards a General Hydride Donor Ability Scale

*Influence of the counterion.* In previous studies it has been demonstrated, that the anionic counterion of the cationic electrophile has no influence on the rate of hydride transfer from organic silanes to benzhydrylium,[15] as well as to tritylium ions (ref. [7e] and Chapter 3). Furthermore, $Na^+BH_4^-$, $K^+BH_4^-$, and $Bu_4N^+BH_4^-$ all exhibited comparable reactivities towards benzhydrylium ions in DMSO.[6] It may, therefore, be assumed that this independence also holds true for reactions with tritylium ions.

*Influence of the solvent.* In reactions of neutral hydride donors with cationic acceptors, charge is neither generated nor destroyed in the rate-determining step. The solvent polarity can therefore be assumed to exert a marginal effect on the rate constant. In line with this expectation, hydride transfers in dichloromethane solution proceed only 6-8 times faster than in acetonitrile solution (Table 5.6). A big difference is only observed for the reaction of N-benzyl-1,4-dihydronicotinamide (**1i**) with $(MeO)_3Tr^+$ in dichloromethane and water. Possibly, this difference is due to nucleophilic assistance by water, similar to the assistance reported for acetic acid in the reaction of silanes with the tris(2,6-dimethoxyphenyl)methyl cation.[7c] It was postulated that the atom which releases the hydride ion is simultaneously attacked by a solvent molecule.

Table 5.6. Second-order rate constants $k$ for hydride transfers to tritylium ions in different solvents, 20 °C.

| donor | acceptor | $T$ [°C] | $k$ [L mol$^{-1}$ s$^{-1}$] [a] | | | $k(CH_2Cl_2)/$ $k(X)$ |
|---|---|---|---|---|---|---|
| | | | $CH_2Cl_2$ | $CH_3CN$ | 90W10AN | |
| HSiEt$_3$ | Tr$^+$ | 20 | $1.36 \times 10^2$ | $2.11 \times 10^{1}$ [b] | - | 6.4 |
| **1a** | Tr$^+$ | 20 | 1.77 | $2.16 \times 10^{-1}$ [c] | - | 8.2 |
| **1c** | Tr$^+$ | 20 | $8.15 \times 10^{-2}$ | $1.08 \times 10^{-2}$ [d] | - | 7.5 |
| **1i** | (MeO)$_3$Tr$^+$ | 20 | $2.40 \times 10^3$ | - | $1.24 \times 10^5$ | 0.02 |

[a] From Table 5.3 if not otherwise noted; [b] this work; [c] calculated from $4.13 \times 10^{-1}$ L mol$^{-1}$ s$^{-1}$ (29.8 °C) in ref. [16]; [d] calculated from $1.65 \times 10^{-2}$ L mol$^{-1}$ s$^{-1}$ (25 °C) in ref. [17].

*Influence of the electrophile.* Table 5.5 shows that in almost all cases the experimental rate constants are smaller than predicted by equation (5.1). This can be explained by the additional aryl ring in tritylium ions compared to benzhydrylium ions, leading to an increased steric repulsion between the two reactants. Obviously, the steric requirements of electrophilic attack at carbocations are more important for hydride donors than primary amines and water, which were used for the evaluation of the *E*-parameters in Table 5.4.

In contrast to most other hydride donors, the borohydride anion and N-benzyl-1,4-dihydronicotinamide (**1i**) react faster with tritylium ions than calculated by equation (5.1) – a factor of 10 is found for the reaction of $Bu_4N^+BH_4^-$ with $(Me_2N)_2Tr^+$. Because of its high reactivity, $Bu_4N^+BH_4^-$ could only be combined with the least reactive tritylium ions $(Me_2N)_2Tr^+$ and $(Me_2N)_3Tr^+$.

When log $k$ for the reactions of $HSiR_3$ with substituted tritylium ions are plotted versus the corresponding electrophilicities $E$, linear correlations with slopes of approximately unity are found (Figure 5.3). By contrast, the nucleophile-specific slope parameters $s_N$ of silanes range from 0.72 to 0.81 (Table 5.3). The higher dependency of the rates on the electrophilicities of the reaction partners in the trityl series result in increasing deviations of the experimental rate constants from the calculated ones when going from top to bottom in the last column of Table 5.5. While the ratios $k_{calc}/k_{exp}$ are rather small for $Tr^+$ reacting with silanes (factor 2-8), they reach higher values for $(MeO)_2Tr^+$ (factor 20-40).

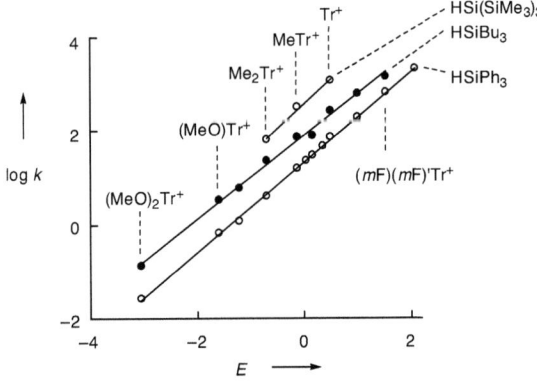

Figure 5.3. Correlation of log $k$ for the reactions between tritylium ions and silanes with the corresponding electrophilicity parameters $E$ of the tritylium ions; correlation equations: log $k$ = 0.97$E$ + 1.32 ($HSiPh_3$), log $k$ = 0.89$E$ + 1.89 ($HSiBu_3$), log $k$ = 1.04$E$ + 2.57 ($HSi(SiMe_3)_3$).

$HSi(SiMe_3)_3$ was intensely studied by Chatgilialoglu,[18] and is primarily known as a hydrogen atom source in radical reactions. It is well-known that the supersilyl group facilitates electron-donation in β-position to π-bonds,[19] but Table 5.3 demonstrates a comparable hydride donating ability towards benzhydrylium ions as normal organic silanes.

Despite the bulkiness of the supersilyl group, good agreement between experimental and calculated rate constants are found for $HSi(SiMe_3)_3$ as hydride donor ($k_{calc}/k_{exp}$ = 2.1 for $Tr^+$).

## 5. Towards a General Hydride Donor Ability Scale 97

On the other hand, Chojnowski reported about the inertness of tri(*tert*-butyl)silane toward Tr$^+$,[7e] indicating that steric effects are indeed present in these reactions. Possibly, the retarding steric effect caused by the supersilyl group in HSi(SiMe$_3$)$_3$ is compensated by its pronounced electronic effect, making its hydride donor ability comparable to HSiEt$_3$.

A good agreement between experimental and calculated rates is also found for HSnBu$_3$ (Table 5.5), which is illustrated by the correlation in Figure 5.4.

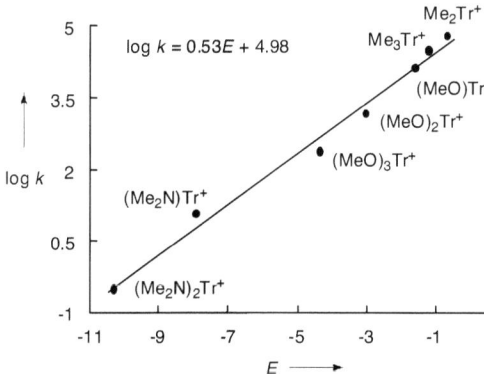

Figure 5.4. Plot of log $k$ for the reactions of HSnBu$_3$ with tritylium ions (CH$_2$Cl$_2$, 20 °C) vs. the corresponding $E$-parameters; $n = 7$, $R^2 = 0.9874$.

Bearing in mind that equation (5.1) covers a reactivity range of 40 orders of magnitude, deviations of factors of 10-100 are usually considered to constitute its tolerance interval. Because all values in the last column of Table 5.5 are well within this interval, equation (5.1) is applicable to a large variety of hydride abstractions by tritylium ions. It cannot be excluded, however, that other classes of hydride donors deviate from the predictions by equation (5.1) to a higher extent.

Nevertheless, these results justify the evaluation of $N$-parameters for hydride donors from the manifold of rate constants for hydride transfers to tritylium ions reported in the literature. The temperature in many investigations differed from 20 °C, to which equation (5.1) refers. In some of these cases activation parameters have been reported, thus offering the possibility to calculate rate constants at 20 °C. In the remaining cases the activation entropy was estimated.

Only rate constants determined in solvents of low Lewis basicity (methylene chloride, acetonitrile, dichloroethane, nitromethane) have been considered in the following. Studies performed in acetic acid have not been taken into account, due to the presumed nucleophilic assistance exerted by this solvent.[7c]

*Literature Survey.* Silanes can be employed in ionic hydrogenation reactions of alkenes, cyclopropanes, ketones, aldehydes, and alcohols,[20] which demonstrates their versatility.

Several alkylsilanes and HGeEt$_3$ have been studied in reactions with Tr$^+$SbF$_6^-$ by Chojnowski.[7d,e] In order to evaluate $N$ parameters, slope parameters of 0.75 for the silanes were assumed, resembling the values in Table 5.3. Table 5.7 summarizes the reported rate constants for the different silanes, and gives an overview of the derived reactivity parameters.

While ordinary alkylsilanes have $N$-parameters of 2-3, trialkoxysilanes are considerably less reactive. The highest $N$ value is found for HGeEt$_3$ ($N$ = 4.0). It may be compared with those of HGeBu$_3$ ($N$ = 5.92) and HGePh$_3$ ($N$ = 3.99), which have been obtained by reactions with benzhydrylium ions.[4]

Table 5.7. Second-order rate constants for hydride transfers from silanes and HGeEt$_3$ to Tr$^+$SbF$_6^-$ (CH$_2$Cl$_2$), and the derived nucleophilicity parameters $N$ ($s_N$ = 0.75).

| donor | $k_{25 °C}^{exp}$ [L mol$^{-1}$ s$^{-1}$] | ref. | $k_{20 °C}$ [L mol$^{-1}$ s$^{-1}$] | $N$ |
|---|---|---|---|---|
| HSiMe$_3$ | 2.23 × 10$^2$ | [7d,e] | 1.85 × 10$^{2\,[a]}$ | 2.6 |
| HSiMe$_2$Et | 1.64 × 10$^{2\,[b]}$ | [7e] | 1.34 × 10$^{2\,[c]}$ | 2.3 |
| HSiMeEt$_2$ | 1.38 × 10$^{2\,[d]}$ | [7e] | 1.18 × 10$^{2\,[e]}$ | 2.3 |
| HSi$^n$Pr$_3$ | 1.75 × 10$^2$ | [7d,e] | 1.43 × 10$^{2\,[f]}$ | 2.4 |
| HSi$^n$Hex$_3$ | 2.56 × 10$^2$ | [7e] | 2.10 × 10$^{2\,[f]}$ | 2.6 |
| HSiMe$_2$(CH$_2$Cl) | 1.20 | [7d] | 8.97 × 10$^{-1\,[f]}$ | –0.6 |
| HSiMePh(CH$_2^t$Bu) | 1.40 × 10$^1$ | [7d] | 1.09 × 10$^{1\,[f]}$ | 0.9 |
| HSiMe$_2$Bn | 8.41 × 10$^1$ | [7e] | 6.76 × 10$^{1\,[f]}$ | 1.9 |
| HSiMe$_2$(*m*-ClBn) | 2.83 × 10$^1$ | [7e] | 2.23 × 10$^{1\,[f]}$ | 1.3 |
| HMeSi—□ (cyclobutyl) | 1.63 × 10$^2$ | [7e] | 1.31 × 10$^{2\,[g]}$ | 2.3 |
| HMeSi—⬠ (cyclopentyl) | 1.39 × 10$^2$ | [7e] | 1.13 × 10$^{2\,[h]}$ | 2.2 |
| HSi(OEt)$_3$ | 1.5 × 10$^{-1}$ | [7d] | 1.08 × 10$^{-1\,[f]}$ | –1.8 |
| HSiMe$_2$(OTMS) | 2.35 × 10$^2$ | [7d] | 1.92 × 10$^{2\,[f]}$ | 2.5 |
| HSiMe$_2$(OPr) | 1.99 × 10$^2$ | [7d] | 1.62 × 10$^{2\,[f]}$ | 2.4 |
| HGeEt$_3$ | 2.78 × 10$^3$ | [7e] | 2.26 × 10$^{3\,[i]}$ | 4.0 |

[a] calculated with $\Delta S^{\neq}$ = –117 J K$^{-1}$ mol$^{-1}$;[7e] [b] the same author reported a value of 2.01 × 10$^2$ L mol$^{-1}$ s$^{-1}$ ten years earlier;[7d] [c] calculated with $\Delta S^{\neq}$ = –113 J K$^{-1}$ mol$^{-1}$;[7e] [d] the same author reported a value of 1.85 × 10$^2$ L mol$^{-1}$ s$^{-1}$ ten years earlier;[7d] [e] calculated with $\Delta S^{\neq}$ = –117 J K$^{-1}$ mol$^{-1}$;[7e] [f] calculated with an estimated activation entropy $\Delta S^{\neq}$ = –110 J K$^{-1}$ mol$^{-1}$; [g] calculated with $\Delta S^{\neq}$ = –105 J K$^{-1}$ mol$^{-1}$;[7e] [h] calculated with $\Delta S^{\neq}$ = –109 J K$^{-1}$ mol$^{-1}$;[7e] [i] calculated with $\Delta S^{\neq}$ = –86 J K$^{-1}$ mol$^{-1}$ (determined for the reactions of HGeBu$_3$ with benzhydrylium ions).[21]

## 5. Towards a General Hydride Donor Ability Scale

A hydride in β-position to the metal may be abstracted from tetralkylated silanes, -stannanes, and -plumbanes. For example, $^{sec}Bu_4Sn$, $^nPr_4Sn$, and $^iBu_4Sn$ were used to reduce a series of substituted tritylium ions,[22] and plots of log $k$ versus the electrophilicities $E$ reveal linear relationships (Figure 5.5). As the slopes of the correlation lines are almost identical to the value obtained for $^nBu_4Sn$ in reactions with benzhydrylium ions ($s_N$ = 1.1, Table 5.3), this $s_N$-parameter was chosen for the evaluation of $N$-parameters for the hydride donors in Table 5.8.

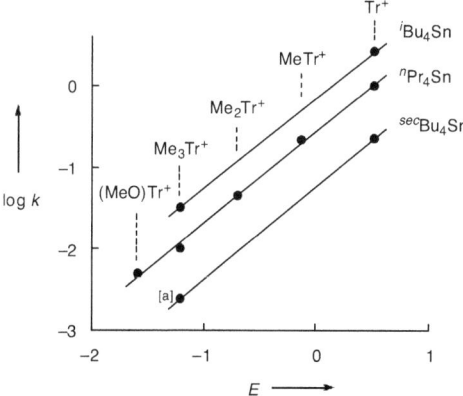

Figure 5.5. Plots of log $k$ for the reactions of $^{sec}Bu_4Sn$, $^nPr_4Sn$, and $^iBu_4Sn$ with tritylium ions (from ref. [22], dichloroethane, 20 °C) vs. $E$. Correlation equations: log $k$ = 1.15$E$ − 1.23 ($^{sec}Bu_4Sn$); log $k$ = 1.13$E$ − 0.56 ($^nPr_4Sn$); log $k$ = 1.11$E$ − 0.15 ($^iBu_4Sn$); [a] rate constant calculated from values at higher temperatures (see Table 5.6 and ref. [22]).

Table 5.8. Second-order rate constants for hydride transfers from peralkylated metals to tritylium ions, and the derived nucleophilicity parameters $N$ ($s_N$ = 1.1).

| donor | acceptor | solvent | $T$ [°C] | $k_T^{exp}$ [L mol$^{-1}$ s$^{-1}$] | ref. | $k_{20°C}$[a] [L mol$^{-1}$ s$^{-1}$] | $N$ |
|---|---|---|---|---|---|---|---|
| Et$_4$Si | Tr$^+$BF$_4^-$ | CH$_3$CN | 29.8 | 6.8 × 10$^{-7}$ | [16] | 2.88 × 10$^{-7}$ | −6.5 |
| BuSiMe$_3$ | Tr$^+$BF$_4^-$ | CH$_2$Cl$_2$ | 22 | 5 × 10$^{-6}$ | [23] | 4 × 10$^{-6}$ | −5.4 |
| PhCH$_2$CH$_2$SiMe$_3$ | Tr$^+$BF$_4^-$ | CH$_2$Cl$_2$ | 22 | 5.30 × 10$^{-4}$ | [23] | 4.66 × 10$^{-4}$ | −3.5 |
| Ph$_2$CHCH$_2$SiEt$_2$Me | Tr$^+$BF$_4^-$ | CH$_2$Cl$_2$ | 22 | 3 × 10$^{-7}$ | [23] | 3 × 10$^{-7}$ | −6.4 |
| Me$_3$Si(CH$_2$)$_6$SiMe$_3$ | Tr$^+$BF$_4^-$ | CH$_2$Cl$_2$ | 22 | 3.00 × 10$^{-5}$ | [23] | 2.58 × 10$^{-5}$ | −4.7 |
| Me$_3$Si(CH$_2$)$_3$SiMe$_3$ | Tr$^+$BF$_4^-$ | CH$_3$CN/ CH$_2$Cl$_2$ (1:1) | 30 | 1.24 × 10$^{-5}$ | [24] | 5.71 × 10$^{-6}$ | −5.3 |
| ⌐SiMe$_2$ (cyclobutyl) | Tr$^+$BF$_4^-$ | CH$_2$Cl$_2$ | 22 | 2.00 × 10$^{-2}$ | [23] | 1.80 × 10$^{-2}$ | −2.1 |
| SiMe$_2$ (cyclopentyl) | Tr$^+$BF$_4^-$ | CH$_2$Cl$_2$ | 22 | 4 × 10$^{-6}$ | [23] | 3 × 10$^{-6}$ | −5.5 |
| SiMe$_2$ (cyclohexyl) | Tr$^+$BF$_4^-$ | CH$_2$Cl$_2$ | 23 | 2.21 × 10$^{-3}$ | [25] | 1.85 × 10$^{-3}$ | −3.0 |

Table 5.8. Continued.

| donor | acceptor | solvent | T [°C] | $k_T^{exp}$ [L mol$^{-1}$ s$^{-1}$] | ref. | $k_{20°C}$ [a] [L mol$^{-1}$ s$^{-1}$] | N |
|---|---|---|---|---|---|---|---|
| (cycloheptyl)SiMe$_2$ | Tr$^+$BF$_4^-$ | CH$_2$Cl$_2$ | 22 | 1.58 × 10$^{-3}$ | [23] | 1.40 × 10$^{-3}$ | −3.1 |
| Et$_4$Ge | Tr$^+$BF$_4^-$ | CH$_3$CN | 29.8 | 4.8 × 10$^{-5}$ | [16] | 2.3 × 10$^{-5}$ | −4.7 |
| Me$_3$Ge(CH$_2$)$_3$SiMe$_3$ | Tr$^+$BF$_4^-$ | CH$_3$CN | 30 | 3.8 × 10$^{-5}$ | [24] | 1.8 × 10$^{-5}$ | −4.8 |
| Me$_3$Ge(CH$_2$)$_3$GeMe$_3$ | Tr$^+$BF$_4^-$ | CH$_3$CN | 30 | 3.6 × 10$^{-4}$ | [24] | 1.9 × 10$^{-4}$ | −3.4 |
| Et$_4$Sn | Tr$^+$ClO$_4^-$ | DCE [b] | 20 | 3.1 × 10$^{-2}$ [c] | [22] | - | −1.9 |
| Pr$_4$Sn | Tr$^+$ClO$_4^-$ | DCE | 20 | 1.0 | [22] | - | −0.5 [d] |
|  | MeTr$^+$ClO$_4^-$ | DCE | 20 | 2.2 × 10$^{-1}$ | [22] | - |  |
|  | Me$_2$Tr$^+$ClO$_4^-$ | DCE | 20 | 4.4 × 10$^{-2}$ | [22] | - |  |
|  | Me$_3$Tr$^+$ClO$_4^-$ | DCE | 20 | 9.9 × 10$^{-3}$ | [22] | - |  |
|  | (MeO)Tr$^+$ClO$_4^-$ | DCE | 20 | 4.9 × 10$^{-3}$ | [22] | - |  |
| $^i$Bu$_4$Sn | Tr$^+$ClO$_4^-$ | DCE | 20 | 2.6 | [22] | - | −0.1 [d] |
|  | Me$_3$Tr$^+$ClO$_4^-$ | DCE | 20 | 3.2 × 10$^{-2}$ | [22] | - |  |
| $^{sec}$Bu$_4$Sn | Tr$^+$ClO$_4^-$ | DCE | 20 | 2.3 × 10$^{-1}$ | [22] | - | −1.1 [d] |
|  | Me$_3$Tr$^+$ClO$_4^-$ | DCE | 30 | 4.3 × 10$^{-3}$ | [22] | 2.4 × 10$^{-3}$ |  |
|  | Me$_3$Tr$^+$ClO$_4^-$ | DCE | 40 | 7.4 × 10$^{-3}$ | [22] | 2.4 × 10$^{-3}$ |  |
|  | Me$_3$Tr$^+$ClO$_4^-$ | DCE | 50 | 1.23 × 10$^{-2}$ | [22] | 2.4 × 10$^{-3}$ |  |
| Me$_3$Sn(CH$_2$)$_3$SiMe$_3$ | Tr$^+$BF$_4^-$ | CH$_3$CN | 30 | 1.18 × 10$^{-3}$ | [24] | 6.34 × 10$^{-4}$ | −3.4 |
| Me$_3$Sn(CH$_2$)$_3$GeMe$_3$ | Tr$^+$BF$_4^-$ | CH$_3$CN | 30 | 1.42 × 10$^{-2}$ | [24] | 8.31 × 10$^{-3}$ | −2.4 |
| Me$_3$Sn(CH$_2$)$_3$SnMe$_3$ | Tr$^+$BF$_4^-$ | CH$_3$CN | 30 | 8.0 × 10$^{-2}$ | [24] | 5.0 × 10$^{-2}$ | −1.7 |
| Me$_3$SnCH$_2$Bn | Tr$^+$BF$_4^-$ | CH$_3$CN | 29.8 | 9.56 × 10$^{-2}$ | [16] | 6.03 × 10$^{-2}$ | −1.6 |
| Me$_3$Sn(CH$_2$)$_3$CMe$_3$ | Tr$^+$BF$_4^-$ | CH$_3$CN/CH$_2$Cl$_2$ (1:1) | 30 | 4 × 10$^{-4}$ | [24] | 2 × 10$^{-4}$ | −3.9 |
| Et$_4$Pb | Tr$^+$BF$_4^-$ | CH$_3$CN | 29.8 | 5.9 | [16] | 4.3 | 0.1 |
| Me$_3$Pb(CH$_2$)$_3$SiMe$_3$ | Tr$^+$BF$_4^-$ | CH$_3$CN | 30 | 2.6 × 10$^{-1}$ | [24] | 1.7 × 10$^{-1}$ | −1.2 |
| Me$_3$Pb(CH$_2$)$_3$GeMe$_3$ | Tr$^+$BF$_4^-$ | CH$_3$CN | 30 | 4 × 10$^{-1}$ | [24] | 3 × 10$^{-1}$ | −1.0 |
| Et$_2$Hg | Tr$^+$BF$_4^-$ | CH$_3$CN | 29.8 | 0.9 | [16] | 6 × 10$^{-1}$ | −0.7 |

[a] Calculated from $k_T^{exp}$ with $\Delta S^{\neq} = -158$ J K$^{-1}$ mol$^{-1}$ (determined for the reaction of $^{sec}$Bu$_4$Sn with Me$_3$Tr$^+$);[22] [b] DCE = dichloroethane; [c] a value of 1.0 × 10$^{-2}$ L mol$^{-1}$ s$^{-1}$ (25 °C, CH$_3$CN) was reported in ref. [16]; [d] calculated by the correlation equation in Figure 5.5.

The series of tetraalkylstannanes offers to analyze the influence of elongation and branching in the alkyl residues on the hydride donor abilities. As the hydride transfer constitutes

## 5. Towards a General Hydride Donor Ability Scale    101

the rate-determining step, the stability of the generated carbenium ion may serve as an indicator for the magnitude of the rate constant. Increasing the stabilization of the resulting carbenium ion leads to an increase of $N$ from $-1.9$ to $-0.1$ in the first line of Figure 5.6, although the number of potential hydrides in the neutral donors decreases from 3 to 1 per alkyl residue. However, keeping the reactive site a secondary carbon, branching leads to pronounced sterical hindrance in reactions with hydride acceptors. The values in Figure 5.6 are based on reactions with tritylium ions, and may be compared with $N = -0.30$ for $^n$Bu$_4$Sn (Table 5.3), which refers to reactions with benzhydrylium ions.

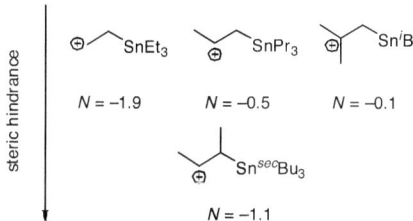

Figure 5.6. Comparison of hydride donor nucleophilicities of different tetraalkylstannanes; the $N$ parameters refer to the neutral, saturated compounds.

Transition metal hydrides have been intensely studied by Bullock,[26,27] who determined the rates of reactions of HMo(CO)$_3$Cp* with a series of substituted tritylium ions. Again, a linear correlation of log $k$ versus $E$ is observed (Figure 5.7). The slope of approximately 0.8 was chosen as an estimate for the evaluation of nucleophilicity parameters for the compounds in Table 5.9.

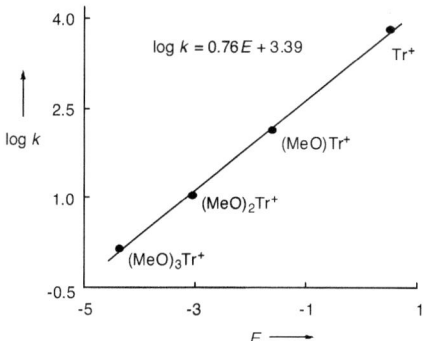

Figure 5.7. Correlation of log $k$ for the reactions of HMo(CO)$_3$Cp* with substituted ritylium ions (data from ref. [27], CH$_2$Cl$_2$, 25 °C) with the corresponding $E$-parameters.

The different rates of reactions (25 °C) of HMo(CO)$_3$Cp* with (MeO)$_2$Tr$^+$ in dichloromethane ($k$ = 1.1 × 10$^1$ L mol$^{-1}$ s$^{-1}$)[27] and acetonitrile ($k$ = 8.10 × 10$^2$ L mol$^{-1}$ s$^{-1}$)[28] can be explained by the fact that the latter solvent serves as a much better ligand to molybdenum than the former, thus accelerating the hydride donation of the transition metal hydride. Therefore, only data referring to dichloromethane have been considered in Table 5.9.

Table 5.9. Second-order rate constants for hydride transfers from transition metal hydrides to tritylium ions (from ref. [27], CH$_2$Cl$_2$), and the derived nucleophilicity parameters $N$ ($s_N$ = 0.8).

| donor | acceptor | $k_{25\,°C}^{exp}$ [L mol$^{-1}$ s$^{-1}$] | $k_{20\,°C}$ [a] [L mol$^{-1}$ s$^{-1}$] | $N$ |
|---|---|---|---|---|
| HW(CO)$_3$(C$_5$H$_4$CO$_2$Me) | Tr$^+$BF$_4^-$ | 7.2 × 10$^{-1}$ | 5.2 × 10$^{-1}$ | −0.9 |
| HMn(CO)$_5$ | Tr$^+$BF$_4^-$ | 5.0 × 10$^1$ | 3.9 × 10$^1$ | 1.5 |
| HCr(CO)$_3$Cp* | Tr$^+$BF$_4^-$ | 5.7 × 10$^1$ | 4.5 × 10$^1$ | 1.6 |
| HW(CO)$_3$Cp | Tr$^+$BF$_4^-$ | 7.6 × 10$^1$ | 6.0 × 10$^1$ | 1.7 |
| cis-HMn(PCy$_3$)(CO)$_4$ | Tr$^+$BF$_4^-$ | 1.7 × 10$^2$ | 1.4 × 10$^2$ | 2.2 |
| cis-HMn(PPh$_3$)(CO)$_4$ | Tr$^+$BF$_4^-$ | 2.3 × 10$^2$ | 1.8 × 10$^2$ | 2.3 |
| HW(CO)$_3$(C$_5$H$_4$Me) | Tr$^+$BF$_4^-$ | 2.5 × 10$^2$ | 2.0 × 10$^2$ | 2.4 |
| HMo(CO)$_3$Cp | Tr$^+$BF$_4^-$ | 3.8 × 10$^2$ | 3.1 × 10$^2$ | 2.6 |
| HW(CO)$_3$Cp* | Tr$^+$BF$_4^-$ | 1.9 × 10$^3$ | 1.6 × 10$^3$ | 3.5 |
| HW(CO)$_3$(indenyl) | Tr$^+$BF$_4^-$ | 2.0 × 10$^3$ | 1.7 × 10$^3$ | 3.5 |
| HRe(CO)$_5$ | Tr$^+$BF$_4^-$ | 2.0 × 10$^3$ | 1.7 × 10$^3$ | 3.5 |
| cis-HRe(PPh$_3$)(CO)$_4$ | Tr$^+$BF$_4^-$ | 1.2 × 10$^4$ | 1.0 × 10$^{4\,[b]}$ | 4.5 |
| HW(NO)$_2$Cp | Tr$^+$BF$_4^-$ | 1.9 × 10$^4$ | 1.6 × 10$^4$ | 4.8 |
| trans-HMo(CO)$_2$(PCy$_3$)Cp | Tr$^+$BF$_4^-$ | 4.3 × 10$^5$ | 3.7 × 10$^{5\,[c]}$ | 6.5 |
| trans-HMo(CO)$_2$(PPh$_3$)Cp | Tr$^+$BF$_4^-$ | 5.7 × 10$^5$ | 5.2 × 10$^5$ | 6.6 |
| trans-HMo(CO)$_2$(PMe$_3$)Cp | Tr$^+$BF$_4^-$ | 4.6 × 10$^6$ | 4.1 × 10$^{6\,[d]}$ | 7.8 |
| HMo(CO)$_3$Cp* | Tr$^+$BF$_4^-$ | 6.5 × 10$^3$ | 5.5 × 10$^{3\,[e]}$ | 4.5 [f] |
|  | (MeO)Tr$^+$BF$_4^-$ | 1.4 × 10$^2$ | 1.1 × 10$^{2\,[e]}$ |  |
|  | (MeO)$_2$Tr$^+$BF$_4^-$ | 1.1 × 10$^1$ | 8.4 [e] |  |
|  | (MeO)$_3$Tr$^+$BF$_4^-$ | 1.4 | 1.0 [e] |  |

[a] Calculated from $k_T^{exp}$ with $\Delta S^{\neq}$ = −100 J K$^{-1}$ mol$^{-1}$ unless otherwise noted; [b] calculated with $\Delta S^{\neq}$ = −84 J K$^{-1}$ mol$^{-1}$;[27] [c] calculated with $\Delta S^{\neq}$ = −75 J K$^{-1}$ mol$^{-1}$;[27] [d] calculated with $\Delta S^{\neq}$ = −75 J K$^{-1}$ mol$^{-1}$;[27] [e] calculated with $\Delta S^{\neq}$ = −100 J K$^{-1}$ mol$^{-1}$;[27] [f] calculated by the correlation equation in Figure 5.7.

The dioxolane 1e is the weakest hydride donor in Table 5.3. Although product studies clearly showed the formation of the dioxolenium ion and triarylmethanes, the tritylium absor-

## 5. Towards a General Hydride Donor Ability Scale

bances did not decrease mono-exponentially during the reactions with **1e**, and second-order rate constants could not be determined. The same problem occurred with the 2-phenyl-substituted dioxolane, the reasons for these phenomena being unknown.

Nevertheless, the literature data of hydride transfers from ethers and acetals to the triphenylmethyl cation have been used to calculate $N$-parameters with the assumption of $s_N = 0.8$ (Table 5.10), the slope-parameter of **1e** in Table 5.3.

As can be seen, the nucleophilicity of 2-methyldioxolane ($N = -3.3$) is significantly lower than that of **1e** ($N = -1.06$, Table 5.3). It is at least questionable, whether the steric hindrance caused by the propyl group in **1e** can be overcompensated by its superior inductive effect compared to the methyl group in 2-methyldioxolane. The $N$-values in Table 5.10 must, therefore, be considered as preliminary and taken with care.

Table 5.10. Second-order rate constants for hydride transfers from ethers and acetals to tritylium ions (CH$_2$Cl$_2$), and the derived nucleophilicity parameters $N$ ($s_N = 0.8$).

| donor | acceptor | $T$ [°C] | $k_T^{exp}$ [L mol$^{-1}$ s$^{-1}$] | ref. | $k_{20\,°C}$ [L mol$^{-1}$ s$^{-1}$] | $N$ |
|---|---|---|---|---|---|---|
| Et$_2$O | Tr$^+$SbCl$_6^-$ | 25 | $3.0 \times 10^{-4}$ | [2b] | $2.2 \times 10^{-4}$ [a] | −5.1 |
| (tetrahydrofuran) | Tr$^+$SbCl$_6^-$ | 25 | $6.3 \times 10^{-3}$ | [2b] | $4.3 \times 10^{-3}$ [b] | −3.5 |
|  | Tr$^+$BF$_4^-$ | 18 | $3.90 \times 10^{-3}$ | [29] | $4.6 \times 10^{-3}$ [b] | −3.4 |
| MeOCH$_2$OMe | Tr$^+$SbCl$_6^-$ | 25 | $6.0 \times 10^{-4}$ | [2b] | $3.3 \times 10^{-4}$ [c] | −4.9 |
| (1,3-dioxolane) | Tr$^+$PF$_6^-$ | 25 | $9.42 \times 10^{-3}$ | [29] | $5.90 \times 10^{-3}$ [d] | −3.3 |
|  | Tr$^+$SbCl$_6^-$ | 22 | $8.74 \times 10^{-3}$ | [29] | $7.24 \times 10^{-3}$ [d] | −3.2 |
|  | Tr$^+$SbCl$_6^-$ | 18 | $7.40 \times 10^{-3}$ [e] | [29] | $8.94 \times 10^{-3}$ [d] | −3.1 |
|  | Tr$^+$SbCl$_6^-$ | 22.5 | $8.3 \times 10^{-3}$ | [30] | $6.6 \times 10^{-3}$ [d] | −3.2 |
|  | Tr$^+$SbCl$_6^-$ | 23 | $7.71 \times 10^{-3}$ | [31] | $5.81 \times 10^{-3}$ [d] | −3.3 |
| (2-methyl-1,3-dioxolane) | Tr$^+$SbCl$_6^-$ | 25 | $7.9 \times 10^{-3}$ | [2b] | $5.4 \times 10^{-3}$ [f] | −3.3 |
| Ph–(2-phenyl-1,3-dioxolane) | Tr$^+$SbCl$_6^-$ | 25 | $1.5 \times 10^{-2}$ | [2b] | $1.3 \times 10^{-2}$ [g] | −2.9 |
|  | Tr$^+$SbCl$_6^-$ | 23 | $4.14 \times 10^{-2}$ | [31] | $3.37 \times 10^{-2}$ [h] | −2.4 |
|  | Tr$^+$SbCl$_6^-$ | 23 | $3.23 \times 10^{-2}$ [i] | [31] | $2.56 \times 10^{-2}$ [h] | −2.5 |
|  | Tr$^+$SbCl$_6^-$ | 23 | $1.29 \times 10^{-2}$ | [31] | $1.06 \times 10^{-2}$ [j] | −3.0 |
|  | Tr$^+$SbCl$_6^-$ | 23 | $1.51 \times 10^{-2}$ | [31] | $1.28 \times 10^{-2}$ [k] | −2.9 |

## 5. Towards a General Hydride Donor Ability Scale

Table 5.10. Continued.

| donor | acceptor | $T$ [°C] | $k_T^{exp}$ [L mol$^{-1}$ s$^{-1}$] | ref. | $k_{20 °C}^{calc}$ [L mol$^{-1}$ s$^{-1}$] | $N$ |
|---|---|---|---|---|---|---|
| | Tr$^+$SbCl$_6^-$ | 23 | $2.31 \times 10^{-2}$ | [31] | $1.96 \times 10^{-2}$ [k] | −2.6 |
| | Tr$^+$SbCl$_6^-$ | 23 | $1.75 \times 10^{-3}$ | [31] | $1.43 \times 10^{-3}$ [k] | −4.1 |
| | Tr$^+$SbCl$_6^-$ | 23 | $9.8 \times 10^{-4}$ | [31] | $8.0 \times 10^{-4}$ [l] | −4.4 |
| | Tr$^+$SbCl$_6^-$ | 23 | $7.35 \times 10^{-2}$ | [31] | $5.93 \times 10^{-2}$ [m] | −2.0 |
| | Tr$^+$SbCl$_6^-$ | 23 | $3.41 \times 10^{-2}$ | [31] | $2.79 \times 10^{-2}$ [n] | −2.5 |
| | Tr$^+$SbCl$_6^-$ | 23 | $3.51 \times 10^{-2}$ | [31] | $2.95 \times 10^{-2}$ [o] | −2.4 |
| | Tr$^+$SbCl$_6^-$ | 23 | $5.5 \times 10^{-2}$ | [31] | $4.4 \times 10^{-2}$ [p] | −2.2 |
| | Tr$^+$BF$_4^-$ | 22 | $2.69 \times 10^{-3}$ | [29] | $2.28 \times 10^{-3}$ [q] | −3.8 |
| | Tr$^+$BF$_4^-$ | 18 | $1.24 \times 10^{-3}$ [e] | [29] | $1.47 \times 10^{-3}$ [q] | −4.1 |

[a] Calculated with $\Delta S^{\neq} = -167$ J K$^{-1}$ mol$^{-1}$;[2b] [h] calculated with $\Delta S^{\neq} = -109$ J K$^{-1}$ mol$^{-1}$;[2b] [c] calculated with $\Delta S^{\neq} = -17$ J K$^{-1}$ mol$^{-1}$;[2b] [d] calculated with $\Delta S^{\neq} = -63$ J K$^{-1}$ mol$^{-1}$;[2b] [e] in dichloroethane; [f] calculated with $\Delta S^{\neq} = -109$ J K$^{-1}$ mol$^{-1}$;[2b] [g] calculated with $\Delta S^{\neq} = -218$ J K$^{-1}$ mol$^{-1}$;[2b] [h] calculated with $\Delta S^{\neq} = -92$ J K$^{-1}$ mol$^{-1}$;[31] [i] in nitromethane; [j] calculated with $\Delta S^{\neq} = -130$ J K$^{-1}$ mol$^{-1}$;[31] [k] calculated with $\Delta S^{\neq} = -151$ J K$^{-1}$ mol$^{-1}$;[31] [l] calculated with $\Delta S^{\neq} = -143$ J K$^{-1}$ mol$^{-1}$;[31] [m] calculated with $\Delta S^{\neq} = -100$ J K$^{-1}$ mol$^{-1}$;[31] [n] calculated with $\Delta S^{\neq} = -118$ J K$^{-1}$ mol$^{-1}$;[31] [o] calculated with $\Delta S^{\neq} = -138$ J K$^{-1}$ mol$^{-1}$;[31] [p] calculated with $\Delta S^{\neq} = -123$ J K$^{-1}$ mol$^{-1}$;[31] [q] calculated with estimated $\Delta S^{\neq} = -100$ J K$^{-1}$ mol$^{-1}$.

Hydrocarbons can also act as hydride donors when the resulting carbenium centers are part of a conjugated electronic π-system. Examples are **1a,b,c,d,f,g,h** in Table 5.3. According to Table 5.5, the rate constants of the reactions of cycloheptatriene (**1a**) with tritylium ions are in good agreement with the calculated values ($k_{exp}/k_{calc} < 6$). On the other hand, discrepancies between experimental and calculated rates are considerable larger in the reactions of 1,4-dihydronaphthalene (**1c**, $k_{exp}/k_{calc} = 35$) or 9,10-dihydroanthracene (**1d**, $k_{exp}/k_{calc} = 33$) with Tr$^+$. These findings can be rationalized by the much bigger steric shielding of the hydride to be transferred in **1c** and **1d** compared to that in **1a**. The steric hindrance probably becomes a serious limitation of equation (5.1) when hydride ions are transferred from triarylmethanes to triarylmethyl cations.

## 5. Towards a General Hydride Donor Ability Scale

Despite this concern, $N$-parameters for the hydride donors in Table 5.11 have been evaluated with the assumption of $s_N = 1.0$, similar to the values of the structurally related compounds in Table 5.3. It should be mentioned that all rate constants in Table 5.11 refer to acetonitrile as solvent, in which hydride transfers have been shown to proceed about 6-8 times slower than in dichloromethane (Table 5.6).

Attempts to determine second-order rate constants for hydride transfers from Hantzsch esters to tritylium ions in dichloromethane have been unsuccessful (see Chapter 7).

Table 5.11. Second-order rate constants for hydride transfers from hydrocarbons to tritylium ions, and the derived nucleophilicity parameters $N$ ($s_N = 1.0$).

| donor | acceptor | solvent | $T$ [°C] | $k_T^{exp}$ [L mol$^{-1}$ s$^{-1}$] | ref. | $k_{20\,°C}^{[a]}$ [L mol$^{-1}$ s$^{-1}$] | $N$ |
|---|---|---|---|---|---|---|---|
| (Me)$_3$TrH | Tr$^+$ClO$_4^-$ | CH$_3$CN | 23 | 2.65 × 10$^{-6}$ | [32] | 1.92 × 10$^{-6}$ | −6.2 |
| (MeO)$_2$TrH | Tr$^+$ClO$_4^-$ | CH$_3$CN | 23 | 6.18 × 10$^{-6}$ | [32] | 4.53 × 10$^{-6}$ | −5.9 |
|  | (Me)$_3$Tr$^+$ ClO$_4^-$ | CH$_3$CN | 23 | 2.51 × 10$^{-7}$ | [32] | 1.78 × 10$^{-7}$ | −5.5 |
| (MeO)$_3$TrH | Tr$^+$ClO$_4^-$ | CH$_3$CN | 23 | 7.20 × 10$^{-5}$ | [32] | 5.41 × 10$^{-5}$ | −4.8 |
|  | Tr$^+$BF$_4^-$ | CD$_3$CN | 22 | 4.63 × 10$^{-5}$ | [33] | 3.82 × 10$^{-5}$ | −4.9 |
|  | (Me)$_3$Tr$^+$ ClO$_4^-$ | CH$_3$CN | 23 | 2.82 × 10$^{-6}$ | [32] | 2.05 × 10$^{-6}$ | −4.5 |
|  | (MeO)Tr$^+$ BF$_4^-$ | CD$_3$CN | 26 | 5.02 × 10$^{-7}$ | [33] | 2.56 × 10$^{-7}$ | −5.0 |
|  | (MeO)Tr$^+$ ClO$_4^-$ | CH$_3$CN | 23 | 1.07 × 10$^{-6}$ | [32] | 7.70 × 10$^{-7}$ | −4.5 |
| (Me$_2$N)$_2$TrH | Tr$^+$ClO$_4^-$ | CH$_3$CN | 23 | 1.31 × 10$^{-3}$ | [32] | 1.01 × 10$^{-3}$ | −3.5 |
| (Me$_2$N)$_3$TrH | Tr$^+$ClO$_4^-$ | CH$_3$CN | 23 | 6.17 × 10$^{-2}$ | [32] | 5.00 × 10$^{-2}$ | −1.8 |
| 1a | Tr$^+$BF$_4^-$ | CH$_3$CN | 29.8 | 4.13 × 10$^{-1}$ | [16] | 2.16 × 10$^{-1}$ | −1.2 |
|  | (Me)$_3$Tr$^+$ ClO$_4^-$ | CH$_3$CN | 23 | 2.68 × 10$^{-2}$ | [32] | 2.14 × 10$^{-2}$ | −0.5 |
|  | (MeO)Tr$^+$ ClO$_4^-$ | CH$_3$CN | 23 | 2.04 × 10$^{-2}$ | [32] | 1.62 × 10$^{-2}$ | −0.2 |
|  | (MeO)$_2$Tr$^+$ ClO$_4^-$ | CH$_3$CN | 23 | 1.09 × 10$^{-3}$ | [32] | 8.42 × 10$^{-4}$ | 0.0 |
|  | (MeO)$_3$Tr$^+$ ClO$_4^-$ | CH$_3$CN | 80 | 1.63 × 10$^{-3}$ | [32] | 9.92 × 10$^{-6}$ | −0.7 |
| Ph H Ph Ph (cyclopropene) | Tr$^+$ClO$_4^-$ | CH$_3$CN | 23 | 9.1 × 10$^{-2}$ | [32] | 7.4 × 10$^{-2}$ | −1.6 |
|  | Me$_3$Tr$^+$ ClO$_4^-$ | CH$_3$CN | 23 | 7.1 × 10$^{-3}$ | [32] | 5.6 × 10$^{-3}$ | −1.0 |
| (dihydronaphthalene) | Tr$^+$BF$_4^-$ | CH$_3$CN | 25 | 2.67 × 10$^{-3}$ | [17] | 1.61 × 10$^{-3\,[b]}$ | −3.3 |

Table 5.11. Continued.

| donor | acceptor | solvent | $T$ [°C] | $k_T^{exp}$ [L mol$^{-1}$ s$^{-1}$] | ref. | $k_{20\,°C}$ [a] [L mol$^{-1}$ s$^{-1}$] | $N$ |
|---|---|---|---|---|---|---|---|
| (structure) | Tr$^+$ClO$_4^-$ | CH$_3$CN | 25 | $1.7 \times 10^3$ | [34] | $1.4 \times 10^3$ | 4.0 [c] |
| 1i | Tr$^+$ClO$_4^-$ | CH$_3$CN | 25 | $8.9 \times 10^4$ | [34] | $7.9 \times 10^4$ | (6.5) [c] |

[a] Calculated from $k_T^{exp}$ with $\Delta S^\neq = -100$ J K$^{-1}$ mol$^{-1}$; [b] calculated with $\Delta S^\neq = -57$ J K$^{-1}$ mol$^{-1}$;[17] [c] calculated with $s_N = 0.7$, in accordance with the values in Table 5.3.

The nucleophilicity of cycloheptatriene (1a) is one order of magnitude smaller in CH$_3$CN [$N = -0.5$ (mean value)] than in CH$_2$Cl$_2$ ($N = 0.52$, Table 5.3), in agreement with the results of Table 5.6.

Because the rates of hydride transfers from the dihydronicotinamide 1i to tritylium ions in CH$_2$Cl$_2$ or H$_2$O were shown to be equal to or even higher than the predictions of equation (5.1) (Table 5.5), the $N$-parameter for 1i in CH$_3$CN (6.5, Table 5.11) appears to be too low, in view of its $N$-value in CH$_2$Cl$_2$ of 8.67 (Table 5.3) and the observation that the rates in CH$_2$Cl$_2$ are only slightly bigger than in CH$_3$CN (Table 5.6). A possible reason might be the $s_N$-parameter of 0.7 chosen for the evaluation of $N$ in Table 5.11. The appropriate choice of $s_N$-parameters is a general problem, as their estimation significantly affects the resulting $N$-parameters. As we have seen before, $s_N$-values do not only depend on the solvent (Table 5.3), but also on the carbocationic reaction partners (see the context of Figure 5.3). A value of $s_N = 0.6$ for 1i in CH$_3$CN would give a nucleophilicity of $N = 7.7$, which appears to be more reasonable.

However, in light of the huge scope of different compounds characterized in this work, and the estimations that were necessary for their characterization, deviations of $\Delta N = \pm 2$ have to be accepted.

*Test of reliability.* In Table 5.12 some of the newly evaluated nucleophilicity parameters are tested with respect to their reliability. For this purpose, the hydride donors have been combined with benzhydrylium ions, i.e., the reference electrophiles of equation (5.1), and the experimental rate constants are compared with those calculated by equation (5.1).

## 5. Towards a General Hydride Donor Ability Scale

Table 5.12. Comparison of experimental ($CH_2Cl_2$, 20 °C) and calculated rates of hydride transfers to acceptors different from tritylium ions.

| donor | $N$ [a] ($s_N$) | acceptor ($E$) [b] | $k_{exp}$ [L mol$^{-1}$ s$^{-1}$] | ref. | $k_{calc}$ [c] [L mol$^{-1}$ s$^{-1}$] | $k_{exp}$/$k_{calc}$ |
|---|---|---|---|---|---|---|
| $HSiMe_3$ | 2.6 (0.75) | (ani)PhCH$^+$ (2.11) | 4.81 × 10$^3$ [d] | [15] | 3.4 × 10$^3$ | 1.4 |
| $HSiMe_2Et$ | 2.3 (0.75) | (ani)PhCH$^+$ (2.11) | 6.14 × 10$^3$ [d] | [15] | 2.0 × 10$^3$ | 3.1 |
| $HSiMeEt_2$ | 2.3 (0.75) | (ani)PhCH$^+$ (2.11) | 7.29 × 10$^3$ [d] | [15] | 2.0 × 10$^3$ | 3.6 |
| $HSi^nPr_3$ | 2.4 (0.75) | (ani)PhCH$^+$ (2.11) | 1.15 × 10$^4$ [d] | [15] | 2.4 × 10$^3$ | 4.8 |
| $HSi^nHex_3$ | 2.6 (0.75) | (ani)PhCH$^+$ (2.11) | 1.65 × 10$^4$ [d] | [15] | 3.4 × 10$^3$ | 4.9 |
| $HSiMe_2(CH_2Cl)$ | –0.6 (0.75) | (ani)PhCH$^+$ (2.11) | 9.20 × 10$^1$ [d] | [15] | 1.4 × 10$^1$ | 6.6 |
| $HSiMe_2Bn$ | 1.9 (0.75) | (ani)PhCH$^+$ (2.11) | 2.57 × 10$^3$ [d] | [15] | 1.0 × 10$^3$ | 2.6 |
| $HSiMe_2(OTMS)$ | 2.5 (0.75) | (ani)PhCH$^+$ (2.11) | 4.54 × 10$^3$ [d] | [15] | 2.9 × 10$^3$ | 1.6 |
| $HGeEt_3$ | 4.0 (0.75) | Trop$^+$ (–3.27) [e] | 1.95 [f] | [7e] | 3.5 | 0.6 |
| $^iBu_4Sn$ | –0.1 (1.1) | (ani)$_2$CH$^+$ (0.00) | 3.05 × 10$^1$ | [35] | 7.8 × 10$^{-1}$ | 39 |
| ⟨SiMe$_2$⟩ (cyclohexyl) | –3.0 (1.1) | (tol)$_2$CH$^+$ (3.63) | 2.23 × 10$^1$ | [35] | 4.9 | 4.6 |
| ⟨SiMe$_2$⟩ (cyclopentyl) | –5.5 (1.1) | Ph$_2$CH$^+$ (5.90) | 5.53 × 10$^1$ | [35] | 2.8 | 20 |
| $HMo(CO)_3Cp$ | 2.6 (0.8) | (fur)$_2$CH$^+$ (–1.36) | 6.65 × 10$^2$ | [35] | 9.8 | 68 |
| dioxolane | –3.2 [g] (0.8) | (tol)$_2$CH$^+$ (3.63) | 2.30 | [35] | 2.2 | 1.0 |
| Ph-dioxolane | –2.9 (0.8) | (ani)(tol)CH$^+$ (1.48) | 2.90 | [35] | 7.3 × 10$^{-2}$ | 40 |
| (Me)$_3$TrH | –6.2 (1.0) | (tol)$_2$CH$^+$ (3.63) | 1.01 | [35] | 2.7 × 10$^{-3}$ | 374 |
| (MeO)$_3$TrH | –4.7 [h] (1.0) | (ani)$_2$CH$^+$ (0.00) | 5.7 × 10$^{-5}$ [i] | [32] | 2.0 × 10$^{-5}$ | 2.9 |
| Ph₂cyclopropyl-H | –1.3 [j] (1.0) | (ani)$_2$CH$^+$ (0.00) | 1.69 × 10$^1$ | [35] | 5.0 × 10$^{-2}$ | 338 |
| acridane | 4.0 (0.7) | (dma)$_2$CH$^+$ (–7.02) | 1.23 × 10$^{-1}$ | [35] | 7.7 × 10$^{-3}$ | 16 |
|  |  | (dpa)$_2$CH$^+$ (–4.72) | 5.44 | [35] | 3.1 × 10$^{-1}$ | 18 |

[a] From Tables 5.7-5.11; [b] from ref. [4]; [c] calculated by equation (5.1), using $N$, $s_N$, and $E$; [d] calculated from rate constants at –70 °C with an estimated activation entropy of $\Delta S^{\neq}$ =

−100 J K$^{-1}$ mol$^{-1}$; [e] from ref. [36]; [f] calculated from 2.63 L mol$^{-1}$ s$^{-1}$ (25 °C) with $\Delta S^{\neq}$ = −100 J K$^{-1}$ mol$^{-1}$; [g] averaged value from Table 5.10; [h] averaged value from Table 5.11; [i] calculated from 7.6 × 10$^{-5}$ L mol$^{-1}$ s$^{-1}$ (23 °C) with $\Delta S^{\neq}$ = −100 J K$^{-1}$ mol$^{-1}$, in CH$_3$CN; [j] averaged value from Table 5.11.

The ratios $k_{exp}/k_{calc}$ for hydrosilanes in the last column of Table 5.12 resemble the corresponding numbers for $k_{calc}/k_{exp}$ in the last column of Table 5.5, thus indicating consistency between the nucleophilicities of hydrosilanes in Table 5.3 and Table 5.7.

The nucleophilicities $N/s_N$ of the last four hydride donors in Table 5.12 refer to CH$_3$CN as solvent, whereas the measurements were mostly performed in CH$_2$Cl$_2$. The two largest deviations from $k_{calc}$ (more than a factor of 100) fall within this group. While the reaction of (MeO)$_3$TrH with (ani)$_2$CH$^+$ in CH$_3$CN is characterized by an almost perfect agreement between experimentally observed and predicted rates ($k_{exp}/k_{calc}$ = 2.9), (Me)$_3$TrH reacts 374 times faster with (tol)$_2$CH$^+$ in CH$_2$Cl$_2$ than expected.

The second significant deviation is found for triphenylcyclopropene, which transfers a hydride 338 times faster to (ani)$_2$CH$^+$ than calculated by equation (5.1). In both cases only part of the deviations can be due to the change of solvents and an error in one of the experimental data appears to be likely.

Apart from these two cases, all other $N$-parameters in Table 5.12 seem to be quite reliable, with $k_{exp}/k_{calc}$ ≤ 68. As already stated above, deviations of this magnitude are within the confidence limit of equation (5.1), considering its reactivity range of 40 powers of ten, that is currently covered.

## 5.4. Conclusion

In summary, we could evaluate nucleophilicity parameters $N$ for numerous hydride donors widely varying in structure by using literature rate constants of hydride transfers to tritylium ions (Scheme 5.2). While ethers and acetals are relatively weak hydride donors with −5 < $N$ < −2, similar to hydrocarbons (−6 < $N$ < −1), trialkylsilanes and -germanes are considerable stronger reducing agents (0 < $N$ < 4). The hydride donating abilities of tetraalkyl-group-14-elements depend on the central atom, with silicon forming the least reactive, and lead the most reactive compounds. The nucleophilicities of transition metal hydrides cover a wide range with −1 < $N$ < 8. Their most reactive representatives contain molybdenum as central metal, and are comparable to dihydropyridines in CH$_3$CN. A comprehensive hydride donor ability scale could thus be established.

## 5. Towards a General Hydride Donor Ability Scale

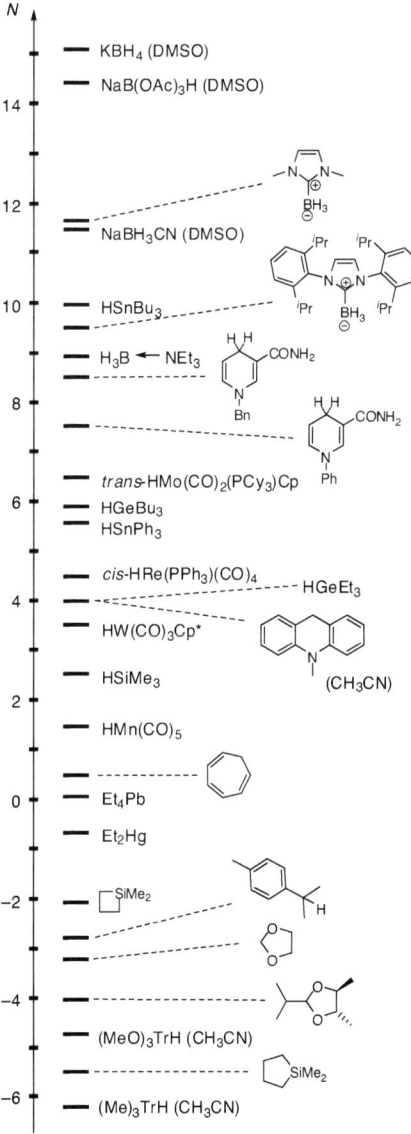

Scheme 5.2. Hydride donor ability scale of donors widely varying in structure (in $CH_2Cl_2$, if not otherwise noted); for $N$-values of compounds not reported in this chapter, see ref. [37] and Chapter 7.

## 5.5. References

[1] a) J. M. Berg, J. L. Tymoczko, L. Stryer, *Biochemistry*, 6th ed., W. H. Freeman, New York, **2008**, p. 510; b) J. W. Bunting, *Bioorg. Chem.* **1991**, *19*, 456-491.

[2] a) R. S. Velichkova, V. D. Toncheva, I. M. Panayotov, *J. Polym. Sci. Part A: Polym. Chem.* **1987**, *25*, 3283-3292; b) S. Penczek, *Makromol. Chem.* **1974**, *175*, 1217-1252.

[3] F. A. Carey, R. J. Sundberg, *Advanced Organic Chemistry, Part B: Reactions and Synthesis*, 5th ed., Springer, **2007**.

[4] H. Mayr, T. Bug, M. F. Gotta, N. Hering, B. Irrgang, B. Janker, B. Kempf, R. Loos, A. R. Ofial, G. Remennikov, H. Schimmel, *J. Am. Chem. Soc.* **2001**, *123*, 9500-9512.

[5] H. Mayr, A. R. Ofial, *J. Phys. Org. Chem.* **2008**, *21*, 584-595.

[6] D. Richter, H. Mayr, *Angew. Chem.* **2009**, *121*, 1992-1995; *Angew. Chem. Int. Ed.* **2009**, *48*, 1958-1961.

[7] a) J. Y. Corey, R. West, *J. Am. Chem. Soc.* **1963**, *85*, 2430-2433; b) F. A. Carey, H. S. Tremper, *J. Am. Chem. Soc.* **1968**, *90*, 2578-2583; c) F. A. Carey, C.-L. Wang Hsu, *J. Organomet. Chem.* **1969**, *19*, 29-41; d) J. Chojnowski, L. Wilczek, W. Fortuniak, *J. Organomet. Chem.* **1977**, *135*, 13-22; e) J. Chojnowski, W. Fortuniak, W. Stanczyk, *J. Am. Chem. Soc.* **1987**, *109*, 7776-7781; f) K.-T. Smith, M. Tilset, *J. Organomet. Chem.* **1992**, *431*, 55-64; g) C. A. Mullen, M. R. Gagné, *J. Am. Chem. Soc.* **2007**, *129*, 11880-11881; h) R. Stewart, *Can. J. Chem.* **1957**, *35*, 766-777.

[8] S. Minegishi, H. Mayr, *J. Am. Chem. Soc.* **2003**, *125*, 286-295.

[9] H. Mayr, M. Roth, R. Faust, *Macromolecules*, **1996**, *29*, 6110-6113.

[10] H. Mayr, G. Lang, A. R. Ofial, *J. Am. Chem. Soc.* **2002**, *124*, 4076-4083.

[11] G. Lang, Ph.D. Thesis, Darmstadt, **1998**.

[12] K.-H. Müller, Ph.D. Thesis, Darmstadt, **1997**.

[13] I. M. Panayotov, N. E. Manolova, R. S. Velichkova, *Polym. Bull.* **1981**, *4*, 653-660.

[14] C. A. Bunton, S. K. Huang, C. H. Paik, *Tetrahedron Lett.* **1976**, *18*, 1445-1448.

[15] H. Mayr, N. Basso, G. Hagen, *J. Am. Chem. Soc.* **1992**, *114*, 3060-3066.

[16] J. M. Jerkunica, T. G. Traylor, *J. Am. Chem. Soc.* **1971**, *93*, 6278-6279.

[17] G. Giese, A. Heesing, *Chem. Ber.* **1990**, *123*, 2737-2380.

[18] a) A. Postigo, S. Kopsov, C. Ferreri, C. Chatgilialoglu, *Org. Lett.* **2007**, *9*, 5159-5162; b) C. Chatgilialoglu, *Chem. Eur. J.* **2008**, *14*, 2310-2320.

[19] H. Bock, J. Meuret, R. Baur, K. Ruppert, *J. Organomet. Chem.* **1993**, *446*, 113-122.

[20]  a) D. N. Kursanov, Z. N. Parnes, N. M. Loim, *Synthesis* **1974**, 633-651; b) E. Keinan, *Pure & Appl. Chem.* **1989**, *61*, 1737-1746.
[21]  H. Mayr, N. Basso, *Angew. Chem.* **1992**, *104*, 1103-1105.
[22]  É. V. Uglova, V. D. Makhaev, O. A. Reutov, *J. Org. Chem. USSR (Engl. Trans.)* **1975**, *11*, 1-4.
[23]  S. S. Washburne, R. Szendroi, *J. Org. Chem.* **1981**, *46*, 691-693.
[24]  T. G. Traylor, G. S. Koermer, *J. Org. Chem.* **1981**, *46*, 3651-3657.
[25]  S. S. Washburne, J. B. Simolike, *J. Organomet. Chem.* **1974**, *81*, 41-44.
[26]  a) T.-Y. Cheng, R. M. Bullock, *Organometallics*, **1995**, *14*, 4031-4033; b) T.-Y. Cheng, R. M. Bullock, *J. Am. Chem. Soc.* **1999**, *121*, 3150-3155; c) T.-Y. Cheng, R. M. Bullock, *Organometallics* **2002**, *21*, 2325-2331.
[27]  T.-Y. Cheng, B. S. Brunschwig, R. M. Bullock, *J. Am. Chem. Soc.* **1998**, *120*, 13121-13137.
[28]  N. Sarker, J. W. Bruno, *J. Am. Chem. Soc.* **1999**, *121*, 2174-2180.
[29]  Kabir-ud-Din, P. H. Plesch, *J. Chem. Soc. Perkin Trans. 2* **1978**, 937-938.
[30]  P. Kubisa, S. Penczek, *Makromol. Chem.* **1971**, *144*, 169-182.
[31]  Z. Jedliński, J. Lukaszczyk, J. Dudek, M. Gibas, *Macromolecules* **1976**, *9*, 622-625.
[32]  L. M. McDonough, Ph.D. Thesis, Washington, **1960**.
[33]  G. J. Karabatsos, M. Tornaritis, *Tetrahedron Lett.* **1989**, *30*, 5733-5736.
[34]  M. Ishikawa, S. Fukuzumi, T. Goto, T. Tanaka, *Bull. Chem. Soc. Jpn.* **1989**, *62*, 3754-3756.
[35]  L. H. Schappele, Ph.D. Thesis, München, **2001**.
[36]  H. Mayr, B. Kempf, A. R. Ofial, *Acc. Chem. Res.* **2003**, *36*, 66-77.
[37]  Reactivity database in the internet: http://www.cup.uni-muenchen.de/oc/mayr/DBintro.html, 17.6.2011.

# 6. Reduction Potentials of Substituted Tritylium Ions

## 6.1. Introduction

Triphenylmethyl compounds and derivatives thereof have previously been studied with respect to their electrochemical properties. Electrochemical oxidations of triarylmethanes to the corresponding carbocations,[1] as well as oxidations of tritylium ions such as crystal violet, malachite green, ethyl violet, and brilliant green to further oxidation products[2] have been reported.

Electrochemical reduction potentials of tritylium ions have been determined in several investigations (Scheme 6.1). Hereby, experimental conditions like electrochemical method, solvent, temperature, conductive salt, electrode material and reference electrode, varied tremendously. Table 6.1 gives an overview of literature data concerning reduction potentials of the triphenylmethyl cation.

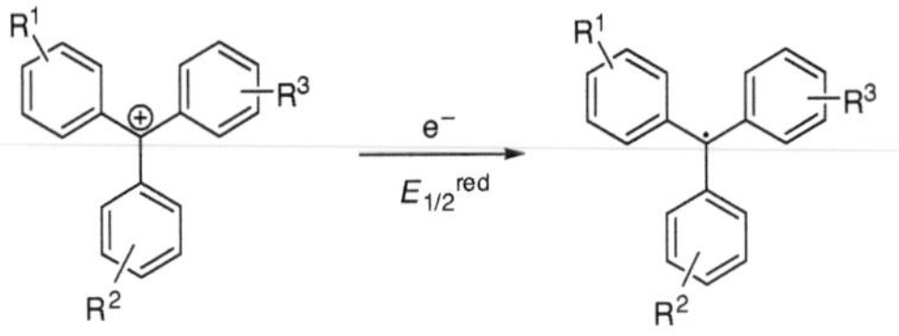

Scheme 6.1.

In some of the studies included in Table 6.1, substituted triaryl carbenium ions were analyzed, too. While Volz and Lotsch reported on the reduction potentials of several donor-substituted tritylium ions in acetonitrile,[5] Arnett determined reduction potentials of donor-substituted tritylium ions in sulfolane and oxidation potentials of substituted trityl anions in DMSO.[9] Some reduction potentials of stabilized tritylium systems in DMSO were given by Breslow.[10] Relative reduction potentials of tritylium ions were reported by Taft.[3]

Most of these studies aimed at using the electrochemical data for the calculation of other thermodynamic parameters, e.g., $pK_a$ values, with the help of thermodynamic cycles.

It becomes obvious from Table 6.1, that $E_{1/2}^{red}$ for $Tr^+$ in acetonitrile and benzonitrile is almost independent of the nature of the counterion, as $PF_6^-$ and $ClO_4^-$ or $SbCl_6^-$ and $ClO_4^-$ gave essentially the same result.

## 6. Reduction Potentials of Substituted Tritylium Ions

Table 6.1. Reduction potentials $E_{1/2}^{red}$ of the triphenylmethyl cation Tr$^+$ under various conditions.

| counterion | solvent | T [°C] | reference electrode | conductive salt | $E_{1/2}^{red}$ [mV] | ref. |
|---|---|---|---|---|---|---|
| PF$_6^-$ | CH$_2$Cl$_2$ | 23 | SCE [a] | 0.1 M Et$_4$NClO$_4$ | 331 | [4] |
| PF$_6^-$ | CH$_3$CN | 23 | SCE [a] | 0.1 M Et$_4$NClO$_4$ | 261 | [4] |
| ClO$_4^-$ | CH$_3$CN | 25 | SCE [a] | 0.1 M Bu$_4$NClO$_4$ | 270 | [5] |
| SbCl$_6^-$ | PhCN | 25 | Ag/AgCl in PhCN | Bu$_4$NClO$_4$ | 467 | [6] |
| ClO$_4^-$ | PhCN | 25 | Ag/AgCl in PhCN | Bu$_4$NClO$_4$ | 474 | [6] |
| n.r. [b] | CH$_2$Cl$_2$ | 22 | AgI in CH$_2$Cl$_2$ | Bu$_4$NBF$_4$ | 465 | [7] |
| TrOH [c] | MeSO$_3$H | 22 | Hg/HgSO$_4$ in 98 % H$_2$SO$_4$ | - | −635 | [7] |
| TrOH [c] | 97 % H$_2$SO$_4$ | 20 | n.r. [b] | - | −780 | [8] |
| BF$_4^-$ | sulfolane | 25 | NHE [d] | 0.1 M Bu$_4$NBF$_4$ | 542 | [9] |
| ClO$_4^-$ | DMSO | n.r. | SCE [a,e] | 0.1 M Bu$_4$NClO$_4$ | 190 | [10] |

[a] standard calomel electrode; [b] not reported; [c] precursor; [d] normal hydrogen electrode, measured against Ag/AgNO$_3$; [e] measured against Ag/AgCl in DMSO.

In the present work, substituted tritylium ions as well as other types of electrophiles were subjected to steady-state cyclic voltammetry in acetonitrile to determine their reduction potentials $E_{1/2}^{red}$. As all substrates were studied under the same conditions, direct comparisons become possible, and the relationship of the obtained reduction potentials with the corresponding electrophilicity parameters $E$ can be analyzed.

## 6.2. Results

The electrochemical window, in which data could be obtained, was limited by the solvent acetonitrile. As shown by Figure 6.1, acetonitrile gets reduced at approximately −1.8 V and oxidized at approximately 3 V with respect to the Ag/Ag$_2$O reference electrode.[11,12] Accurate values could only be obtained within this interval.

Because an array of 8 platinum ultramicroelectrodes was used as working electrode (each disk had a radius of 10 $\mu$m), the detected currents $i$ were so small that the $iR$ drop caused by the resistance of the solution was assumed to be negligible. Consequently, no conductive salt was added.

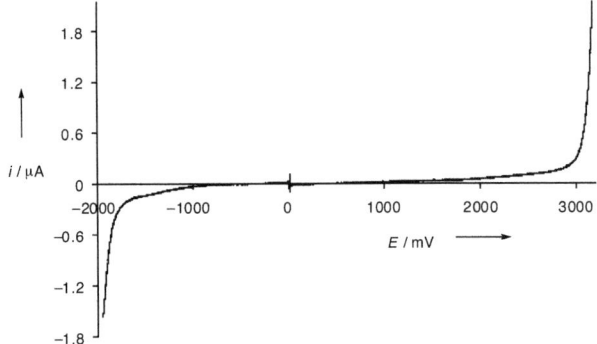

Figure 6.1: The electrochemical window of the solvent acetonitrile (25 °C), vs. Ag/Ag$_2$O.

Even for reversible, diffusion-controlled processes, i.e., when the rate constant for the heterogeneous electron transfer at the electrode is fast, the actually measured half-wave reduction potentials $E_{1/2}^{\text{red}}$ are not identical with the thermodynamic standard reduction potentials $E^0$. The difference between the two values is small, however, when the diffusion coefficients $D$ and the activity coefficients $\gamma$ of the reduced and oxidized forms are similar (equation 6.1).

$$E_{1/2}^{\text{red}} = E^0 + \frac{RT}{nF} \ln\left(\frac{D_R}{D_O}\right) + \frac{RT}{nF} \ln\left(\frac{\gamma_O}{\gamma_R}\right) \tag{6.1}$$

Figure 6.2 illustrates, that forward and backward scan were not exactly superimposed, but the deviations were negligible.

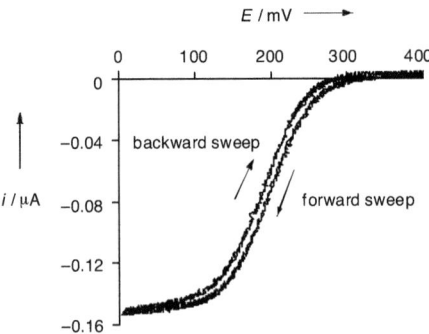

Figure 6.2. Forward and backward scan for the reduction of Me$_3$Tr$^+$BF$_4^-$, CH$_3$CN, 25 °C, vs. Ag/Ag$_2$O.

## 6. Reduction Potentials of Substituted Tritylium Ions 115

This phenomenon can either be explained by the fact, that the solutions were not stirred during the experiments, causing small depletions of the concentrations of the electroactive species near the electrodes, or by non-faradayic charging currents. The following presentation and the discussion of results are based on the forward scans.

The scan rate was usually adjusted to 20 mV/s. This small rate and the employment of ultramicroelectrodes gave rise to steady-state shapes of the recorded curves as depicted in Figure 6.2 and Figure 6.3a. When the scan rates were increased (Figure 6.3b-d) the shapes of the curves approached that obtained in experiments using electrodes of bigger size, where the current is governed by linear diffusion. As can be seen in Figure 6.3 (in b-d both, forward and backward scans, are shown) a current peak develops (in contrast to steady-state currents) when the scan rate is increased from 20 to 200 to 500 to 1000 mV/s. The larger gradients lead to pronounced transient diffusion, meaning that the reduction process at the electrode is faster than the transport of the cationic species to the electrode.

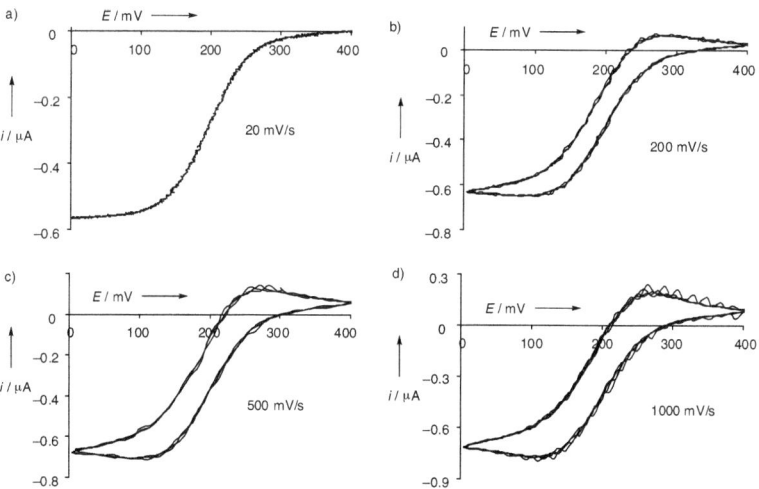

Figure 6.3. Influence of the scan rate on the shape of the waves for the reduction of $Me_3Tr^+BF_4^-$ ($CH_3CN$, 25 °C), vs. $Ag/Ag_2O$.

As expected, higher concentrations of the substrate lead to increased steady-state currents (limiting currents $i_l$), as is shown in Figure 6.4 for the reduction of $(MeO)Tr^+BF_4^-$.

When the applied potentials were further decreased (beyond −50 mV in Figure 6.4), reduction waves corresponding to the conversion radical → anion could be observed. As reduction potentials of the tritylium ions (not the radicals) were of interest, and constant limiting currents were not reached in the second waves due to irreversibilities caused by the instabilities of the tritylium anions, these waves have not been evaluated. It was previously shown that the reduction of $Tr^+$ in the aprotic solvent benzonitrile proceeds reversibly, whereas the reduction of the trityl radical to the anion is irreversible.[6a]

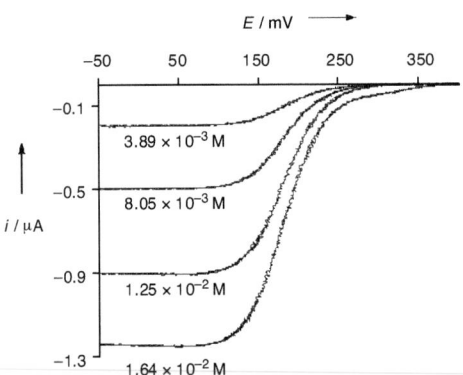

Figure 6.4. Reduction waves for $(MeO)Tr^+BF_4^-$ (20 mV/s, $CH_3CN$, 25 °C, vs. $Ag/Ag_2O$) in dependence of the substrate concentration.

Figure 6.4 also demonstrates the independence of the half-wave reduction potential $E_{1/2}^{red}$ (the potential at which $i = i_l/2$) of the substrate concentration.

For a disk shaped electrode the limiting current $i_l$ is given by equation (6.2),[13] where $n$ refers to the stoichiometric number of electrons transferred, $D$ is the diffusion coefficient, $c$ the substrate concentration and $r$ the radius of the disk.

$$i_l = 4nFDcr \qquad (6.2)$$

Therefore, plots of $i_l$ versus $c$ result in straight lines (Figure 6.5) with the slopes $4nFDr$, from which the diffusion coefficients $D$ can be derived (the fact that an array of 8 microdisc electrodes was used in the present work had to be taken into account).

## 6. Reduction Potentials of Substituted Tritylium Ions

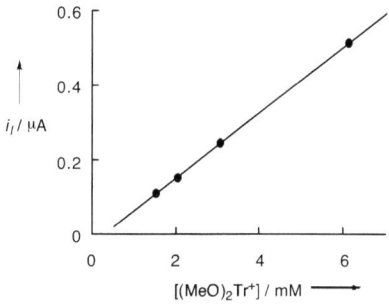

Figure 6.5. Correlation of the limiting currents $i_l$ for the reductions of $(MeO)_2Tr^+BF_4^-$ with $[(MeO)_2Tr^+BF_4^-]$ $(CH_3CN, 25\ °C)$.

When reduction processes are considered without reduced species initially present in the solution (as it was the case in this work, as only tritylium ions were present), equation (6.3) is valid, and plotting the potential $E$ against $\log\left(\dfrac{i_l - i}{i}\right)$ should give rise to straight lines.

$$E = E_{1/2}^{\text{red}} + \frac{RT}{nF}\ln\left(\frac{i_l - i}{i}\right) \qquad (6.3)$$

The slopes of the lines drawn in Figure 6.6 are near the theoretical value of 59 mV for a reversible one-electron transfer. It is shown later that not all species could be reduced ideally reversibly. The intercepts of the lines in Figure 6.6 refer to the reduction half-wave potentials $E_{1/2}^{\text{red}}$.

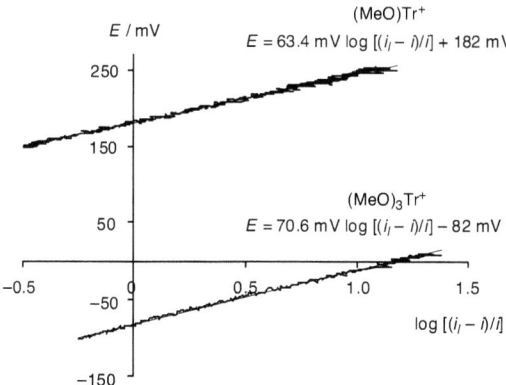

Figure 6.6. Plot of potential $E$ vs. $\log[(i_l - i)/i]$ for the scans of $(MeO)Tr^+$ ($3.89 \times 10^{-3}$ mol $L^{-1}$) and $(MeO)_3Tr^+$ ($1.87 \times 10^{-3}$ mol $L^{-1}$) in $CH_3CN$ at 25 °C.

To avoid liquid junction potentials by working in two different solvents with a reference electrode in aqueous solution, the quasi-reference electrode Ag/Ag$_2$O in acetonitrile was used. Because quasi-reference electrodes are known to be not very stable, each compound was studied with a small amount of ferrocene as internal standard.

Due to the robustness of its reversibility at experimentally accessible potentials, the ferrocene/ferrocenium (Fc/Fc$^+$) redox couple has been established as one of the standards for the calibration of electrochemical measurements.[14,15,16] However, it has been reported that the oxidation potential of ferrocene is strongly dependent on the dielectric properties of the solvent. A shift of 160 mV was observed when the concentration of tetrabutylammonium tetrafluoroborate was varied from 1 to 500 mM in acetonitrile.[14]

The observation of slightly differing oxidation potentials for ferrocene in Table 6.2 might therefore be due to either slightly different ionic strengths in the experiments, or the instability of the reference electrode.

Furthermore, when destabilized tritylium ions were combined with ferrocene, chemical interactions seemed to occur, leading to a significant current between the two half-waves (not shown). Only for stabilized systems like (Me$_2$N)(MeO)Tr$^+$, no chemical interaction was observed, the two half-waves being well separated from each other with no current flow at potentials in between (Figure 6.7).

In consequence, ferrocene was not suited as an internal reference for the determination of reduction potentials of tritylium ions in general, and the obtained data have not been adjusted respectively.

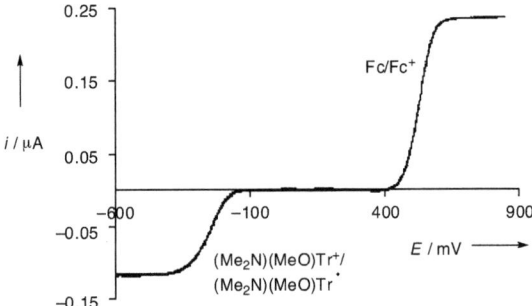

Figure 6.7. Detection of ferrocene oxidation and (Me$_2$N)(MeO)Tr$^+$ reduction in a single experiment (20 mV/s, CH$_3$CN, 25 °C, vs. Ag/Ag$_2$O).

## 6. Reduction Potentials of Substituted Tritylium Ions

Table 6.2 gives an overview of all obtained reduction potentials $E_{1/2}^{red}$, diffusion coefficients $D$ and the oxidation potentials $E_{1/2}^{ox}$ of ferrocene. For comparison, the diffusion coefficient of ferrocene in acetonitrile has been reported to be $2.2 \times 10^{-9}$ m$^2$/s.[12]

All cationic species have been employed as either tetrafluoroborate or chloride salts. Only the *meta*-fluoro-substituted tritylium ion has been generated in solution by mixing the corresponding bromide with a small excess of the Lewis acid GaCl$_3$.

It has to be mentioned that the values for Tr$^+$ and less stabilized derivatives have to be taken with care, as often no stable limiting currents were reached. The systems (*p*F)Tr$^+$BF$_4^-$ and (*m*F)(*m*F)'TrBr/GaCl$_3$ did not give satisfying results at all.

### 6.3. Discussion

In case of the neutral compounds, i.e., the dinitrothiophene **1b**, the tetrazolopyridine **1e**, and the benzofuroxan **1g**, the oxidation potentials $E_{1/2}^{ox}$ of ferrocene are significantly high (Table 6.2). It has been reported that $E_{1/2}^{ox}$ of ferrocene increases with decreasing ionic strength of the solution.[14] As a conductive salt has not been used in the present work, the ionic strengths of the solutions were especially low in the experiments involving **1b**, **1e**, and **1g**, thus explaining the results in Table 6.2.

The *S*-values (last column of Table 6.2) theoretically approach 59 mV for reversible one-electron reductions. The only system, where this value is observed, is (*m*$^t$Bu)$_6$Tr$^+$, possibly due to the shielding effect of the six bulky groups, preventing the generated radicals to undergo subsequent reactions. All other species display slightly or significantly larger values of *S*. Particularly high numbers are again found for **1b**, **1e** and **1g**, which is another indication that that the conductivities of the solutions have been too low in these experiments to reliably evaluate reduction potentials. Unfortunately, a dry supporting electrolyte was not available for increasing the conductivities of the solutions.

The tritylium ions (*m*F)Tr$^+$, (*p*F)$_3$Tr$^+$, and (Me$_2$N)Tr$^+$ also exhibit large *S*-parameters, the corresponding radicals obviously being particularly unstable. The reason for the big deviation of *S* for (Me$_2$N)Tr$^+$ is not understood at the moment.

As it is not trivial to evaluate reduction potentials from half-waves with *S*-values significantly different from 59 mV, the corresponding values in Table 6.2 have to be considered as preliminary. In these cases $E_{1/2}^{red}$ do not reflect the thermodynamic reduction potentials $E^0$.

Table 6.2: Electrophilicity parameters $E$, reduction potentials $E_{1/2}^{red}$, and diffusion coefficients $D$ of the electrophilic species studied in this work, as well as the oxidation potentials $E_{1/2}^{ox}$ of the internal standard ferrocene, $CH_3CN$, 25 °C.

| compound [a] | counterion | $E$ [b] | $E_{1/2}^{red}$ [c] [mV] | $D$ [m$^2$ s$^{-1}$] | $E_{1/2}^{ox}$ [c] [mV] | $S$ [d] [mV] |
|---|---|---|---|---|---|---|
| ($m$F)Tr$^+$ | R-Br/GaCl$_3$ | 1.01 | (424) [e] | - | - | 113 |
| ($p$F)$_3$Tr$^+$ | BF$_4^-$ | 0.05 | (195) [e] | - | (542) [e] | 121 |
| ($m^t$Bu)$_6$Tr$^+$ | BF$_4^-$ | - | (249) [e] | - | (544) [e] | 59 |
| Tr$^+$ | BF$_4^-$ | 0.51 | (381) [e] | - | - | 73 |
| MeTr$^+$ | BF$_4^-$ | −0.13 | 312 | 2.31 × 10$^{-9}$ | - | 70 |
| Me$_2$Tr$^+$ | BF$_4^-$ | −0.70 | 262 | 1.72 × 10$^{-9}$ | 535 | 63 |
| Me$_3$Tr$^+$ | BF$_4^-$ | −1.21 | 197 | 2.57 × 10$^{-9}$ | 518 | 78 |
| (MeO)Tr$^+$ | BF$_4^-$ | −1.59 | 182 | 2.76 × 10$^{-9}$ | 503 | 63 |
| (MeO)$_2$Tr$^+$ | BF$_4^-$ | −3.04 | 52 | 2.85 × 10$^{-9}$ | 537 | 66 |
| (MeO)$_3$Tr$^+$ | BF$_4^-$ | −4.35 | −82 | 3.47 × 10$^{-9}$ | 534 | 71 |
| (Me$_2$N)Tr$^+$ | BF$_4^-$ | −7.93 | (−216) [e] | 2.54 × 10$^{-9}$ | 550 | 107 |
| (Me$_2$N)(MeO)Tr$^+$ | BF$_4^-$ | −7.98 | −306 | 2.90 × 10$^{-9}$ | 527 | 72 |
| (Me$_2$N)$_2$Tr$^+$ | BF$_4^-$ | −10.29 | −406 | 6.06 × 10$^{-9}$ | 478 | 82 |
| (Me$_2$N)$_3$Tr$^+$ | Cl$^-$ | −11.26 | −548 | 4.28 × 10$^{-9}$ | 521 | 78 |
| 1a | BF$_4^-$ | −3.72 [f] | −72 | 5.96 × 10$^{-9}$ | 526 | 68 |
| 1b | - | −12.33 [g] | (−433) [e] | 1.35 × 10$^{-9}$ | 590 | 104 |
| 1c | BF$_4^-$ | - | −796 | 3.43 × 10$^{-9}$ | 547 | 83 |
| 1d | BF$_4^-$ | −5.90 [h] | −135 | 3.76 × 10$^{-9}$ | - | 72 |
| 1e | - | - | (−755) [e] | 3.63 × 10$^{-9}$ | 597 | 133 |
| 1f | BF$_4^-$ | −7.30 [h] | −169 | 4.57 × 10$^{-9}$ | 523 | 80 |
| 1g | - | - | −6.41 [g] | (−262) [e] | 1.55 × 10$^{-9}$ | 525 | 127 |

[a] For the substitution pattern of tritylium ions, see Scheme 6.1; [b] from Chapters 2 and 3, if not otherwise noted; [c] vs. Ag/Ag$_2$O in CH$_3$CN; [d] $S = dE/d\log[(i_l - i)/i]$; [e] these values have to be taken with care, as no stable limiting currents were observed, or because of the high $S$-values; [f] from ref. [17]; [g] from ref. [18]; [h] not published.

## 6. Reduction Potentials of Substituted Tritylium Ions 121

A linear correlation between the empirical electrophilicity parameters $E$ of benzhydrylium ions with their one-electron reduction potentials $E_{1/2}^{red}$ in acetonitrile was previously reported.[19]

While the reduction potentials determined in the present work have been measured with respect to the Ag/Ag$_2$O reference electrode, the values for benzhydrylium ions refer to the standard calomel electrode (SCE).

A direct comparison between the two classes of compounds is complicated by the fact, that the Ag/Ag$_2$O electrode is commonly used as a quasi-reference electrode, and therefore its potential against established references, like the SCE, is not known.

However, if the redox potential of the Fc/Fc$^+$ couple in acetonitrile is averaged to 530 mV vs. Ag/Ag$_2$O (cf. Table 6.2), this value can be combined with the Fc/Fc$^+$ potential of 380 mV vs. the SCE,[16] to calculate a potential of the Ag/Ag$_2$O reference electrode in acetonitrile against the SCE of –150 mV. The new value for Me$_3$Tr$^+$ of 47 mV may be compared with a literature value of 50 mV.[5]

In Figure 6.8, which compares benzhydrylium and tritylium ions, all reduction potentials refer to the SCE. It can be seen, that a linear correlation also exists for tritylium ions, although slight scattering must be admitted. The largest deviation exhibits the system $(pF)_3Tr^+$, for which $E_{1/2}^{red}$ is not reliable.

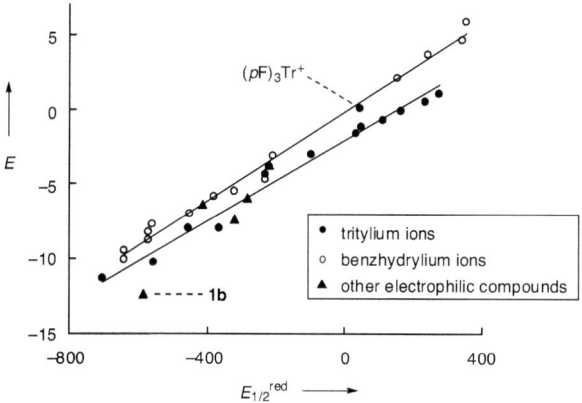

Figure 6.8. Correlation of empirical electrophilicity parameters $E$ with reduction potentials $E_{1/2}^{red}$ (vs. SCE, in mV, CH$_3$CN, 25 °C); data for benzhydrylium ions from ref. [19]. The two correlation lines refer to the benzhydryl and the trityl series, respectively.

Despite a clear common trend comprising the benzhydryl and the trityl series, two diverging correlations are observed for the two series, with the gap between the lines increasing for decreasing stabilization of the carbenium ions.

The tropylium ion (**1a**), the tetrazoles **1c-f** and the benzofuroxan **1g** also fit the correlation, their scattering being larger than the difference between the two correlation lines. The relatively high deviation of the thiophene **1b** may be due to the sulfur, which possibly alters the surface of the electrode during the reduction process and therefore falsifies the reduction potential.

The fact that tritylium ions are slightly less electrophilic than benzhydrylium ions of the same reduction potential can be explained by the higher steric demand of the reactions with tritylium ions. It is not clear, however, why the differences increase when more reactive systems are considered.

When the reduction potentials of benzhydrylium and tritylium ions are plotted against the gas phase LUMO energies, which have been calculated on the B3LYP/6-31G(d,p) level of theory, the amino-substituted tritylium ions follow a correlation line different to that of the other tritylium ions (filled circles in Figure 6.9). Although $(Me_2N)Tr^+$ has a lower LUMO energy in the gas phase, it also has a lower reduction potential than $(MeO)_3Tr^+$.

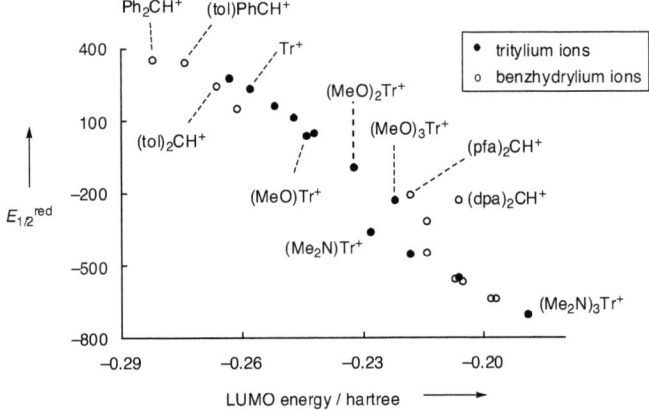

Figure 6.9. Correlation of reduction potentials $E_{1/2}^{red}$ (vs. SCE, in mV, $CH_3CN$, 25 °C) with the corresponding LUMO energies [B3LYP/6-31G(d,p)], data for benzhydrylium ions from ref. [19] and [20].

## 6. Reduction Potentials of Substituted Tritylium Ions 123

This situation resembles the picture given in Figure 2.7, where it was shown that although these two systems are of equal stability in the gas phase, the stability of $(Me_2N)Tr^+$ exceeds that of $(MeO)_3Tr^+$ by 18 kJ mol$^{-1}$ in aqueous solution. Exceptionally high solvation enthalpies of the mono- and diamino substituted tritylium ions, which have been used to rationalize the results of Figure 2.7, may also explain the results in Figure 6.9.

Less data are available in the benzhydryl series, and considerable scattering exists for amino-substituted benzhydrylium ions. Although a thorough analysis, therefore, appears to be difficult, a general common trend comprising both series of compounds becomes obvious from Figure 6.9. A good linear correlation of $E_{1/2}^{red}$ values with LUMO energies has also been reported for substituted benzofurazanes.[21]

A mechanistic question always controversially discussed, is whether hydride transfer reactions proceed in a stepwise or a concerted manner.[19,22] The stepwise reaction involves an initial single electron transfer (SET) leading to a radical pair, followed by a subsequent hydrogen atom transfer. In contrast, the polar mechanism proceeds via a one step hydride transfer (Scheme 6.2). Whether one or the other mechanism applies, is strongly dependent on the reaction partners as well as the reaction conditions. It was argued previously that hydride transfers from silanes to tritylium ions are characterized by SET.[23]

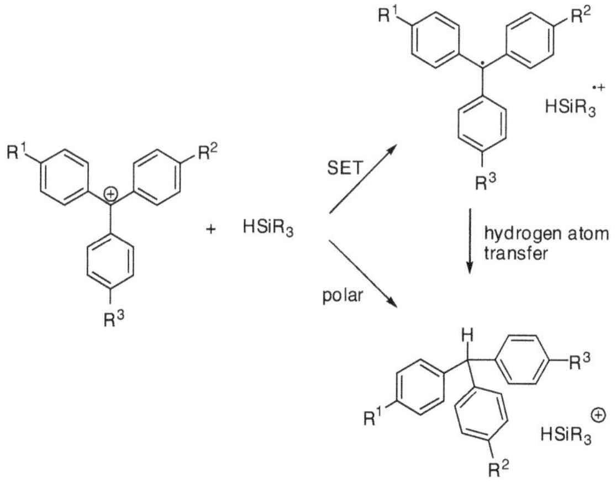

Scheme 6.2. Two mechanistic pathways for the hydride transfer from a silane to a tritylium ion.

In Chapters 3 and 5 of this thesis, hydride transfers to tritylium ions have been investigated kinetically. With the reduction potentials of tritylium ions at hand, it is now possible to analyze the underlying mechanism of these reactions in a more detailed fashion.

The free energy of the SET process $\Delta G^0_{SET}$ can be calculated by equation (6.4). As the corresponding free energy of activation, $\Delta G^{\ddagger}_{SET}$, cannot be smaller, the maximum rate of hydride transfer according to the SET can be obtained by substituting $\Delta G^0_{SET}$ into the Eyring equation.

$$\Delta G^0_{SET} = F\Delta E^0 = F(E^{ox} - E^{red}) \qquad (6.4)$$

The oxidation potentials of dimethylphenylsilane and phenylsilane in acetonitrile have been reported as 2.2 V vs. SCE and 2.1 V vs. Ag/AgNO$_3$ (0.01 M), respectively.[24] The latter value can be converted to the SCE as reference by addition of 0.3 V.[16] It is reasonable to expect the oxidation potentials of other trialkylsilanes to be of similar magnitude, i.e., approximately 2 V vs. SCE.

Figure 6.10 exemplifies the improbability for a stepwise process in the reactions of tritylium ions with dimethylphenylsilane.

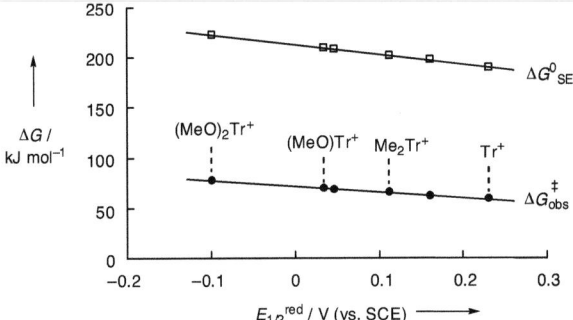

Figure 6.10. Reactions of substituted tritylium ions with dimethylphenylsilane. Plots of $\Delta G^0_{SET}$ [calculated by equation (6.4), CH$_3$CN, 25 °C] and $\Delta G^{\ddagger}_{obs}$ (from the experimentally determined rate constants in Chapter 5, CH$_2$Cl$_2$, 20 °C) against the reduction potentials of substituted tritylium ions.

The free energies $\Delta G^0_{SET}$ have been calculated according to equation (6.4) with $E^{ox} = 2.2$ V and the reduction potentials of tritylium ions against the SCE as $E^{red}$. $\Delta G^{\ddagger}_{obs}$ are the free energies of activation, which have been obtained by substitution of the experimentally determined rate constants of hydride transfers from dimethylphenylsilane to tritylium ions (see Chapter 5) into the Eyring equation.

Although the values for $\Delta G^0_{SET}$ and $\Delta G^{\ddagger}_{obs}$ refer to acetonitrile and dichloromethane, respectively, it is unlikely that the different solvents can account for a change in $\Delta G$ of more than 100 kJ mol$^{-1}$. As a matter of fact, it was shown in Table 5.6, that hydride transfers from silanes to tritylium ions proceed approximately 10 times faster in dichloromethane than in acetonitrile. As a deceleration of the rate by a factor of 10 leads to an increase of $\Delta G^{\ddagger}_{obs}$ by only 5.6 kJ mol$^{-1}$ (at 20 °C), the large gap of more than 100 kJ mol$^{-1}$ between the two lines in Figure 6.10 strongly suggests a polar one-step hydride shift in the reactions of silanes with tritylium ions.

## 6.4. References

[1]  J. E. Kuder, W. W. Limburg, M. Stolka, S. R. Turner, *J. Org. Chem.* **1979**, *44*, 761-766.
[2]  Z. Galus, R. Adams, *J. Am. Chem. Soc.* **1964**, *86*, 1666–1671.
[3]  W. D. Jenson, R. W. Taft, *J. Am. Chem. Soc.* **1964**, *86*, 116-117.
[4]  M. F. Asaro, G. S. Bodner, J. A. Gladysz, S. R. Cooper, N. J. Cooper, *Organometallics* **1985**, *4*, 1020-1024.
[5]  H. Volz, W. Lotsch, *Tetrahedron Lett.* **1969**, 2275-2278.
[6]  a) G. Kothe, W. Sümmermann, H. Baumgärtel, H. Zimmermann, *Tetrahedron Lett.* **1969**, 2185-2188; b) G. Kothe, W. Sümmermann, H. Baumgärtel, H. Zimmermann, *Tetrahedron* **1972**, *28*, 5949-5955.
[7]  a) P. H. Plesch, *J. Chem. Soc., Perkin Trans. 2* **1989**, 1139-1142; b) P. H. Plesch, F. G. Thomas, *J. Chem. Soc., Perkin Trans. 2* **1975**, 1532-1535.
[8]  M. I. James, P. H. Plesch, *Chem. Commun.* **1967**, 508-510.
[9]  E. M. Arnett et al., *J. Phys. Org. Chem.* **1997**, *10*, 499-513.
[10] R. Breslow, W. Chu, *J. Am. Chem. Soc.* **1972**, *95*, 411-418.
[11] S. S. Wang, D. W. Chilcott, *US Patent* 5385652.
[12] A.-M. Gonçalves, C. Mathieu, M. Herlem, A. Etcheberry, *J. Electroanal. Chem.* **1999**, 140-145.
[13] A. J. Bard, L. R. Faulkner, *Electrochemical Methods, Fundamentals and Applications*, 2nd ed. **2001**, Wiley.

[14]  D. Bao, B. Millare, W. Xia, B. G. Steyer, A. A. Gerasimenko, A. Ferreira, A. Contreras, V. I. Vullev, *J. Phys. Chem. A* **2009**, *113*, 1259-1267.

[15]  M. V. Mirkin, T. C. Richards, A. J. Bard, *J. Phys. Chem.* **1993**, *97*, 7672-7677.

[16]  V. V. Pavlishchuk, A. W. Addison, *Inorg. Chim. Acta* **2000**, 97-102.

[17]  H. Mayr, B. Kempf, A. R. Ofial, *Acc. Chem. Res.* **2003**, *36*, 66-77.

[18]  F. Terrier, S. Lakhdar, T. Boubaker, R. Goumont, *J. Org. Chem.* **2005**, *70*, 6242-6253.

[19]  A. R. Ofial, K. Ohkubo, S. Fukuzumi, R. Lucius, H. Mayr, *J. Am. Chem. Soc.* **2003**, *125*, 10906-10912.

[20]  T. Singer, *PhD Thesis*, **2008**, Munich.

[21]  W. R. Fawcett, P. A. Forte, R. O. Loutfy, J. M. Prokipcak, *Can. J. Chem.* **1971**, *50*, 263-269.

[22]  a) Y. Apeloig, O. Merin-Aharoni, D. Danovich, A. Ioffe, S. Shaik, *Israel J. Chem.* **1993**, *33*, 387-402; b) R. J. Klingler, K. Mochida, J. K. Koch, *J. Am. Chem. Soc.* **1979**, *101*, 6626-663; c) H. Mayr, N. Basso, G. Hagen, *J. Am. Chem. Soc.* **1992**, *114*, 3060-3066.

[23]  J. Chojnowski, W. Fortuniak, W. Stańczyk, *J. Am. Chem. Soc.* **1987**, *109*, 7776-7781.

[24]  A. Kunai, T. Kawakami, E. Toyoda, T. Sakurai, M, Ishikawa, *Chem. Lett.* **1993**, 1945-1948.

# 7. Miscellaneous Experiments

## 7.1. Nucleophilicitiy Parameters for N-Heterocyclic Carbene Boranes

N-heterocyclic carbene boranes have recently been employed as hydride donors in synthetic applications.[1] In order to include these substances in the comprehensive nucleophilicity scale of hydride donors in Chapter 5, the reactions of 1,3-bis(2,6-diisopropylphenyl)-imidazol-2-ylidene borane (**1**) and 1,3-dimethylimidazol-2-ylidene borane (**2**) with benzhydrylium ions were analyzed kinetically.

The rates of these reactions have been measured in dichloromethane solution at 20 °C. When a high excess of **1** or **2** was added to a benzhydrylium tetrafluoroborate, the absorbance of the carbenium ion decreased mono-exponentially, according to a second-order rate law. Plots of the observed rate constants $k_{obs}$ versus the concentrations of the carbene boranes were linear with the slopes of the correlation lines representing the second-order rate constants $k$ for the hydride transfers (Figure 7.1 and Table 7.1).

Table 7.1. Second-order rate constants for hydride transfers from carbene boranes **1** and **2** to substituted benzhydrylium ions ($CH_2Cl_2$, 20 °C), and derived nucleophilicity parameters $N/s_N$.

| donor | acceptor [a] | electrophilicity $E$ [b] | $k$ / L mol$^{-1}$ s$^{-1}$ | $N$ ($s_N$) |
|---|---|---|---|---|
| 1 | (dma)$_2$CH$^+$ | −7.02 | 1.16 × 10$^2$ | 9.55 (0.81) |
|   | (thq)$_2$CH$^+$ | −8.22 | 1.04 × 10$^1$ |   |
|   | (jul)$_2$CH$^+$ | −9.45 | 1.27 |   |
| 2 | (thq)$_2$CH$^+$ | −8.22 | 4.05 × 10$^2$ | 11.77 (0.84) |
|   | (jul)$_2$CH$^+$ | −9.45 | 5.05 × 10$^1$ |   |

[a] dma = 4-(dimethylamino)phenyl; thq = 1,2,3,4-tetrahydroquinoline-6-yl; jul = julolidine-4-yl; [b] from ref. [4].

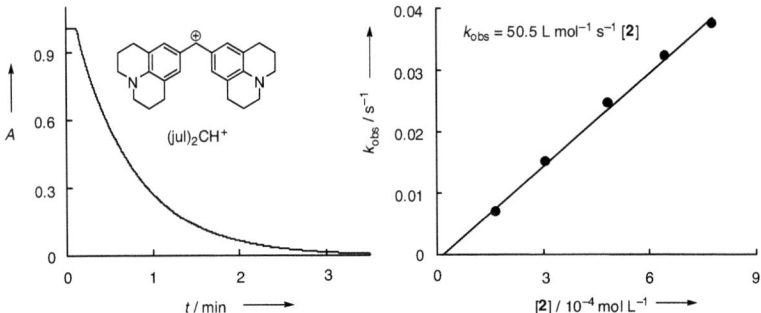

Figure 7.1. Left: Absorbance decay for the reaction of (jul)$_2$CH$^+$BF$_4^-$ ($c_0$ = 1.38 × 10$^{-5}$ mol L$^{-1}$) with **2** (c = 4.84 × 10$^{-4}$ mol L$^{-1}$), $k_{obs}$ = 2.45 × 10$^{-2}$ s$^{-1}$, CH$_2$Cl$_2$, 20 °C; right: plot of $k_{obs}$ versus [**2**], $k$ = 5.05 × 10$^1$ L mol$^{-1}$ s$^{-1}$.

According to equation (7.1), plots of log $k$ against the known electrophilicity parameters $E$ of the benzhydrylium ions (Table 7.1) give rise to straight lines, from which the nucleophilicity parameters $N$ and $s_N$ of the carbene boranes can be derived (Table 7.1).

$$\log k = s_N(E + N) \qquad (7.1)$$

The linear correlations in Figure 7.2 reveal a 100-fold higher reactivity of the methyl-substituted system **2** compared with the aryl-substituted system **1**.

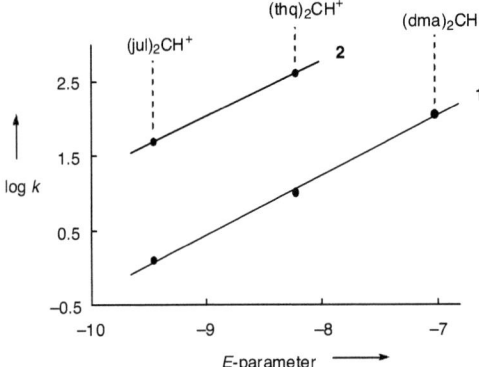

Figure 7.2. Plots of log $k$ against the electrophilicity parameters $E$ of benzhydrylium ions.

# 7. Miscellaneous Experiments

Recently, quantum chemical calculations showed, that while the two phenyl rings in diphenylimidazole-2-ylidene are only slightly distorted out of the plane of the heterocycle, the two mesityl groups in dimesitylimidazole-2-ylidene are almost perpendicular to the 5-membered ring.[2]

|  | Ph | Mes |  |
|---|---|---|---|
| diehdral angle | 27 ° | ca. 90 ° | Scheme 7.1. |

As the steric demand of the 2,6-bis(isopropyl)phenyl group in **1** is even larger than that of the mesityl residue, a perpendicular orientation of the aryl rings can also be assumed for compound **1**. Consequently, no mesomeric interaction between the heterocycle and the aryl-substituent can occur, the influence of the latter being purely inductive. Both, the smaller electron-donating inductive effect as well as the considerable steric demand of the two bis(isopropyl)phenyl rings in **1** compared to the methyl groups in **2**, may explain the lower reactivity of the former compound.

With $N$-parameters between 9 and 12 (Table 7.1), the carbene boranes are slightly less nucleophilic than the borohydride anion $BH_4^-$ in DMSO ($N \approx 15$).[3] They do, however, possess similar nucleophilicities to the cyanoborohydride anion $BH_3CN^-$ in DMSO ($N \approx 11.5$),[3] and are even slightly better hydride donors than the triethylamine-borane complex in $CH_2Cl_2$ ($N \approx 8.9$).[4]

## 7.2. Hydride Transfers from Dihydropyridines to Tritylium Ions

The redox couple 1,4-dihydropyridine/pyridinium ion plays an important role in biology (NADH/NAD$^+$). The 1,4-dihydropyridine moiety is prone to hydride loss, as this conversion leads to a stable aromatic system (Scheme 7.2). A series of different 1,4-dihydropyridines have previously been studied in reactions with substituted benzhydrylium ions in order to evaluate their nucleophilicity parameters $N$, according to equation (7.1).[3]

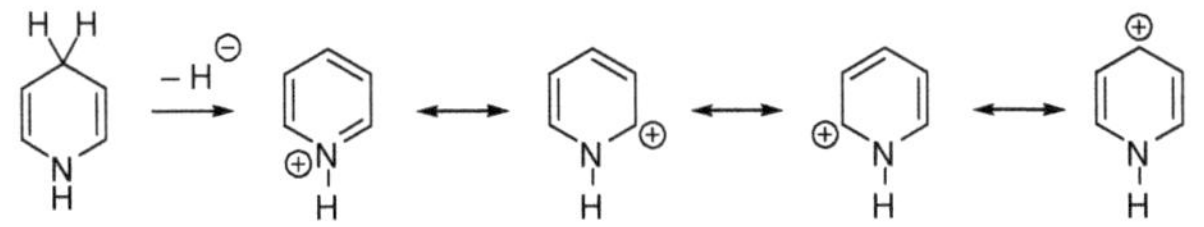

# 7. Miscellaneous Experiments

Scheme 7.2.

It was furthermore found that the experimentally determined rate constants of hydride transfers from N-benzyldihydronicotinamide (**3a**, Scheme 7.3) to tritylium ions agreed well with those calculated by equation (7.1).[3,5] In the present work, reactions of the Hantzsch esters **3b** and **3c** with tritylium ions were studied in dichloromethane solution at 20 °C.

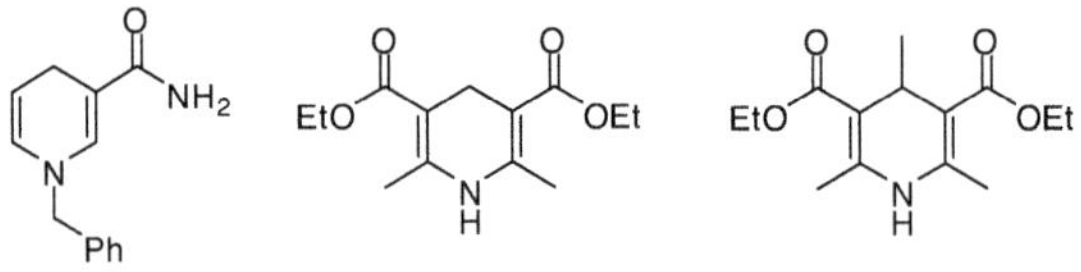

**3a**    **3b**    **3c**    Scheme 7.3.

Product studies revealed the exclusive formation of triarylmethanes and pyridinium ions, when tritylium tetrafluoroborates were combined with equimolar amounts of **3b** and **3c**. Figure 7.3 shows the general down-field shift of the protons in **3c** when it becomes oxidized. Especially the 9 protons b and f of the 3 methyl groups of the collidine moiety are affected by the developing adjacent positive charge (Scheme 7.2). Proton d disappears in the course of the reaction, while the dublet of the protons f transforms into a singlet.

That a lutidinium ion is formed from **3b** is demonstrated in Figure 7.4. The remaining hydrogen at the position f is quite deshielded, and resonates at 9.22 ppm.

However, the time-dependent absorbances of the tritylium ions during these reactions indicated reaction pathways which were not trivial. Hydride transfers with second-order kinetics would require mono-exponential decays when the nucleophiles are used in high excess over the electrophiles. As depicted in Figure 7.5, neither the reaction of $(Me_2N)Tr^+$ with **3b**, nor the reaction of $(MeO)Tr^+$ with **3c** fulfilled this requirement.

It is noteworthy that although $(MeO)Tr^+$ was immediately consumed ($t < 1$ s) when it was combined with **3b**, its concentration did not reach zero even after 40 min when it was combined with **3c** (Figure 7.5b). These results can not be explained satisfactorily at the moment, and because no mechanistic model was available, an evaluation of the data appeared to be difficult.

# 7. Miscellaneous Experiments 131

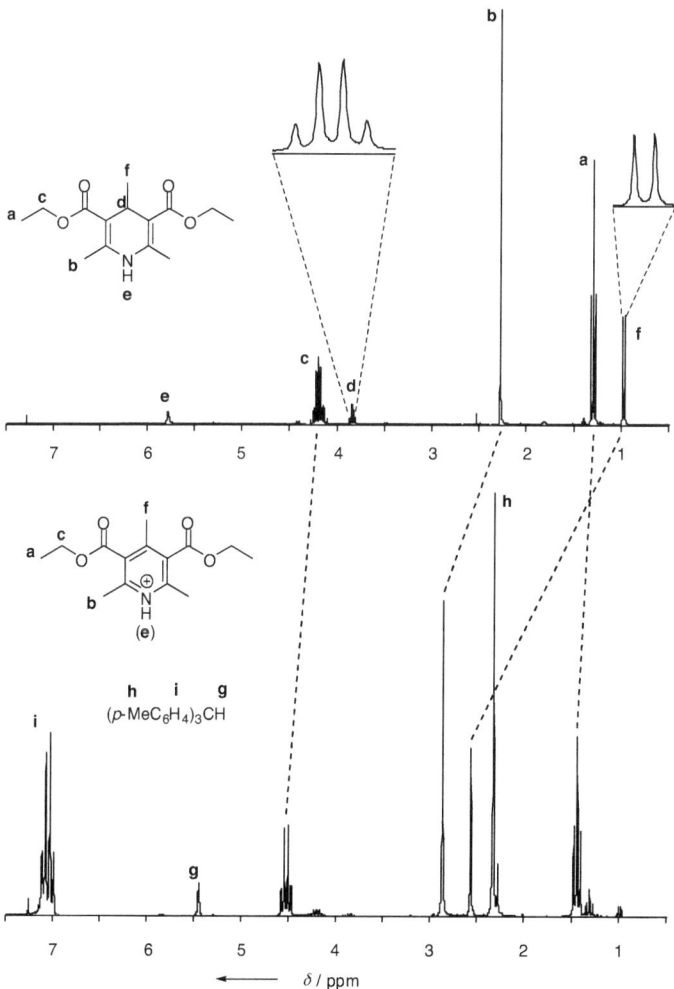

Figure 7.3. $^1$H-NMR spectrum for **3c** (top), and after the addition of an equimolar amount of Me$_3$Tr$^+$BF$_4^-$ (bottom) (200 MHz, CDCl$_3$, 27 °C).

A problem might be the fact, that the dihydropyridines **3** contain more than one reactive site (ambident nucleophiles). Apart from the hydride transfer, fast and reversible attack at oxygen or nitrogen is conceivable. Furthermore, compounds **3** might also react as enamines (Scheme 7.4).

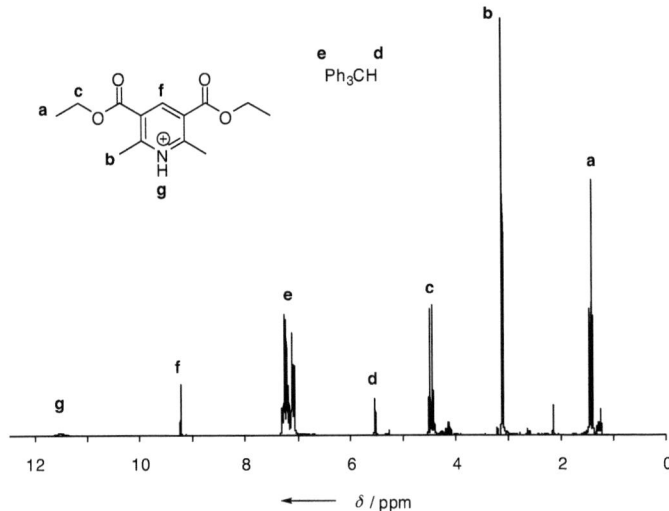

Figure 7.4. $^1$H-NMR spectrum after the reaction of Tr$^+$BF$_4^-$ with **3b** (200 MHz, CDCl$_3$).

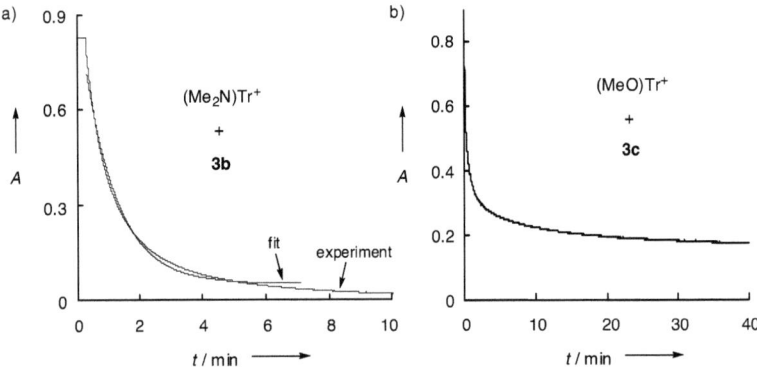

Figure 7.5. Time-dependent absorbances of tritylium ions during their reactions with Hantzsch esters (CH$_2$Cl$_2$, 20 °C). a) (Me$_2$N)Tr$^+$BF$_4^-$ ($c_0$ = 4.13 × 10$^{-5}$ M) with **3b** ($c_0$ = 6.63 × 10$^{-4}$ M), mono-exponential least-squares-fit according to $A = A_0 e^{kt} + C$; b) (MeO)Tr$^+$BF$_4^-$ ($c_0$ = 2.69 × 10$^{-5}$ M) with **3c** ($c_0$ = 1.10 × 10$^{-3}$ M).

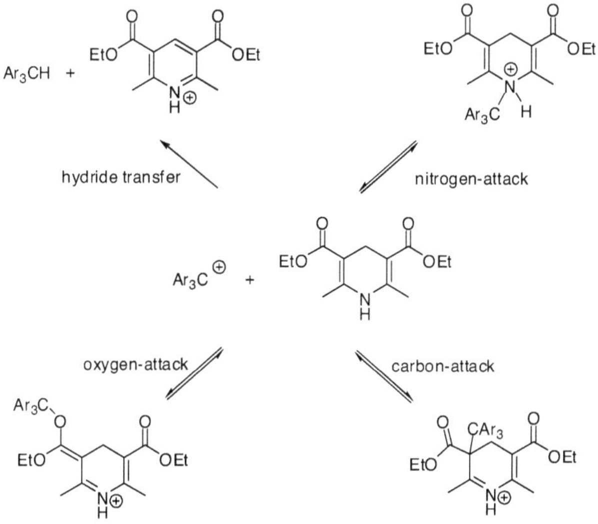

# 7. Miscellaneous Experiments 133

Scheme 7.4.

While the equilibria in Scheme 7.4 may be established relatively fast, the actual hydride transfer proceeds more slowly, thus explaining the rapid decays in Figure 7.5, followed by the shallow parts of the curves.

Whether this hypothesis is correct, and why problems similar to those outlined in Figure 7.5 did not occur in reactions of N-benzyldihydronicotinamide (**3a**) with tritylium ions,[3] are questions which have to be addressed in future work.

## 7.3. Reactivities of Tritylium Ions toward Imidazoles

The rate constants for the reactions of the tritylium ions in Table 7.2 with imidazole (**4a**) and 2-methylimidazole (**4b**) have been determined in acetonitrile at 20 °C.

Table 7.2. $E$-parameters of substituted tritylium ions.[6]

| electrophile | $E$ |
| --- | --- |
| $(MeO)_2Tr^+$ | −3.04 |
| $(MeO)_3Tr^+$ | −4.35 |
| $(Me_2N)Tr^+$ | −7.93 |
| $(Me_2N)(MeO)Tr^+$ | −7.98 |

# 7. Miscellaneous Experiments

The tritylium ions have been introduced as tetrafluoroborate salts, and stopped-flow and conventional photospectrometry have been used for monitoring the time-dependent tritylium ion absorbances.

Nucleophilicity parameters $N/s_N$ for **4a** and **4b** in acetonitrile according to equation (7.1) have been reported previously,[7] thus offering the possibility of a comparison between experimental and calculated rate constants.

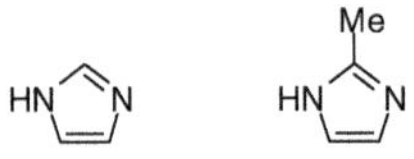

|       | **4a** | **4b** |
|-------|--------|--------|
| $N$   | 11.47  | 11.74  |
| $s_N$ | 0.79   | 0.76   |

The possible reaction mechanisms are depicted in Scheme 7.5: Apart from the simple, reversible attack of one imidazole at a tritylium ion ($k_2$ and $k_{-2}$) forming a positively charged adduct, general base catalysis ($k_{cat}$) might serve as a second pathway to the products on the lower right side. The latter pathway is characterized by the attack of an imidazole with simultaneous abstraction of the proton by a second imidazole. The neutral adduct and the imidazolium ion (lower right side) are in equilibrium with the positively charged adduct and the neutral imidazole (lower left side) through proton transfer ($k_p$).

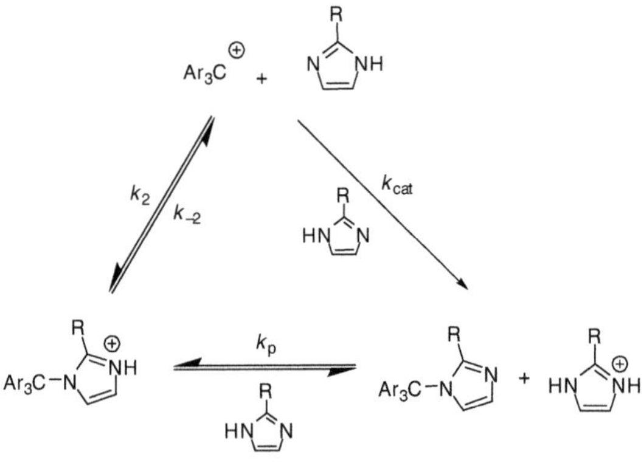

Scheme 7.5. Reactions of tritylium ions with imidazole (R = H) and 2-methylimidazole (R = Me).

# 7. Miscellaneous Experiments 135

In the kinetic experiments, the nucleophiles have been employed in large excess so that their concentrations could be assumed to stay constant during the reactions. In the case of imidazole (**4a**), mono-exponential decays of the carbocation absorbances according to a pseudo-first-order rate law have been observed. The absorbances always reached zero, implying a complete consumption of the tritylium ions in these reactions. Plots of first-order rate constants versus the concentrations of **4a** were linear with the slopes of the correlation lines representing the second-order rate constants $k_2$ (Figure 7.6). Because the kinetics of the reactions with **4a** were found to be of first order in both, the electrophile and the nucleophile, general base catalysis could be ruled out. The results thus indicate the direct addition of imidazole at the carbocation with a fast subsequent proton transfer.

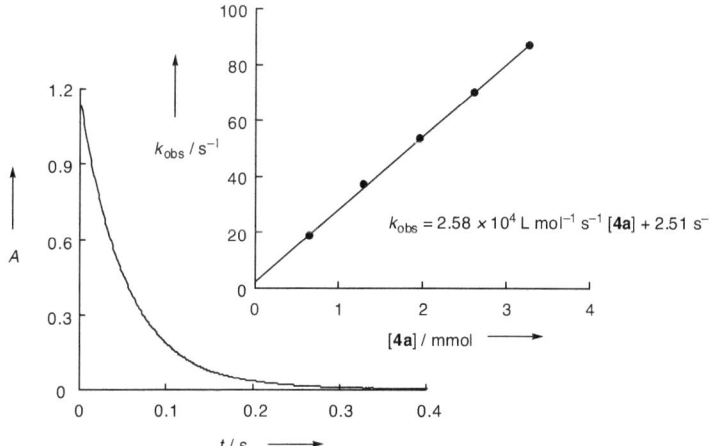

Figure 7.6. Decay of the absorbance of $(MeO)_3Tr^+BF_4^-$ ($c_0 = 3.24 \times 10^{-5}$ mol L$^{-1}$) during its reaction with **4a** ($c_0 = 6.56 \times 10^{-4}$ mol L$^{-1}$); stopped-flow photospectrometry, CH$_3$CN, 20 °C. Inset: plot of $k_{obs}$ vs. [**4a**]; $k = 2.58 \times 10^4$ L mol$^{-1}$ s$^{-1}$.

In contrast, the reactions of 2-methylimidazole (**4b**) proved to be more complicated. When it was combined with $(MeO)_2Tr^+$ or $(MeO)_3Tr^+$, the electrophile absorbances did not decrease mono-exponentially. At the moment, no satisfying explanation can be given for this phenomenon.

# 7. Miscellaneous Experiments

In the case of $(Me_2N)Tr^+$ and $(Me_2N)(MeO)Tr^+$, the recorded curves followed mono-exponential laws, but they did not reach zero. These observations indicated significant equilibria $K = k_2/k_{-2}$ or $K = k_{cat}k_p/k_{-2}$.

For $(Me_2N)Tr^+$, neither the plot of $k_{obs}$ against [**4b**], nor that against [**4b**]$^2$ is linear (Figure 7.7), suggesting that both pathways ($k_2$ and $k_{cat}$) are followed simultaneously in this reaction.

The reaction with $(Me_2N)(MeO)Tr^+$ gave rise to a linear plot of $k_{obs}$ vs. [**4b**]$^2$ (Figure 7.8), indicating a second-order dependence on the nucleophile concentration, in agreement with the pathway $k_{cat}$. However, too many open questions remained in the reactions with 2-methylimidazole as nucleophile, so that a deeper analysis and interpretation of the kinetic data has not been undertaken.

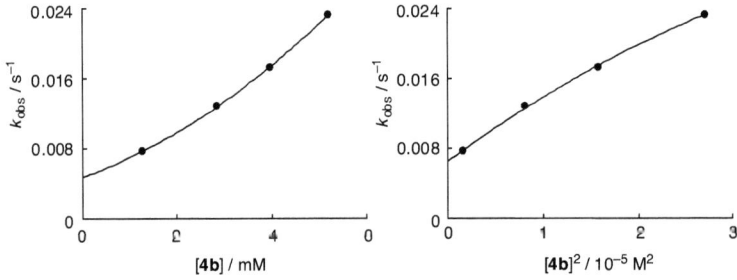

Figure 7.7. Plot of $k_{obs}$ vs. [**4b**] (left) and vs. [**4b**]$^2$ (right) for the reaction of $(Me_2N)Tr^+BF_4^-$ with **4b**; $CH_3CN$, 20 °C.

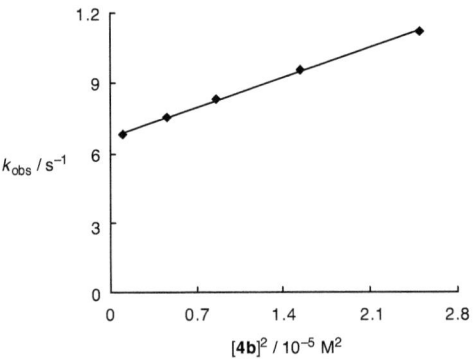

Figure 7.8. Plot of $k_{obs}$ vs [**4b**]$^2$ for the reaction of $(Me_2N)(MeO)Tr^+BF_4^-$ with **4b**; $CH_3CN$, 20 °C. Correlation equation: $k_{obs} = 181 \; L^2 \, mol^{-2} \, s^{-1}$ [**4b**]$^2 + 6.72 \times 10^{-3} \, s^{-1}$.

In summary, three rate constants $k_2$ could be determined for the reactions of substituted tritylium ions with imidazole (**4a**). Table 7.3 shows that the experimental values $k_{exp}$ are only slightly smaller than the rate constants calculated by substitution of the corresponding reactivity parameters into equation (7.1). This result can be rationalized by the pronounced steric hindrance inherent to tritylium ions when compared to benzhydrylium ions, with which the nucleophilicity parameters of **4a** have been evaluated.

Table 7.3. Comparison of calculated and experimental rate constants for the reactions between tritylium ions and imidazole (**4a**), $CH_3CN$, 20 °C.

| electrophile | $k_{exp}$ [L mol$^{-1}$ s$^{-1}$] | $k_{calc}$ [L mol$^{-1}$ s$^{-1}$] | $k_{calc}/k_{exp}$ |
|---|---|---|---|
| $(MeO)_2Tr^+$ | $1.64 \times 10^5$ | $4.57 \times 10^6$ | 28 |
| $(MeO)_3Tr^+$ | $2.58 \times 10^4$ | $4.22 \times 10^5$ | 16 |
| $(Me_2N)Tr^+$ | $5.19 \times 10^1$ | $6.26 \times 10^2$ | 12 |

## 7.4. References

[1] Q. Chu, M. M. Brahmi, A. Solovyev, S.-H. Ueng, D. P. Curran, M. Malacria, L. Fensterbank, E. Lacôte, *Chem. Eur. J.* **2009**, *15*, 12937-12940.

[2] B. Maji, M. Breugst, H. Mayr, *Angew. Chem.* **2011**, accepted.

[3] D. Richter, H. Mayr, *Angew. Chem.* **2009**, *121*, 1992-1995; *Angew. Chem. Int. Ed.* **2009**, *48*, 1958-1961.

[4] H. Mayr, T. Bug, M. F. Gotta, N. Hering, B. Irrgang, B. Janker, B. Kempf, R. Loos, A. R. Ofial, G. Remennikov, H. Schimmel, *J. Am. Chem. Soc.* **2001**, *123*, 9500-9512.

[5] C. A. Bunton, S. K. Huang, C. H. Paik, *Tetrahedron Lett.* **1976**, 1445-1448.

[6] S. Minegishi, H. Mayr, *J. Am. Chem. Soc.* **2003**, *125*, 286-295.

[7] M. Baidya, F. Brotzel, H. Mayr, *Org. Biomol. Chem.* **2010**, *8*, 1929-1935.

# Appendix

The following two chapters are not related to tritylium ions. They contain studies which were supported by quantum chemical calculations performed by the author of this thesis. Data concerning the experimental work of these two chapters are not included in the Experimental Part.

Appendix A was published in: L. Shi, M. Horn, S. Kobayashi, H. Mayr, *Chem. Eur. J.* **2009**, *15*, 8533-8541. The experimental work has been done by Dr. Lei Shi.

Appendix B was published in: M. Baidya, M. Horn, H. Zipse, H. Mayr, *J. Org. Chem.* **2009**, *74*, 7157–7164. The experimental work has been done by Dr. Mahiuddin Baidya.

# A. Carbocationic n-endo-trig Cyclizations

## A.1. Introduction

Whereas Baldwin's rules proved to be a useful guide for a large variety of cyclization reactions,[1] their applicability to carbocationic π-cyclizations[2] appears to be limited. Thus, we have found that cationic 5-endo-trig cyclizations[3] as well as 4-exo-dig cyclizations[4] provide a straightforward access to five- and four-membered carbocycles, though both cyclization modes were termed as "disfavored" by Baldwin's rules.

Correlation equation (A.1) has been found to provide a reliable estimate for the rates of intermolecular reactions of carbocations with π-systems.[5]

$$\log k_2 \,(20\,°C) = s_N(N + E) \qquad (A.1)$$

In order to apply equation (A.1) also to π-cyclizations, i.e., to intramolecular reactions of carbocations with π-systems, the knowledge of effective molarities[6] is needed, i.e., a correction term, which adjusts the predictions of equation (A.1) to intramolecular processes. For this purpose, we have compared the rates of inter- and intramolecular reactions of the 1-(p-methoxyphenyl)ethyl cation **1a**, its homologue **1b**, and the unsaturated derivatives **1c-e** with different nucleophiles (Scheme A.1).

|  |  |  |  |
|---|---|---|---|
| **1a** | **1b** | $n =$  2   3   4 | |
|  |  | **1c 1d 1e** | Scheme A.1. |

## A.2. Kinetic Investigations

The carbocations **1a-e** were generated by laser irradiation (266 nm, 7 ns, 40-60 mJ) of the corresponding triphenylphosphonium tetrafluoroborates in $CH_3CN$ (Scheme A.2). The resulting UV-vis absorptions were in the range $\lambda_{max}$ = 340-360 nm. When the benzyl cations **1a-e** were generated in the presence of a high excess of the π-systems **2a-d** (Scheme A.3), exponential decays of the absorbances of the carbocations were observed from which the first-order rate constants $k_{obs}$ were obtained. These were then plotted against the variable concentrations of **2a-d**.

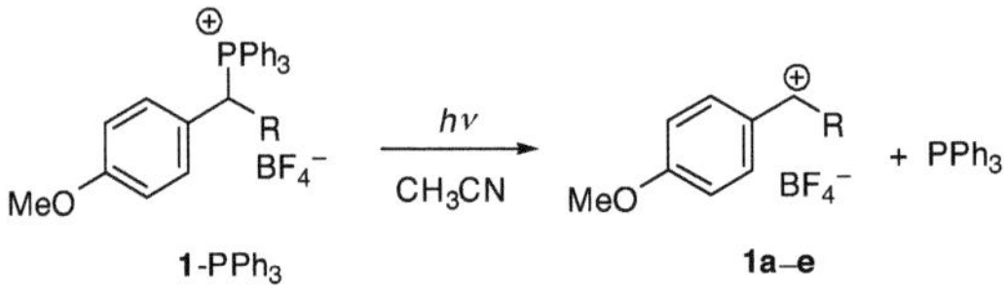

Scheme A.2.

For the reactions of the carbocations **1a-c** and **1e**, the intercepts of the $k_{obs}$ vs. [**2**] plots were small ($< 10^6$ s$^{-1}$) compared to $k_{obs}$, and the second-order rate constants for the reactions of these carbocations with the π-systems were derived from the slopes of these plots, as shown for a typical example in Figure A.1.

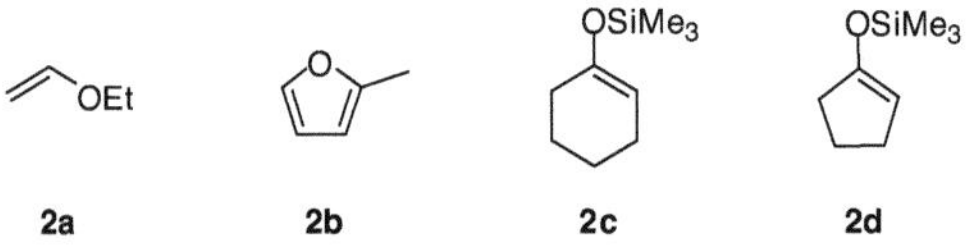

Scheme A.3.

A different behavior was found for carbocation **1d**. As shown in Figure A.2, $k_{obs}$ for the reactions of **1d** with the π-nucleophiles **2** increases strongly with [**2**] at low concentrations of **2** and transforms into linear plots with smaller slopes at higher concentrations, from which the second-order rate constants for the reactions of **1d** with **2** (Table A.1) were derived.

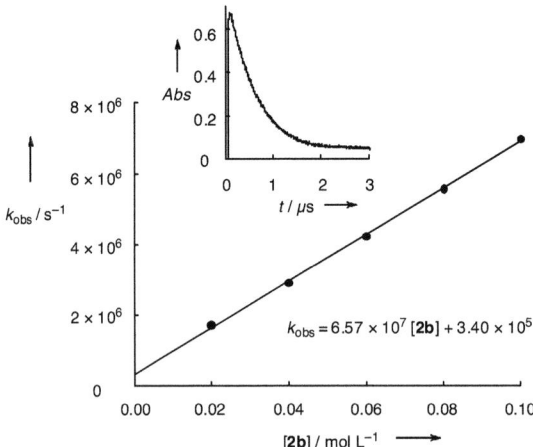

Figure A.1. Plot of $k_{obs}$ vs. [**2b**] for the reaction of **1b** with **2b** in CH$_3$CN (20 °C). Inset: absorbance at 350 nm as a function of time, [**2b**] = 0.02 mol L$^{-1}$.

# Appendix: A. Carbocationic n-endo-trig Cyclizations

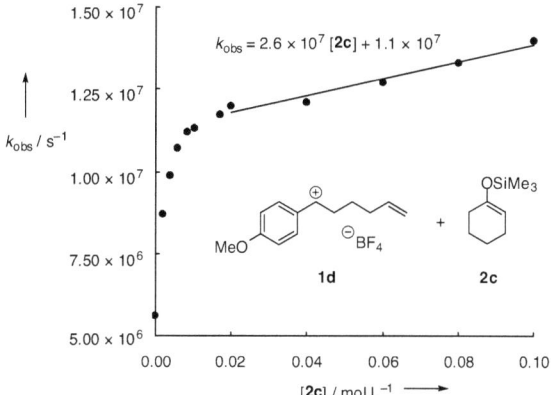

Figure A.2. Dependence of $k_{obs}$ on the concentration of the nucleophile (CH$_3$CN, 20 °C) for the reaction of **1d** with **2c**. The linear part of the curve was used to determine the second-order rate constant.

In trifluoroethanol (TFE), all carbocations **1a-e** decayed according to first-order rate laws, with **1d** reacting considerably faster than all other systems (last column of Table A.1).

## A.3. Product Studies

In order to investigate the products of the reactions of the benzyl cations **1** with the π-nucleophiles **2a-d**, the carbocations **1a,b** were generated in situ by treatment of the benzyl alcohols (**1a,b**)-OH with Bi(OTf)$_3$ in CH$_3$NO$_2$. The alcohol **1a**-OH reacted with the silyl enol ethers **2c** and **2d** in the presence of 2.5 mol% Bi(OTf)$_3$ to give the 2-substituted cycloalkanones **3ac** and **3ad**, respectively, as 1:1 mixtures of diastereoisomers (Scheme A.4).

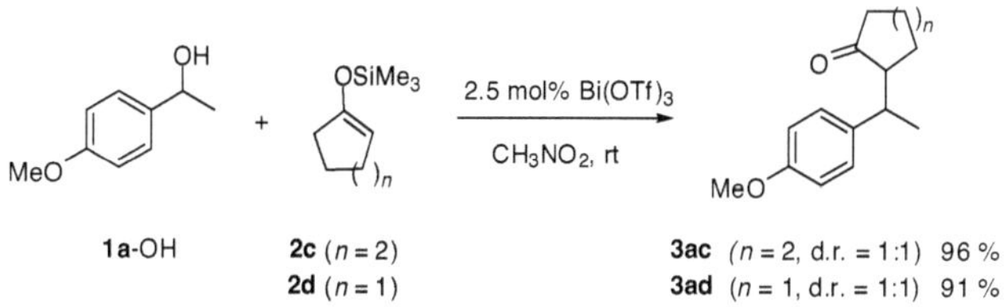

Scheme A.4.

Table A.1. Second-order rate constants (L mol$^{-1}$ s$^{-1}$) for the reactions of carbocations **1a-e** with the π-systems **2a-d** in acetonitrile and with trifluoroethanol (s$^{-1}$) at 20 °C.

| Carbocation | 2a (OEt) | 2b | 2c (OSiMe$_3$) | 2d (OSiMe$_3$) | CF$_3$CH$_2$OH [a] |
|---|---|---|---|---|---|
| **1a** (R = CH$_3$) | $8.1 \times 10^6$ | $8.2 \times 10^7$ | $6.3 \times 10^7$ | - | $4.3 \times 10^5$ |
| **1b** (R = n-C$_4$H$_9$) | $4.3 \times 10^6$ | $6.6 \times 10^7$ | $3.4 \times 10^7$ | $1.2 \times 10^8$ | $2.4 \times 10^5$ |
| **1c** (R = (CH$_2$)$_2$CH=CH$_2$) | - | $6.3 \times 10^7$ | - | - | $2.4 \times 10^5$ |
| **1d** [b] (R = (CH$_2$)$_3$CH=CH$_2$) | $5.0 \times 10^6$ | $5.6 \times 10^7$ | $2.6 \times 10^7$ | $1.2 \times 10^8$ | $2.0 \times 10^6$ |
| **1e** (R = (CH$_2$)$_4$CH=CH$_2$) | $5.7 \times 10^6$ | $6.8 \times 10^7$ | $3.3 \times 10^7$ | $1.3 \times 10^8$ | $2.6 \times 10^5$ |

[a] First-order rate constants (s$^{-1}$), [b] $k_2$ for the reactions of **1d** with **2a-d** from the slope $k_{obs}$ vs. [**2**] at high nucleophile concentrations.

Under the same conditions, **1b**-OH reacted with **2a-d** to give **3ba-3bd** in 51-85 % yield (Scheme A.5).

**3ba** 51 %   **3bb** 79 %   **3bc** (*n* = 2, d.r. = 1:1) 85 %
                             **3bd** (*n* = 1, d.r. = 1:1) 81 %   Scheme A.5.

When the Bi(OTf)$_3$-catalyzed reaction of **1d**-OH with **2b** in CH$_3$NO$_2$ was carried out at high substrate concentrations ([**1d**-OH] = 0.25 M, [**2b**] = 0.5 M), the non-cyclized product **3db** was isolated in 69 % yield, and formation of a cyclization product was not observed (Scheme A.6).

**1d**-OH   **2b**   2.5 mol% Bi(OTf)$_3$, CH$_3$NO$_2$, rt   **3db** 69 %   Scheme A.6.

At low substrate concentrations ([**1d**-OH] = 0.005 mol L$^{-1}$, [**2b**] = 0.011 mol L$^{-1}$), the reaction became more complicated; while a separation of the products was not achieved, NMR and GC-MS analysis of the product mixture indicated the presence of dehydration products of **1d**-OH as well as of cyclized products in addition to **3db**.

Solvolysis in trifluoroethanol (**3e**) of benzyl chloride **1d**-Cl in the presence of 2-chloropyridine as a proton trap yielded the cyclohexyl ether **3d'e** exclusively (Scheme A.7).

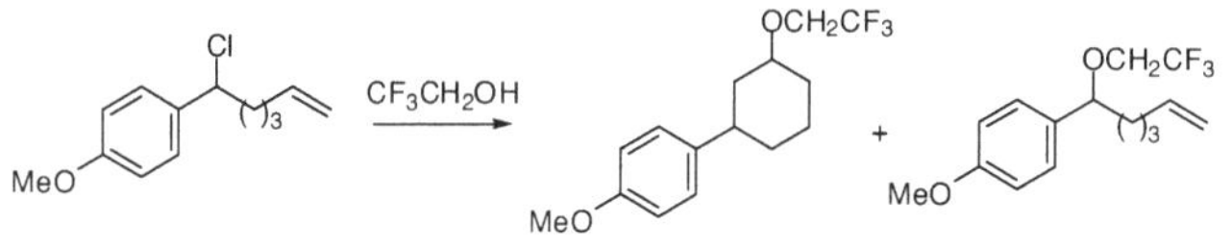

Scheme A.7.

| | 3d'e | 3de |
|---|---|---|
| 2-chloropyridine | > 95 % | - |
| 2,6-lutidine | 50 % | 50 % |

However, when the trifluoroethanolysis of **1d**-Cl was performed in the presence of the stronger base 2,6-lutidine, **3d'e** and **3de** were obtained in equal amounts. In contrast, the solvolyses of **1c**-Cl and **1e**-Cl led to the exclusive formation of the non-cyclized benzyl ethers **3ce** and **3ee**, independent of the nature of the proton trap (Scheme A.8).

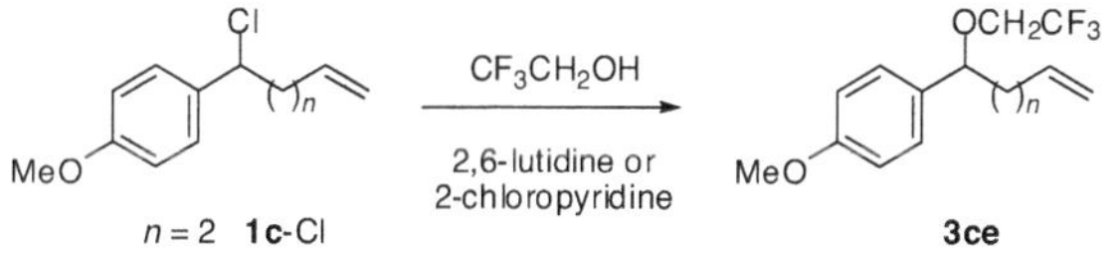

Scheme A.8.

## A.4. Discussion

The carbocations **1a** and **1b** differ only slightly in reactivity, and the somewhat smaller rate constants for **1b** can be explained by steric effects. From the exclusive formation of the non-cyclized ethers **3ce** and **3ee** in the trifluoroethanolyses of the benzyl chlorides **1c**-Cl and **1e**-Cl, even in the presence of 2,6-lutidine (Scheme A.8), one can derive that under these conditions neither the 5-endo-trig cyclization of **1c** nor the 7-endo-trig cyclization of **1e** does occur. As a consequence, the first-order rate constants of the reactions of the cations **1c** and **1e** with trifluoroethanol are almost identical to that of the saturated counterpart **1b** (last column of Table A.1). Furthermore, the second-order rate constants for the reactions of the saturated benzyl cation **1b** and of the unsaturated analogues **1c** and **1e** with the π-systems **2a-d** are closely similar indicating that the carbocationic center in **1c** and **1e** does not interact with the terminal CC-double bond. This observation is in line with the almost identical UV-vis spectra of **1b**, **1c**, and **1e**.

Carbocation **1d** behaves differently, as revealed by the concomitant formation of the benzyl ether **3de** and the cyclohexyl ether **3d'e** during the trifluoroethanolysis of **1d**-Cl (Scheme A.7). According to Scheme A.9, carbocation **1d** either reacts directly with the nucleophiles to give the non-cyclized products **3d** or undergoes a reversible cyclization with formation of **1d'** which then reacts with the nucleophile to give the cyclohexane derivative **3d'**.

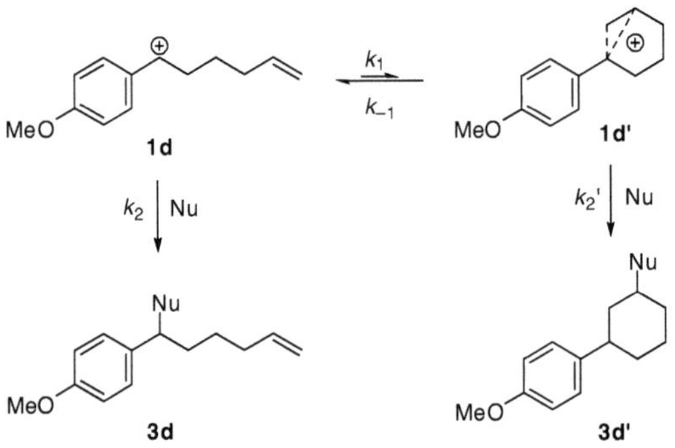

Scheme A.9. Fast reversible cyclization of carbocation **1d**.

Assuming a low equilibrium concentration of **1d'** the kinetics of the consumption of **1d** (Scheme A.9) can be described by equation (A.2).

$$-\frac{d[\mathbf{1d}]}{dt} = k_2[\mathbf{1d}][\text{Nu}] + k_1[\mathbf{1d}]\frac{k_2'[\text{Nu}]}{k_{-1} + k_2'[\text{Nu}]} \qquad (A.2)$$

While the first term of equation (A.2) represents the formation of **3d**, the second term reflects the rate of formation of **3d'**, i.e., $k_1[\mathbf{1d}]$ multiplied with the partitioning ratio [forward reaction ($k_2'[\text{Nu}]$) divided by the sum of backward ($k_{-1}$) and forward reaction].

Rearrangement of equation (A.2) yields equation (A.3) which simplifies to equation (A.4) if [Nu] is used in high excess and thus remains almost constant.

$$-\frac{d[\mathbf{1d}]}{dt} = \left(k_2[\text{Nu}] + k_1\frac{k_2'[\text{Nu}]}{k_{-1} + k_2'[\text{Nu}]}\right)[\mathbf{1d}] \qquad (A.3)$$

For [Nu] >> [1d]:   $-\dfrac{d[\mathbf{1d}]}{dt} = k_{obs}[\mathbf{1d}]$   (A.4)

with   $k_{obs} = k_2[\text{Nu}] + k_1 \dfrac{k_2'[\text{Nu}]}{k_{-1} + k_2'[\text{Nu}]}$   (A.5)

At high concentrations of the nucleophiles, $k_{-1} \ll k_2'[\text{Nu}]$, and equation (A.5) transforms to equation (A.6).

$k_{obs} = k_2[\text{Nu}] + k_1$   (A.6)

According to equation (A.6), the slopes of the linear parts of the $k_{obs}$ vs. [Nu] plots yield the second-order rate constants $k_2$, and the intercepts of these straight lines give the cyclization rate constant $k_1 \approx 1.1 \times 10^7$ s$^{-1}$ which is independent of the nature of the nucleophile.

Table A.1 shows that the second-order rate constants for the direct reaction of **1d** with π-nucleophiles (formation of **3d**) are similar to those for the corresponding reactions of carbocations **1a-c** and **1e** though the overall rate constants $k_{obs}$, which include the formation of **3d'**, are much larger for **1d**. At lower nucleophile concentration [with condition (A.4) still fulfilled], equation (A.5) does not transform into the linear dependence (A.6), and the nonlinear relationship between $k_{obs}$ and [Nu] as expressed by equation (A.5) is observed in the left parts of Figure A.2.

From the observation that the trifluoroethanolysis of **1d–Cl** in the presence of the weakly basic 2-chloropyridine yields the cyclohexyl ether **3d'e** exclusively (Scheme A.7), one can derive that the direct trapping of **1d** by CF$_3$CH$_2$OH is outstripped by the fast cyclization of **1d**.

This conclusion is in line with the fact that the decay of **1d** in CF$_3$CH$_2$OH (Table A.1, right column) is much faster than that of the other carbocations (**1a-c, 1e**). For the direct trapping of **1d** by CF$_3$CH$_2$OH a rate constant close to 2.4 × 10$^5$ s$^{-1}$ (as for **1b**) would be expected.

In previous work it was shown that the rates of reactions of carbocations with alkenes depend only slightly on solvent polarity.[7] If we now assume that the cyclization of **1d** occurs with a similar rate in CF$_3$CH$_2$OH as in acetonitrile ($k_1 = 1.1 \times 10^7$ s$^{-1}$), one can calculate the partitioning coefficient from the second term of equation (A.5), because direct trapping of **1d** [first term of equation (A.5)] does not occur in the absence of 2,6-lutidine (Scheme A.7). In equation (A.7), $k_2'$ [Nu] from equation (A.5) is replaced by $k'_{TFE}$, i.e., the first-order rate constant for the reaction of **1d'** with trifluoroethanol.

$$2.0 \times 10^6 = 1.1 \times 10^7 \frac{k'_{TFE}}{k_{-1} + k'_{TFE}} \qquad (A.7)$$

From equation (A.7) one calculates $k'_{TFE}/(k_{-1} + k'_{TFE}) = 0.18$ or $k_{-1}/k'_{TFE} = 4.5$, i.e., the backward reaction $k_{-1}$ is 4.5 times faster than the reaction of **1d'** with trifluoroethanol. When **1d** is generated in trifluoroethanol in the presence of 2,6-lutidine, which partially deprotonates $CF_3CH_2OH$, a part of **1d** is directly trapped by $CF_3CH_2O^-$, and a mixture of **3d'e** and **3de** is obtained.

## A.5. Quantum Chemical Calculations

In order to elucidate the structures and relative energies of the cations **1d** and **1d'**, theoretical calculations have been performed. For comparison, the unsubstituted parent structures **1h** and **1h'** as well as the *meta*-fluoro substituted structures **1f** and **1f'** have also been investigated (Scheme A.10).

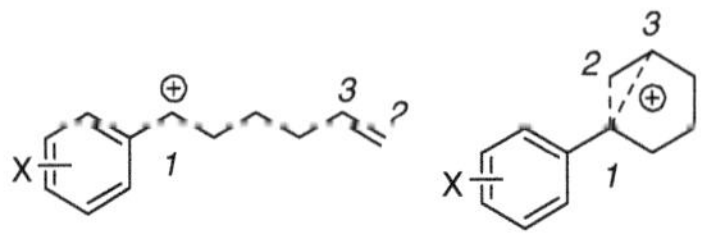

| X = $p$-OMe | **1d** | **1d'** |
| X = H | **1h** | **1h'** |
| X = $m$-F | **1f** | **1f'** |

Scheme A.10.

Conformational analyses of the acyclic structures **1d**, **1h** and **1f** on the basis of the MM3 force field yielded 152, 89 and 175 minima, respectively. The four energetically most favorable conformers of each species were then subjected to geometry optimizations on the B3LYP/G-311G(d,p) level (Tables A.2-4, entries 1-4).

Because of the reduced conformational flexibility of the cyclohexane ring, chair structures of **1d'**, **1h'** and **1f'** with the aryl substituent in equatorial position were used as input with a bond-length of 1.54 Å for C(1)–C(2). Geometry optimization at the B3LYP/G-311G(d,p) level led to the lengthening of the C(1)–C(2) bond in all cases. The methoxy substituted cation **1d'** opened up fully until an energy minimum at C(1)–C(2) = 3.33 Å was reached, i.e., a clearly acyclic structure was formed (Table A.2, last entry).

Appendix: A. Carbocationic n-endo-trig Cyclizations  147

Table A.2. Geometry optimizations of **1d** (4 conformers) and **1d'** (B3LYP/6-311G(d,p)).

| conformer | $\Delta E_{tot}$ / kJ mol$^{-1}$ [a] | bond length / Å | | |
|---|---|---|---|---|
| | | 1-2 | 2-3 | 1-3 |
| **1d** - 1 | 4.12 | 4.33 | 1.33 | 3.91 |
| **1d** - 2 | 6.43 | 5.64 | 1.33 | 4.50 |
| **1d** - 3 | 4.90 | 4.32 | 1.33 | 3.89 |
| **1d** - 4 | 7.40 | 5.27 | 1.33 | 4.55 |
| **1d'** | 0 | 3.33 | 1.33 | 3.03 |

[a] Relative to **1d'**.

In contrast, optimization of **1h'** and **1f'** led to shallow minima with C(1)–C(2) = 1.80 Å and 1.76 Å, respectively, corresponding to non-classical structures (Table A.3 and A.4, last entries).

Table A.3. Geometry optimizations of **1h** (4 conformers) and **1h'** (B3LYP/6-311G(d,p)).

| conformer | $\Delta E_{tot}$ / kJ mol$^{-1}$ [a] | bond length / Å | | |
|---|---|---|---|---|
| | | 1-2 | 2-3 | 1-3 |
| **1h** - 1 | 8.52 | 5.62 | 1.33 | 4.50 |
| **1h** - 2 | 7.21 | 4.33 | 1.33 | 3.90 |
| **1h** - 3 | 10.2 | 5.27 | 1.33 | 4.55 |
| **1h** - 4 | 3.30 | 5.13 | 1.33 | 4.43 |
| **1h'** | 0 | 1.80 | 1.40 | 2.40 |

[a] Relative to **1h'**.

Table A.4. Geometry optimizations of **1f** (4 conformers) and **1f'** (B3LYP/6-311G(d,p)).

| conformer | $\Delta E_{tot}$ / kJ mol$^{-1}$ [a] | bond length / Å | | |
|---|---|---|---|---|
| | | 1-2 | 2-3 | 1-3 |
| **1f** - 1 | 15.7 | 4.35 | 1.33 | 3.92 |
| **1f** - 2 | 18.7 | 5.28 | 1.33 | 4.56 |
| **1f** - 3 | 16.6 | 5.63 | 1.33 | 4.50 |
| **1f** - 4 | 17.4 | 3.11 | 1.33 | 4.37 |
| **1f'** | 0 | 1.76 | 1.41 | 2.39 |

[a] Relative to **1f'**.

As shown in Tables A.2-4, the structures obtained by optimizations of the cyclohexyl systems were slightly more stable than the extended conformers with long C(1)–C(2) distances. Optimizations of the cyclohexyl structures **1d'**, **1h'** and **1f'** with fixed C(1)–C(2) bond lengths show that in all cases classical cyclohexyl cations do not correspond to minima on the B3LYP/6-311G(d,p) level (Figure A.3).

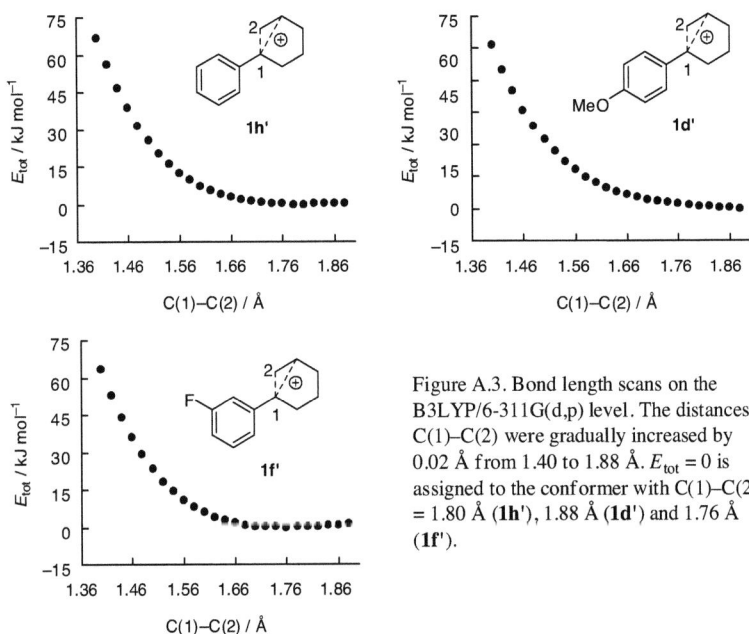

Figure A.3. Bond length scans on the B3LYP/6-311G(d,p) level. The distances C(1)–C(2) were gradually increased by 0.02 Å from 1.40 to 1.88 Å. $E_{tot} = 0$ is assigned to the conformer with C(1)–C(2) = 1.80 Å (**1h'**), 1.88 Å (**1d'**) and 1.76 Å (**1f'**).

*Ab initio* geometry optimizations on the MP2/6-31+G(2d,p) level, which proved to be a good compromise between performance and accuracy for medium sized organic cations,[8] have been performed for the methoxy- and the unsubstituted system, using the preoptimized structures (Tables A.5 and A.6).

In the case of the parent compound, a structure with a hydrogen bridging the atoms 2 and 3 resulted (**1h''** in Table A.5, Figure A.4), when an input bond length C(1)–C(2) = 1.54 Å was used. Optimization starting with C(1)–C(2) = 1.88 Å led to a minimum with C(1)–C(2) = 1.83 Å (**1h'** in Table A.5). The latter structure is 31.6 kJ mol$^{-1}$ lower in energy than the hydrogen bridged structure.

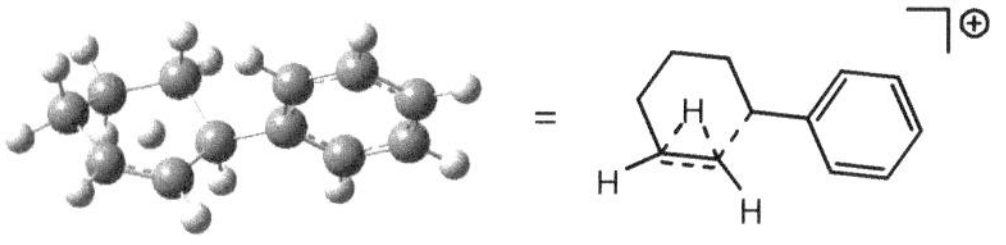

Figure A.4. MP2/6-31+G(2d,p) optimized structure **1h″** of the 3-phenylcyclohexyl cation, when starting with a bond length C(1)–C(2) of 1.54 Å.

Table A.5. Geometry optimizations of **1h** (4 conformers) and **1h′** (MP2/6-31+G(2d,p)).

| conformer | $\Delta E_{tot}$ / kJ mol$^{-1}$ [a] | bond length / Å | | |
|---|---|---|---|---|
| | | 1-2 | 2-3 | 1-3 |
| **1h** - 1 | 52.6 | 5.53 | 1.34 | 4.42 |
| **1h** - 2 | 43.2 | 3.96 | 1.34 | 3.60 |
| **1h** - 3 | 52.6 | 5.15 | 1.34 | 4.43 |
| **1h** - 4 | 45.6 | 4.92 | 1.34 | 4.31 |
| **1h′** | 0 | 1.83 | 1.39 | 2.27 |
| **1h″** | 31.6 | 1.45 | 1.40 | 2.55 |

[a] Relative to **1h′**.

Optimization of the 3-(4-methoxyphenyl)cyclohexyl cation (**1d′**) at the MP2/6-31+G(2d,p) level starting with a C(1)-C(2) distance of 1.54 Å, yielded a structure with C(1)-C(2) = 1.95 Å. In contrast, starting at 3.33 Å – the minimum in the DFT calculations – gave a minimum with C(1)-C(2) = 2.91 Å (**1d″** in Table A.6), which is 1.31 kJ mol$^{-1}$ higher in energy than the one with 1.95 Å.

Table A.6. Geometry optimizations of **1d** (4 conformers) and **1d′** (MP2/6-31+G(2d,p)).

| conformer | $\Delta E_{tot}$ / kJ mol$^{-1}$ [a] | bond length / Å | | |
|---|---|---|---|---|
| | | 1-2 | 2-3 | 1-3 |
| **1d** - 1 | 4.12 | 3.94 | 1.34 | 3.60 |
| **1d** - 2 | 13.9 | 5.55 | 1.34 | 4.42 |
| **1d** - 3 | 4.59 | 3.95 | 1.34 | 3.60 |
| **1d** - 4 | 13.8 | 5.15 | 1.34 | 4.44 |
| **1d′** | 0 | 1.95 | 1.38 | 2.37 |
| **1d″** | 1.31 | 2.91 | 1.35 | 2.83 |

[a] Relative to **1d′**.

In summary, the cyclized structures **1h'**, **1d'** and **1f'** do not correspond to classical cyclohexyl cations. For the former two species *ab initio* calculations predict non-classical carbocations as minima on the energy hypersurface. While structure **1h'** of the unsubstituted system is much more stable than its open chain isomers **1h** (Table A.5), the experimentally studied *p*-methoxyphenyl substituted carbocation **1d'** is only slightly stabilized by interaction of the carbocationic center with the terminal CC-double bond (4 kJ mol$^{-1}$, Table A.6).

## A.6. Consequences for π-Participation in Solvolysis Reactions

Extensive kinetic investigations by Borčić et al. have shown that all 4-methoxy substituted derivatives **4U-Cl** and **4S-Cl** listed in Table A.7 solvolyze with approximately the same rates,[9] indicating that the breaking of the C–Cl bond is not nucleophilically assisted by the π-bond. As a consequence, the ratio $k_U/k_S$ is close to 1 for all compounds in the first line of Table A.7. When the 6-double bond is unsubstituted ($R^1 = R^2 = H$), nucleophilic π-participation was not even observed when the arene ring was bearing an electron-withdrawing substituent (e.g., 3-Br), and all $k_U/k_S$ ratios in the column H, H of Table A.7 are close to 1. The ratio $k_U/k_S$ increases with decreasing electron releasing ability of the substituent X of the aromatic ring and increasing number of methyl groups at the double bond until $k_U/k_S = 58$ is reached on bottom right of Table A.7.

Table A.7. Relative solvolysis rates ($k_U/k_S$) of benzyl chlorides **4U-Cl** and **4S-Cl** in different solvents at 25 °C (data from ref. [9b]).

| X | Solvent [a] | $R^1, R^2$ | | | |
|---|---|---|---|---|---|
| | | H, H | CH$_3$, H | H, CH$_3$ | CH$_3$, CH$_3$ |
| 4-OMe | 95 EtOH | 0.49 [b] | 1.07 | 0.89 | 1.08 |
| 4-Me | 95 EtOH | 0.96 [c] | 1.50 | 2.42 | 3.22 |
| | 80 EtOH | 1.04 | 1.13 | 1.43 | 2.23 |
| H | 80 EtOH | 0.92 [c] | 2.57 | 5.93 | 16.1 |
| 4-Br | 97 TFE | 1.15 [d] | 3.05 | 6.18 | 18.9 |
| 3-Br | 97 TFE | 1.37 [c] | 5.73 | 10.3 | 58.2 |

[a] In vol-%, the residue is water; [b] 5 °C; [c] 50 °C; [d] 30 °C.

Why does the unsubstituted terminal double bond not even assist the heterolysis of the C–Cl bond in compounds **4U**-Cl, which do not carry electron-donating substituents X, though MP2 calculations showed that in the case of **1h** (= **4U**, X = $R^1$ = $R^2$ = H) cyclization is associated with a stabilization of more than 40 kJ mol$^{-1}$ (see Table A.5)?

The observed cyclization rate constant of $1.1 \times 10^7$ s$^{-1}$ for cation **1d** (= **4U**, X = 4-OMe, $R^1$ = $R^2$ = H) indicates that there is still a small energy barrier for the intramolecular attack of the carbenium center of **1d** at the monosubstituted terminal CC-double bond. Using the azide clock method, Jencks and Richard[10] have shown that the unsubstituted 1-phenylethyl cation **1g** reacts $2 \times 10^3$ times faster and the 3-bromo-substituted 1-phenylethyl cation **1i** reacts $2 \times 10^4$ times faster with the solvent (50:50 trifluoroethanol/water) than the 4-methoxy substituted analogue **1a** (Scheme A.11). Exactly the same reactivity ratio has been observed for the reactions of the laser-flash photolytically generated carbocations **1a** and **1g** with hexafluoroisopropanol (HFIP).[11]

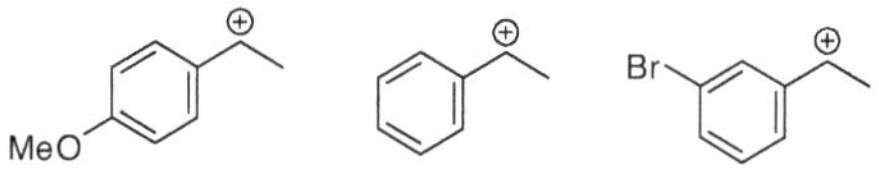

|  | 1a | 1g | 1i |
|---|---|---|---|
| TFE/H$_2$O = 50/50 | 5 x 10$^7$ | 1 x 10$^{11}$ | 1 x 10$^{12}$ |
| HFIP | 2 x 10$^2$ | 4 x 10$^5$ |  |

Scheme A.11. Pseudo-first-order rate constants $k$ (s$^{-1}$) for the reactions of 1-arylethyl cations with 50:50 trifluoroethanol/water (25 °C)[10] and hexafluoroisopropanol (20 °C).[11]

With the assumption that similar substituent effects hold for the reaction with the CC-double bond, one can estimate cyclization rate constants of $2 \times 10^{10}$ s$^{-1}$ ($1.1 \times 10^7 \times 2{,}000$) for **4U** (X = $R^1$ = $R^2$ = H) and of $2 \times 10^{11}$ s$^{-1}$ ($1.1 \times 10^7 \times 20{,}000$) for **4U** (X = 3-Br, $R^1$ = $R^2$ = H).

According to Jencks and Richard,[12] intermediates can only be formed if their lifetimes exceed the duration of a bond-vibration, typically 10$^{-13}$ s. The preceding calculations show that all benzyl cations **4U** with $R^1$ = $R^2$ = H listed in Table A.7 have cyclization rate constants < 10$^{13}$ s$^{-1}$ which implies that π-participation for breaking the C–Cl bond is not needed, in line with $k_U/k_S \approx 1$ in the column H, H of Table A.7.

Typically, 1,1-dialkylethylenes and trialkylethylenes react $10^4$-$10^5$ times faster with carbocations than monoalkylated ethylenes.[13] Multiplication of the cyclization rate constant $k$ = 1.1 × $10^7$ $s^{-1}$ of **1d** (= **4U**, X = 4-OMe, $R^1$ = $R^2$ = H) with this factor again yields cyclization rate constants < $10^{13}$ $s^{-1}$. Accordingly, the first line of Table A.7 shows that none of the solvolyses of the 4-methoxy substituted benzyl chlorides **4U-Cl** is assisted by π-participation. When carbocations **4U** with weaker electron donors X and more nucleophilic double bonds in the side chain are employed, cyclization rate constants > $10^{13}$ $s^{-1}$ can be estimated, implying that carbocations **4U** corresponding to the lower right corner of Table A.7 should not be accessible as intermediates, and the solvolyses of the benzyl chlorides **4U-Cl** in this part of Table A.7 are assisted by π-participation. The fact that a methyl group at $R^2$ position of **4U-Cl** generally accelerates the solvolyses slightly more than a methyl group at $R^1$ position, can be assigned to a non-classical structure of these intermediates in line with previous suggestions by Borčić.

## A.7. Effective Molarities

Because of the lack of a sufficient number of rate constants for the reactions of the carbocations **1** with nucleophiles, it is not possible to determine their electrophilicity parameters $E$. With the assumption that the relative reactivities of 1-(trimethylsiloxy)-cyclohexene (**2c**) and 1-hexene toward benzhydrylium ions (3 × $10^7$) also hold for carbocation **1b**, one can estimate a second-order rate constant of $k$ = 3.4 × $10^7$ L $mol^{-1}$ $s^{-1}$ / 3 × $10^7$ = 1.1 L $mol^{-1}$ $s^{-1}$ for the intermolecular reaction depicted in Scheme A.12.

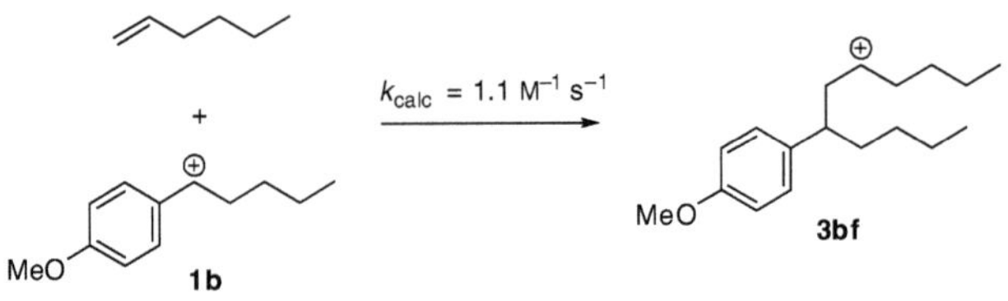

Scheme A.12. Estimated second-order rate constant (L $mol^{-1}$ $s^{-1}$) for the reaction of carbocation **1b** with 1-hexene.

When this rate constant (1.1 L $mol^{-1}$ $s^{-1}$) is compared with the rate constant of the analogous intramolecular reaction in Scheme A.9, one arrives at an effective molarity EM =

$k_{cycl}/k_{acycl} = 1.1 \times 10^7 \text{ s}^{-1}/ 1.1 \text{ L mol}^{-1} \text{ s}^{-1} = 1 \times 10^7 \text{ mol L}^{-1}$ for the 6-endo-trig cyclization depicted in Scheme A.9. With the assumption that similar effective molarities hold for related π-cyclizations, one arrives at the rule of thumb that cationic 6-endo-trig cyclizations are roughly 10 million times faster than calculated for the corresponding intermolecular process by equation (A.1).

Kinetic and product studies indicate that the effective molarities for the Baldwin-disfavored 5-endo-trig and the Baldwin-favored 7-endo-trig cyclizations are significantly smaller. Cyclization experiments with other π-donors are needed in order to examine whether the effective molarities are specific for certain cyclization modes and thus allow introducing a correction term in order to apply equation (A.1) also for cyclization reactions.

## A.8. References

[1] a) J. E. Baldwin, *J. Chem. Soc. Chem. Commun.* **1976**, 734–736; b) J. E. Baldwin, J. Cutting, W. Dupont, L. Kruse, L. Silberman, R. C. Thomas, *J. Chem. Soc. Chem. Commun.* **1976**, 736–738; c) J. E. Baldwin, *J. Chem. Soc. Chem. Commun.* **1976**, 738–741; d) J. E. Baldwin, M. J. Lusch, *Tetrahedron* **1982**, *38*, 2939–2947; e) J. E. Baldwin, R. C. Thomas, L. I. Kruse, L. Silberman, *J. Org. Chem.* **1977**, *42*, 3846–3852; f) C. D. Johnson, *Acc. Chem. Res.* **1993**, *26*, 476–482.

[2] a) P. A. Bartlett, in *Asymmetric Synthesis*, Vol. 3 (Ed: J. D. Morrison), Academic Press, Orlando (FL), **1984**, pp. 341–409; b) J. K. Sutherland, in *Comprehensive Organic Synthesis*, Vol. 3 (Eds: B. M. Trost, I. Fleming, G. Pattenden), Pergamon, Oxford, **1991**, pp. 341–377.

[3] H. Klein, H. Mayr, *Angew. Chem.* **1981**, *93*, 1069–1070; *Angew. Chem. Int. Ed.* **1981**, *20*, 1027–1029.

[4] H. Mayr, B. Seitz, I. K. Halberstadt-Kausch, *J. Org. Chem.* **1981**, *46*, 1041–1043.

[5] a) H. Mayr, M. Patz, *Angew. Chem.* **1994**, *106*, 990–1010; *Angew. Chem. Int. Ed.* **1994**, *33*, 938–957; b) H. Mayr, M. Patz, M. F. Gotta, A. R. Ofial, *Pure. Appl. Chem.* **1998**, *70*, 1993–2000; c) H. Mayr, O. Kuhn, M. F. Gotta, M. Patz, *J. Phys. Org. Chem.* **1998**, *11*, 642–654; d) H. Mayr, T. Bug, M. F. Gotta, N. Hering, B. Irrgang, B. Janker, B. Kempf, A. R. Ofial, G. Remennikov, H. Schimmel, *J. Am. Chem. Soc.* **2001**, *123*, 9500–9512; e) H. Mayr, B. Kempf, A. R. Ofial, *Acc. Chem. Res.* **2003**, *36*,

66–77; f) H. Mayr, A. R. Ofial, *Pure Appl. Chem.* **2005**, *77*, 1807–1827; g) H. Mayr, A. R. Ofial, *Nachr. Chem.* **2008**, *56*, 871–877; H. Mayr, A. R. Ofial, *J. Phys. Org. Chem.* **2008**, *21*, 584–595; H. Mayr, A. R. Ofial, in *Carbocation Chemistry* (Eds: G. A. Olah, G. K. S. Prakash), Wiley, Hoboken (N.J.), **2004**, Chapt. 13, pp. 331–358.

[6]  a) G. Illuminati, L. Mandolini, *Acc. Chem. Res.* **1981**, *14*, 95–102; b) F. M. Menger, *Acc. Chem. Res.* **1985**, *18*, 128–134; c) R. Cacciapaglia, S. Di Stefano, L. Mandolini, *Acc. Chem. Res.* **2004**, *37*, 113–122; d) A. J. Kirby, *Adv. Phys. Org. Chem.* **1980**, *17*, 183–278; e) B. Capon, S. P. McManus, *Neighboring Group Participation*, Vol. 1, Plenum Press, New York, **1976**, p. 15.

[7]  a) J. Bartl, S. Steenken, H. Mayr, R. A. McClelland, *J. Am. Chem. Soc.* **1990**, *112*, 6918–6928; b) H. Mayr, M. Patz, *Macromol. Symp.* **1996**, *107*, 99–110.

[8]  a) Y. Wei, T. Singer, H. Mayr, G. N. Sastry, H. Zipse *J. Comp. Chem.* **2008**, *29*, 291–297; b) W. Koch, B. Liu, D. J. DeFrees, D. E. Sunko, H. Vančik, *Angew. Chem.* **1990**, *102*, 198-200; *Angew. Chem. Int. Ed.* **1990**, *29*, 183-185.

[9]  a) E. Polla, S. Borčić, D. E. Sunko, *Tetrahedron Lett.* **1975**, *16*, 799–802; b) I. Mihel, M. Orlović, E. Polla, S. Borčić, *J. Org. Chem.* **1979**, *44*, 4086–4090; c) M. Orlović, E. Polla, S. Borčić, *J. Org. Chem.* **1983**, *48*, 2278–2280; d) S. Jurić, A. Filipović, O. Kronja, *J. Phys. Org. Chem.* **2003**, *16*, 900–904; e) It has been claimed, however, that absence of a rate effect does not rigorously exclude π-participation: S. Jurić, O. Kronja, *J. Phys. Org. Chem.* **2008**, *21*, 108–111.

[10]  J. P. Richard, M. E. Rothenberb, W. P. Jencks, *J. Am. Chem.Soc.* **1984**, *106*, 1361–1372.

[11]  F. L. Cozens, V. M. Kanagasabapathy, R. A. McClelland, S. Steenken, *Can. J. Chem.* **1999**, *77*, 2069–2082.

[12]  a) W. P. Jencks, *Chem. Soc. Rev.* **1981**, *10*, 345–375; b) J. P. Richard, W. P. Jencks, *J. Am. Chem. Soc.* **1982**, *104*, 4689–4691; c) J. P. Richard, W. P. Jencks, *J. Am. Chem. Soc.* **1982**, *104*, 4691–4692.

[13]  Because of the different nucleophile-specific slope parameters $s$, this ratio depends strongly on the electrophilicity of the carbocation: H. Mayr, R. Schneider, U. Grabis, *J. Am. Chem. Soc.* **1990**, *112*, 4460–4467.

# B. Organocatalytic Activity of Cinchona Alkaloids: Which Nitrogen is more Nucleophilic?

## B.1. Introduction

Since the beginning of the 20$^{th}$ century, alkaloids, such as quinine or quinidine, have been used as catalysts for asymmetric syntheses.[1a,b] A breakthrough were Pracejus' alcoholyses of disubstituted ketenes in the presence of cinchona alkaloids.[1c] Though numerous other classes of tertiary amines have since been investigated with respect to their catalytic efficiencies,[1] the naturally occurring cinchona alkaloids **1a,b** and derivatives thereof (Scheme B.1) have remained in the focus of interest.[2] Though Adamczyk and Rege reported that 1,3-propane sultone reacts with quinine selectively at the $N_{sp2}$ center of the quinoline ring,[3] it is generally assumed that the catalytic activity of the cinchona alkaloids is due to the nucleophilicity of the $N_{sp3}$ center of the quinuclidine ring. During the efforts to characterize the nucleophilic reactivity of cinchona alkaloids in comparison with other organocatalysts it was noticed that in contrast to most other electrophiles, benzhydrylium ions attack selectively at the $N_{sp2}$ center of the quinoline ring. This observation prompted us to systematically investigate the nucleophilic reactivity of the two basic positions in cinchona alkaloids.

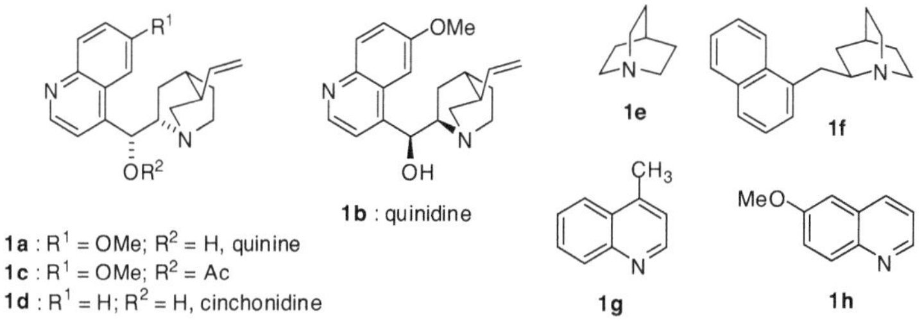

**1a** : R$^1$ = OMe; R$^2$ = H, quinine
**1c** : R$^1$ = OMe; R$^2$ = Ac
**1d** : R$^1$ = H; R$^2$ = H, cinchonidine

**1b** : quinidine

Scheme B.1. Cinchona alkaloids and related compounds.

## B.2. Product Identification

In agreement with earlier reports[4] compounds **1a-d** react with benzyl bromide at the quinuclidine ring (Scheme B.2). The quaternary ammonium salts resulting from benzylation of **1a** and **1d** are commercially available and are used as phase-transfer catalysts.

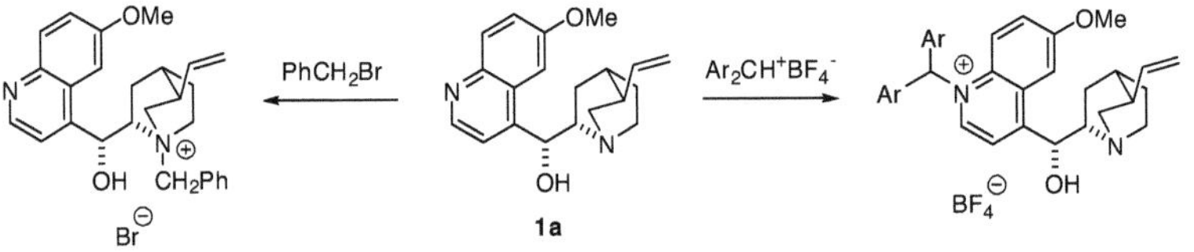

# Appendix: B. Organocatalytic Activity of Cinchona Alkaloids

Scheme B.2. Reactions of quinine (**1a**) with benzyl bromide and benzhydrylium salts.

In contrast, benzhydrylium ions (Table B.1) attack cinchona alkaloids at the quinoline nitrogen. Comparison of the $^1$H- and $^{13}$C-NMR chemical shifts of the adducts of quinuclidine (**1e**), 6-methoxyquinoline (**1h**), and O-acetylquinine (**1c**) with (mfa)$_2$CH$^+$ and (ani)$_2$CH$^+$ reveals exclusive attack of benzhydrylium ions at the N$_{sp2}$ center of cinchona alkaloids.

Table B.1. Abbreviations and electrophilicity parameters $E$ of benzhydrylium ions.

|  | X | $E$ [a] |
|---|---|---|
| Ph$_2$CH$^+$ | H | 5.90 |
| (tol)$_2$CH$^+$ | Me | 3.63 |
| (ani)$_2$CH$^+$ | OMe | 0.00 |
| (pfa)$_2$CH$^+$ | N(Ph)CH$_2$CF$_3$ | −3.14 |
| (mfa)$_2$CH$^+$ | N(CH$_3$)CH$_2$CF$_3$ | −3.85 |
| (dpa)$_2$CH$^+$ | NPh$_2$ | −4.72 |
| (mor)$_2$CH$^+$ | N(CH$_2$CH$_2$)$_2$O | −5.53 |
| (mpa)$_2$CH$^+$ | N(Ph)CH$_3$ | −5.89 |
| (dma)$_2$CH$^+$ | N(CH$_3$)$_2$ | −7.02 |
| (pyr)$_2$CH$^+$ | N(CH$_2$)$_4$ | −7.69 |
| (thq)$_2$CH$^+$ | 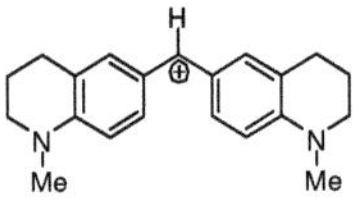  | −8.22 |
| (ind)$_2$CH$^+$ | 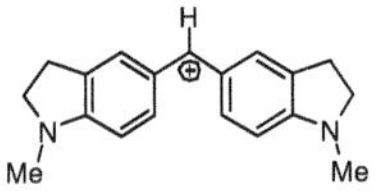  | −8.76 |

[a] From ref. [5].

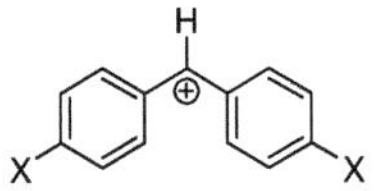

While the reactions of quinuclidine (**1e**) ($\delta$NCH$_2$ = 2.78 ppm) with (mfa)$_2$CH$^+$ and (ani)$_2$CH$^+$ are accompanied by a 0.6-0.8 ppm deshielding of the NCH$_2$-protons, the chemical shifts of the quinuclidine protons remained unaffected when O-acetylquinine (**1c**) was combined with these benzhydrylium ions. On the other hand, benzhydrylation of **1c** led to distinct changes in the chemical shifts of the quinoline moiety similar to that observed upon treatment of 6-methoxyquinoline (**1h**) with benzhydrylium salts.

A further argument for the attack of benzhydrylium ions at the quinoline ring of **1c** comes from the comparison of the chemical shifts of 9-H and C-9 of the adducts in Scheme B.3, which have been assigned by COSY and HSQC. The NMR chemical shifts of the benzhydryl proton 9-H and the benzhydryl carbon C-9 in the adducts obtained from benzhydrylium ions and O-acetylquinine (**1c**) are very similar to those of the corresponding adducts with 6-methoxyquinoline (**1h**), indicating the same environment of the benzhydryl center in both pairs of adducts.

In contrast, the corresponding chemical shifts of the adducts with quinuclidine (**1e**) differ significantly. While the benzhydryl proton resonates at much higher field ($\Delta\delta$ = 2-2.6 ppm), the benzhydryl carbon is more deshielded ($\Delta\delta$ = 7-10 ppm).

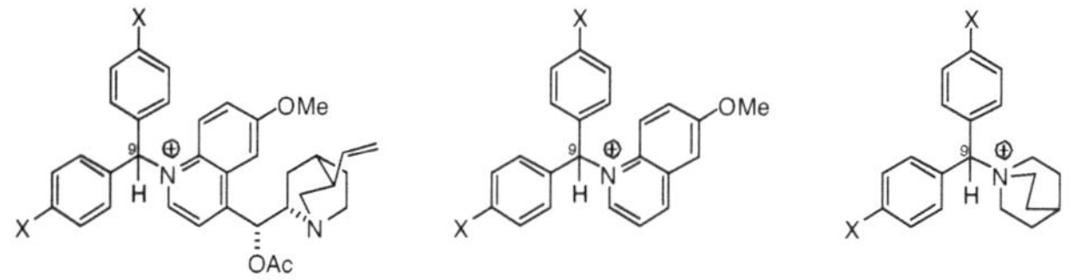

| X | $\delta$(9-H) | $\delta$(C-9) | $\delta$(9-H) | $\delta$(C-9) | $\delta$(9-H) | $\delta$(C-9) |
|---|---|---|---|---|---|---|
| N(CH$_3$)CH$_2$CF$_3$ | 7.77 | 72.8 | 7.80 | 73.5 | 5.20 | 83.5 |
| OMe | 8.20 | 72.8 | 8.20 | 72.9 | 6.28 | 80.1 |

Scheme B.3. Comparison of the $^1$H- and $^{13}$C-NMR chemical shifts of the benzhydryl center in different adducts with amines (in ppm, solvent: CD$_3$CN).

## B.3. Kinetic Investigation

In order to rationalize the opposing selectivities of different electrophiles, the kinetics of the reactions of benzhydrylium ions and benzyl bromide with quinine (**1a**) were studied and compared with the corresponding reactions of related compounds (Scheme B.1).

The decay of the benzhydrylium absorbances has been followed photometrically after combining benzhydrylium tetrafluoroborates with variable excesses of the amines. Pseudo-first-order rate constants $k_{obs}$ were obtained by fitting the decay of the absorbances to the mono-exponential function $A = A_0\,e^{-k_{obs}t} + C$. Plots of $k_{obs}$ vs. the concentrations of the amines were linear (Figure B.1), with the second-order rate constants (Table B.2) being the slopes of the correlation lines. Because of solubility problems, different solvents had to be used for the different reaction series. Comparison of rate constants in $CH_3CN$ and $CH_2Cl_2$ reveals a 3-4 times higher reactivity in $CH_2Cl_2$.

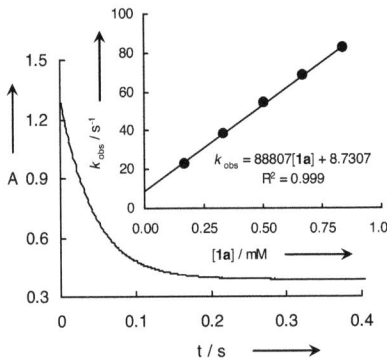

Figure B.1. Exponential decay of the absorbance $A$ at 590 nm and linear correlation of the pseudo-first order rate constants $k_{obs}$ with [**1a**] for the reaction of $(mfa)_2CH^+BF_4^-$ ($c_0 = 1.8 \times 10^{-5}$ mol L$^{-1}$) with amine **1a** ($CH_2Cl_2$, 20 °C); as the reaction is reversible, the final absorbance is not zero.

Table B.2. Second-order rate constants for the reactions of the amines **1a-h** with benzhydrylium tetrafluoroborates (20 °C).

| amine | $N / s_N$ | $Ar_2CH^+$ | $k\,/\,M^{-1}\,s^{-1}$ | |
|---|---|---|---|---|
| | | | $CH_2Cl_2$ | $CH_3CN$ |
| **1a** | 10.46 / 0.75 | $(mfa)_2CH^+$ | $8.88 \times 10^4$ | |
| | ($CH_2Cl_2$) | $(dpa)_2CH^+$ | $1.76 \times 10^4$ | |
| | | $(mor)_2CH^+$ | $4.98 \times 10^3$ | |
| | | $(dma)_2CH^+$ | no rxn | |
| **1b** | 10.54 / 0.74 | $(mfa)_2CH^+$ | $9.36 \times 10^4$ | |
| | ($CH_2Cl_2$) | $(dpa)_2CH^+$ | $1.74 \times 10^4$ | |
| | | $(mor)_2CH^+$ | $5.38 \times 10^3$ | |
| | | $(dma)_2CH^+$ | no rxn | |
| **1c** | | $(mfa)_2CH^+$ | $8.23 \times 10^4$ | $2.68 \times 10^4$ |
| **1e** | 20.54 / 0.60 [a] | $(mfa)_2CH^+$ | | $9.97 \times 10^{8\,[a]}$ |
| | ($CH_3CN$) | $(mor)_2CH^+$ | | $3.34 \times 10^{8\,[a]}$ |
| | | $(dma)_2CH^+$ | | $1.18 \times 10^{8\,[a]}$ |
| | | $(pyr)_2CH^+$ | | $5.22 \times 10^{7\,[a]}$ |
| | | $(ind)_2CH^+$ | | $1.08 \times 10^{7\,[a]}$ |

Table B.2. Continued.

| amine | $N / s_N$ | $Ar_2CH^+$ | $k / M^{-1} s^{-1}$ | |
|---|---|---|---|---|
| | | | $CH_2Cl_2$ | $CH_3CN$ |
| 1f | 15.66 / 0.62 | $(dma)_2CH^+$ | | $1.84 \times 10^5$ |
| | $(CH_3CN)$ | $(pyr)_2CH^+$ | | $1.13 \times 10^5$ |
| | | $(thq)_2CH^+$ | | $4.12 \times 10^4$ |
| | | $(ind)_2CH^+$ | | $1.63 \times 10^4$ |
| | | $(jul)_2CH^+$ | | no rxn |
| | | $(lil)_2CH^+$ | | no rxn |
| 1g | 11.60 / 0.62 | $(pfa)_2CH^+$ | | $1.78 \times 10^5$ |
| | $(CH_3CN)$ | $(mfa)_2CH^+$ | $1.23 \times 10^5$ | $4.22 \times 10^4$ |
| | | $(dpa)_2CH^+$ | | $3.46 \times 10^4$ |
| | | $(mor)_2CH^+$ | | $3.34 \times 10^3$ |
| | | $(mpa)_2CH^+$ | | $4.04 \times 10^3$ |
| | | $(pyr)_2CH^+$ | | no rxn |
| 1h | 10.86 / 0.66 | $(pfa)_2CH^+$ | | $1.37 \times 10^5$ |
| | $(CH_3CN)$ | $(mfa)_2CH^+$ | $7.96 \times 10^4$ | $2.16 \times 10^4$ |
| | | $(dpa)_2CH^+$ | | $2.10 \times 10^4$ |
| | | $(mor)_2CH^+$ | | $2.33 \times 10^3$ |
| | | $(dma)_2CH^+$ | | no rxn |

[a] From ref. [6].

Plots of log $k$ vs. the electrophilicity parameters $E$ of the benzhydrylium ions (Figure B.2) are linear as required by equation (B.1), where $k$ is the second-order rate constant, $E$ the electrophilicity parameter, $N$ the nucleophilicity parameter, and $s_N$ the nucleophile-specific slope parameter.[7]

$$\log k = s_N(N + E) \qquad (B.1)$$

From the slopes and intercepts on the abscissa we can derive the nucleophile-specific parameters $s_N$ and $N$, respectively, which are listed in the second column of Table B.2.

The kinetics of the reactions of benzyl bromide with quinine (**1a**) and several of its substructures have been followed conductimetrically. In all cases, pseudo-first-order conditions were employed with the amines **1** in high excess, giving rise to an exponential increase of the conductance $G$ (equation B.2). The second-order rate constants (Table B.3) were again obtained from plots of $k_{obs}$ vs. [**1**].

$$G_t = G_{max}(1 - e^{-k_{obs}t}) + C \qquad (B.2)$$

# Appendix: B. Organocatalytic Activity of Cinchona Alkaloids

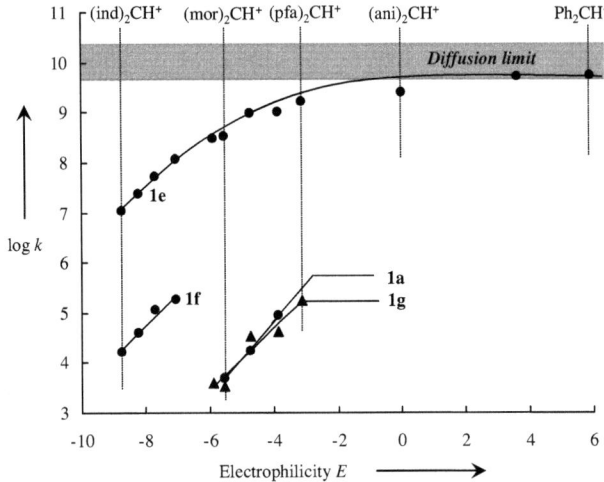

Figure B.2. Plots of log $k$ vs. $E$ for the reactions of amines with benzhydrylium ions.

Table B.3. Second-order rate constants (at 20 °C) for the reactions of amines with benzyl bromide.

| amine | $k$ / $M^{-1}$ $s^{-1}$ | |
|---|---|---|
| | DMSO | CH$_3$CN |
| 1a | 2.88 × 10$^{-2}$ | |
| 1d | 3.68 × 10$^{-2}$ | |
| 1e | 17.3 | 6.32 |
| 1f | | 6.16 × 10$^{-2}$ |
| 1g | | 1.7 × 10$^{-4}$ |

## B.4. Discussion

The similarity of the slope parameters $s_N$ in Table B.2 implies that the relative reactivities of the different amines depend only slightly on the nature of the benzhydrylium ion. For that reason, the relative reactivities toward (mfa)$_2$CH$^+$ given in Table B.4 can be considered to be representative also for reactions with other benzhydrylium ions. Obviously, the quinoline ring in quinine (**1a**) has a similar reactivity towards benzhydrylium ions as the 4-methyl- and 6-methoxy-substituted quinolines **1g** and **1h**. Quinuclidine (**1e**) reacts more than four orders of magnitude faster with benzhydrylium ions, and the 50-fold reduced reactivity of **1f** can be assigned to the steric shielding by the neighboring naphthylmethyl group.

Table B.4. Relative reactivities of different amines toward $(mfa)_2CH^+$ and benzyl bromide in $CH_3CN$ at 20 °C.

|  | 1g | 1h | 1a | 1e | 1f | 1a |
|---|---|---|---|---|---|---|
| $k_{rel}$ $(mfa)_2CH^+$ | 1.0 | 0.51 | 0.70 [a] | $2.4 \times 10^4$ | $5.0 \times 10^{2}$ [b] | << 0.7 |
| $k_{rel}$ $(PhCH_2Br)$ | 1.0 |  | << $6.2 \times 10^1$ | $3.7 \times 10^4$ | $3.6 \times 10^2$ | $6.2 \times 10^{1}$ [c] |

[a] Value in $CH_2Cl_2$ divided by 3, as for **1c** (from Table B.1); [b] calculated by equation (B.1) from $E$, $N$, and $s_N$; [c] value in DMSO divided by 2.7, as for **1e** (from Table B.2).

The additional hydroxy group in the naphthylmethyl group of **1a** must be responsible for a further > $10^3$-fold reduction of reactivity which is derived from the exclusive attack of benzhydrylium ions at the quinoline ring of **1a** (with $k_{rel}$ = 0.70). Though unlikely, one can not rigorously exclude a faster, highly reversible electrophilic attack of benzhydrylium ions at the $N_{sp3}$ center of **1a** and **1b**. The observed mono-exponential decays of the benzhydrylium absorbances in the reactions of $Ar_2CH^+BF_4^-$ with an excess of **1a** or **1b** (pseudo-first-order conditions) allow us, however, to exclude the appearance of noticeable concentrations of intermediate ammonium ions, where the diarylmethyl group is located at the $N_{sp3}$ center. The observed rate constants for the reactions of **1a,b** with $Ar_2CH^+$ can, therefore, unequivocally be assigned to the reactions at the $N_{sp2}$ center.

Comparison of the nucleophilic reactivities of **1e-g** toward benzhydrylium ions and benzyl bromide shows common features. Quinuclidine (**1e**) is four orders of magnitude more reactive than **1g** toward both types of electrophiles, and the attack of both electrophiles is 100-fold retarded by the naphthylmethyl group in **1f**. The additional hydroxy group in **1a,b** reduces the reactivities towards the sterically less demanding benzyl group much less (by a factor of 6) than towards the diarylmethylium ions (> $10^3$).

## B.5. Computational Analysis

Benzhydryl and benzyl cation affinities of quinine and its building blocks were calculated at the MP2(FC)/6-31+G(2d,p)//B3LYP/6-31G(d) level in order to rationalize why benzyl

## Appendix: B. Organocatalytic Activity of Cinchona Alkaloids

bromide reacts selectively at the $N_{sp3}$ center of cinchona alkaloids while benzhydrylium ions react selectively at the $N_{sp2}$ center. As illustrated by the reaction enthalpies $\Delta H_{298}$ in Figure B.3, the substituent effects on the quinoline ring affect its benzhydryl and benzyl cation affinities almost equally.

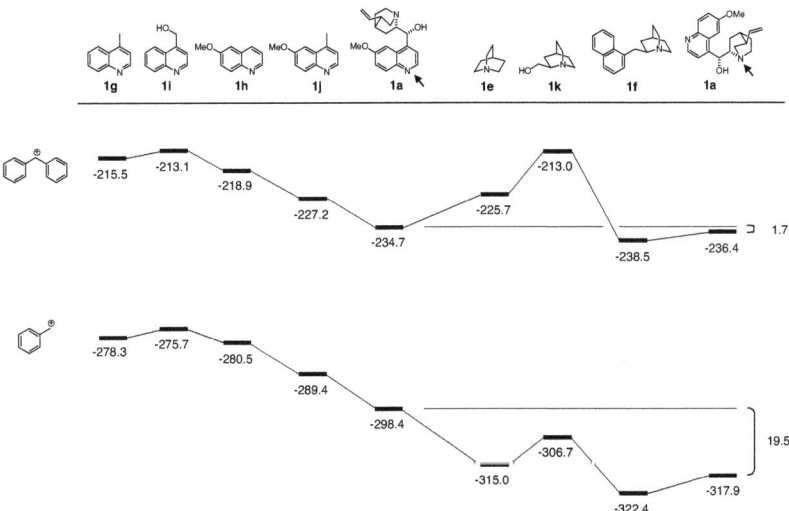

Figure B.3. Comparison of gas-phase benzyl and benzhydryl affinities $\Delta H_{298}$ (kJ mol$^{-1}$) of quinine and several substructures (MP2(FC)/6-31+G(2d,p)//B3LYP/6-31G(d)).

Replacement of CH$_3$ in lepidine (**1g**) by CH$_2$OH (→ **1i**) reduces the carbocation affinities by 2.5 ± 0.1 kJ mol$^{-1}$ while replacement of the 4-CH$_3$ group in **1g** by 6-OCH$_3$ (→ **1h**) raises the cation affinities by 2.8 ± 0.6 kJ mol$^{-1}$. Introduction of the 6-methoxy group into lepidine (**1g** → **1j**) increases the cation affinities by 11.4 ± 0.3 kJ mol$^{-1}$, and benzhydrylation or benzylation of the N$_{sp2}$ center of quinine (**1a**) is 19.6 ± 0.5 kJ mol$^{-1}$ more exothermic than that of lepidine (**1g**).

In contrast to the similar trends of the Ph$_2$CH$^+$ and PhCH$_2^+$ affinities of the differently substituted quinolines, large differences were calculated for the relative benzhydrylation and benzylation enthalpies of the quinuclidines. While the benzylation of quinuclidine (**1e**) is 37 kJ mol$^{-1}$ more exothermic than the benzylation of lepidine (**1g**), this difference shrinks to

10 kJ mol$^{-1}$ for benzhydrylation, which can be explained by two additional gauche-interactions.

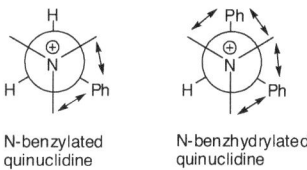

N-benzylated quinuclidine    N-benzhydrylated quinuclidine

The lower carbocation affinity of 2-hydroxymethylquinuclidine **1k** (compared with quinuclidine **1e**) can be assigned to the loss of an intramolecular hydrogen bridge by the quaternization. Surprisingly, the introduction of side chains into **1e** to give **1a** or **1f** increases the affinity toward benzhydryl cations more than towards benzyl cations.

Eventually, Figure B.4 (left) shows that $\Delta H_{298}$ (g) is almost identical for the benzhydrylation of both nitrogens of quinine (**1a**), while $\Delta H_{298}$ (g) for benzylation and methylation are considerably more negative for attack at the $N_{sp3}$ center. When $\Delta G_{298}$ (g) values are compared, a shift in favor of $N_{sp2}$ alkylation is observed (Figure B.4, middle), which is enhanced when solvation is included (Figure B.4, right). As a result, the preferred attack of benzyl bromide at $N_{sp3}$, and of benzhydryl cations at $N_{sp2}$ are in line with the relative thermodynamic stabilities of the reaction products. From these results, one can extrapolate that thermodynamic effects will direct sterically demanding electrophiles to the $N_{sp2}$ center of the cinchona alkaloids, while small electrophiles are directed to the $N_{sp3}$ center.

## B.6. Intrinsic Barriers

Relative reactivities are, however, not exclusively controlled by the relative stabilities of the products. The Marcus equation (B.3) expresses the activation free energy of a reaction ($\Delta G^{\ddagger}$) by a combination of the reaction free energy ($\Delta G^0$) and the intrinsic barrier ($\Delta G_0^{\ddagger}$). The latter term ($\Delta G_0^{\ddagger}$) corresponds to the activation free energy ($\Delta G^{\ddagger}$) of a reaction without thermodynamic driving force (i.e. for $\Delta G^0 = 0$).[8]

$$\Delta G^{\ddagger} = \Delta G_0^{\ddagger} + 0.5\Delta G^0 + [(\Delta G^0)^2/16\Delta G_0^{\ddagger}] \qquad (B.3)$$

# Appendix: B. Organocatalytic Activity of Cinchona Alkaloids

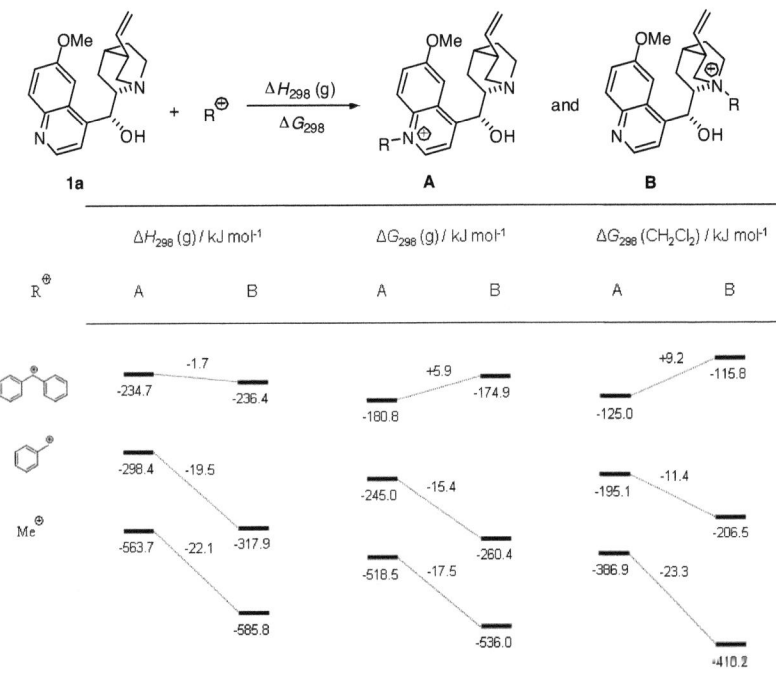

Figure B.4. Benzhydryl, benzyl and methyl cation affinities of the different nitrogen atoms of quinine (**1a**) (MP2(FC)/6-31+G(2d,p)).

In previous work, it was shown that quinuclidine is a much stronger nucleophile than DMAP, although the Lewis basicities, i.e., the equilibrium constants for the generation of the ammonium ions, are the other way around (Scheme B.4).[6]

The fact that quinuclidine (**1e**) reacts $10^3$ times faster with benzhydrylium ions than DMAP, but also departs 50,000 times faster from the benzhydrylium fragment, has been assigned to the lower intrinsic barrier $\Delta G_0^{\ddagger}$ for the reaction of quinuclidine ($\Delta G_0^{\ddagger}$ = 43 kJ mol$^{-1}$) compared with the corresponding reaction of DMAP ($\Delta G_0^{\ddagger}$ = 65 kJ mol$^{-1}$).

Presumably, a large portion of the higher reorganization energy ($\lambda = 4\Delta G_0^{\ddagger}$) in the reaction with DMAP comes from the reorganization of solvent molecules during the formation of the pyridinium ions.

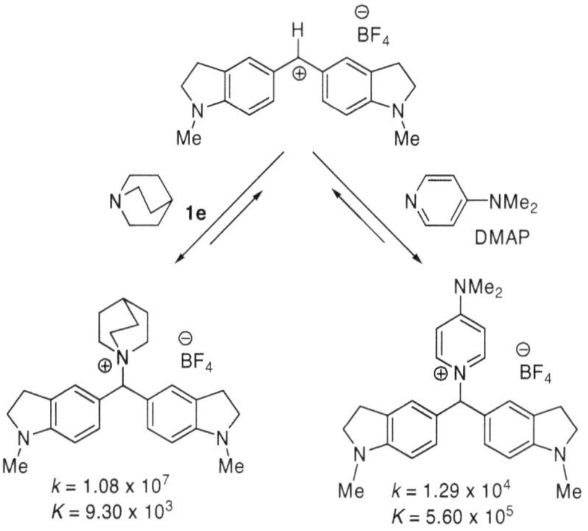

Scheme B.4. Comparison of second-order rate constants $k$ (L mol$^{-1}$ s$^{-1}$) and equilibrium constants $K$ (L mol$^{-1}$) for the reactions of quinuclidine and DMAP with (ind)$_2$CH$^+$BF$_4^-$ in CH$_3$CN at 20 °C (from ref. [6]).

In order to examine whether differences in intrinsic barriers also affect the different nucleophilicities of the two nitrogen atoms in **1a**, the intrinsic barriers for the reactions of benzhydrylium ions with some cinchona alkaloids have been determined. The combinations of (mfa)$_2$CH$^+$ with **1a-c** in CH$_2$Cl$_2$ (equation B.4) do not proceed quantitatively, and the corresponding equilibrium constants have been evaluated by UV/Vis spectroscopy. Assuming the validity of Lambert-Beer's law, the equilibrium constants for reactions B.4 can be expressed by the absorbances of the benzhydrylium ions before (A$_0$) and after (A) the addition of the amines **1a-c** (equation B.5).

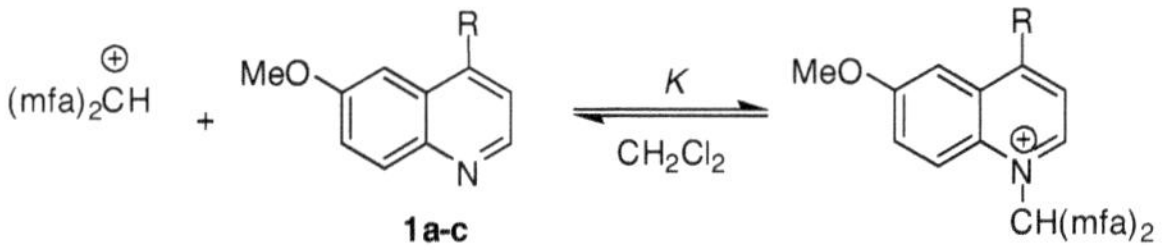

$$K = \frac{[(mfa)_2CH-NR_3^+]}{[(mfa)_2CH^+] \, [1]} = \frac{A_0 - A}{A \, [1]} \quad (B.5)$$

The equilibrium constants $K$ (Table B.5) and rate constants $k$ (Table B.2) were then converted into $\Delta G^0$ and $\Delta G^\ddagger$, respectively, and inserted into the Marcus equation (B.3) to give the intrinsic barriers ($\Delta G_0^\ddagger$) listed in Table B.5.

Table B.5. Equilibrium constants ($K$), reaction free energies ($\Delta G^0$), activation free energies ($\Delta G^\ddagger$), and intrinsic barriers ($\Delta G_0^\ddagger$) for the reactions of $(mfa)_2CH^+$ with the amines **1a-c** in $CH_2Cl_2$ at 20°C.

| amine | $K$ / L mol$^{-1}$ | $\Delta G^0$ / kJ mol$^{-1}$ [a] | $\Delta G^\ddagger$ / kJ mol$^{-1}$ [b] | $\Delta G_0^\ddagger$ / kJ mol$^{-1}$ |
|---|---|---|---|---|
| **1a** | $1.55 \times 10^4$ | -23.5 | 44.0 | 55.1 |
| **1b** | $1.79 \times 10^4$ | -23.9 | 43.8 | 55.1 |
| **1c** | $4.98 \times 10^3$ | -20.7 | 44.1 | 54.0 |

[a] $\Delta G^0 = -RT\ln K$; [b] from $k$ in Table B.2, using the Eyring equation.

With the $\Delta G_0^\ddagger \approx 55$ kJ mol$^{-1}$, the intrinsic barriers for the reactions of $(mfa)_2CH^+$ with **1a**, **1b**, and **1c** in $CH_2Cl_2$ are of similar magnitude as the intrinsic barriers for the reactions of pyridine and of *p*-substituted pyridines with benzhydrylium ions in the same solvent.[9]

Because the intrinsic barriers for the reactions of benzhydrylium ions with quinuclidine have previously been reported to be approximately 20 kJ mol$^{-1}$ smaller then those for the corresponding reactions with pyridines,[6] we can conclude that a similar difference should also hold for the electrophilic attack at the two different nucleophilic sites of the cinchona alkaloids. As a consequence, electrophilic attack at the $N_{sp3}$ center can be expected if the thermodynamic stabilities of the two different products are similar.

## B.7. Conclusion

Quinuclidinium ions arising from $N_{sp3}$ attack of primary alkylating agents at cinchona alkaloids are more stable than the isomeric quinolinium ions arising from the corresponding $N_{sp2}$ attack. In contrast, quinolinium ions are more stable than the isomeric quinuclidinium ions when sterically more demanding alkylating agents are used. Because more reorganization energy is needed for the electrophilic attack at the $N_{sp2}$ than at the $N_{sp3}$ center, kinetically controlled quinuclidine alkylation cannot only be expected when the $N_{sp3}$ adduct is thermodynamically favored, but also when the $N_{sp}^2$ adduct is slightly favored by thermodynamics, i.e., when the less negative $\Delta G^0$ term for $N_{sp3}$ attack of equation (B.3) is overcompensated by the smaller intrinsic barrier $\Delta G_0^\ddagger$.

## B.8. References

[1] a) G. Bredig, P. S. Fiske, *Biochem. Z.* **1913**, *46*, 7-32; b) G. Bredig, M. Minaeff, *Biochem. Z.* **1932**, *249*, 241-244; c) H. Pracejus, *Fortschr. Chem. Forsch.* **1967**, *8*, 493-553; d) K. Kacprzak, J. Gawronski, *Synthesis* **2001**, 961-998; e) Y. Chen, P. McDaid, L. Deng, *Chem. Rev.* **2003**, *103*, 2965-2983; f) S. France, D. J. Guerin, S. J. Miller, T. Lectka, *Chem. Rev.* **2003**, *103*, 2985-3012; g) L. Atodiresei, I. Schiffers, C. Bolm, *Chem. Rev.* **2007**, *107*, 5683-5712; h) M. J. Gaunt, C. C. C. Johansson, *Chem. Rev.* **2007**, *107*, 5596-5605; i) B. Lygo, B. I. Andrews, *Acc. Chem. Res.* **2004**, *37*, 518–525; j) M. J. O'Donnell, *Acc. Chem. Res.* **2004**, *37*, 506-517; k) S. E. Denmark, G. L. Beutner, *Angew. Chem., Int. Ed.* **2008**, *47*, 1560-1638; l) H. M. R. Hoffmann, J. Frackenpohl, *Eur. J. Org. Chem.* **2004**, 4293-4312; m) D. H. Paull, C. J. Abraham, M. T. Scerba, E. Alden-Danforth, T. Lectka, *Acc. Chem. Res.* **2008**, *41*, 655–663; n) T. P. Yoon, E. N. Jacobsen, *Science* **2003**, *299*, 1691-1693; o) T. S. Kaufman, E. A. Ruveda, *Angew. Chem., Int. Ed.* **2005**, *44*, 854-885.

[2] For examples see: a) H. Wack, A. E. Taggi, A. M. Hafez, W. J. Drury III, T. Lectka, *J. Am. Chem. Soc.* **2001**, *123*, 1531-1532; b) S. France, H. Wack, A. E. Taggi, A. M. Hafez, T. R. Wagerle, M. H. Shah, C. L. Dusich, T. Lectka, *J. Am. Chem. Soc.* **2004**, *126*, 4245-4255; c) A. E. Taggi, A. M. Hafez, H. Wack, B. Young, D. Ferraris, T. Lectka, *J. Am. Chem. Soc.* **2002**, *124*, 6626-6635; d) G. S. Cortez, S. H. Oh, D. J. Romo, *Synthesis* **2001**, 1731-1736; e) J. Wolfer, T. Bekele, C. J. Abraham, C. D. Isonagie, T. Lectka, *Angew. Chem., Int. Ed.* **2006**, *45*, 7398-7400; f) S. K. Tian, L. Deng, *J. Am. Chem. Soc.* **2001**, *123*, 6195-6196.

[3] M. Adamczyk, S. Rege, *Tetrahedron Lett.* **1998**, *39*, 9587-9588.

[4] a) K. Maruoka, T. Ooi, *Chem. Rev.* **2003**, *103*, 3013-3028; b) K. Maruoka, T. Ooi, *Angew. Chem., Int. Ed.* **2007**, *46*, 4222-4266; c) K. Maruoka, *Asymmetric Phase Transfer Catalysis*; Wiley-WCH: Weinheim, **2008**; d) S. Santoro, T. B. Poulsen, K. A. Jørgensen, *J. Chem Soc., Chem. Commun.* **2007**, 5155-5157; e) P. I. Dalko, *Enantioselective Organocatalysis*; Wiley-VCH: Weinheim, **2007**; f) A. Berkessel, H. Gröger, *Asymmetric Organocatalysis*; Wiley-VCH: Weinheim, **2005**; g) P. Elsner, L. Bernardi, G. D. Salla, J. Overgaard, K. A. Jørgensen, *J. Am. Chem. Soc.* **2008**, *130*, 4897-4905.

[5] H. Mayr, T. Bug, M. F. Gotta, N. Hering, B. Irrgang, B. Janker, B. Kempf, R. Loos, A. R. Ofial, G. Remennikov, H. Schimmel, *J. Am. Chem. Soc.* **2001**, *123*, 9500-9512.

[6] M. Baidya, S. Kobayashi, F. Brotzel, U. Schmidhammer, E. Riedle, H. Mayr, *Angew. Chem., Int. Ed.* **2006**, *46*, 6176-6179.

[7] a) H. Mayr, B. Kempf, A. R. Ofial, *Acc. Chem. Res.* **2003**, *36*, 66-77; b) H. Mayr, A. R. Ofial, *Pure Appl. Chem.* **2005**, *77*, 1807-1821; c) H. Mayr, A. R. Ofial, *J. Phys. Org. Chem.* **2008**, *21*, 584-595.

[8] a) R. A. Marcus, *J. Phys. Chem.* **1968**, *72*, 891-899; b) W. J. Albery, *Annu. Rev. Phys. Chem.* **1980**, *31*, 227-263.

[9] F. Brotzel, B. Kempf, T. Singer, H. Zipse, H. Mayr, *Chem. Eur. J.* **2007**, *13*, 336-345.

# Experimental Part

## 1. General Information

### 1.1. Methods

*Analytics.* $^1$H-, $^{13}$C- and $^{19}$F-NMR spectra were recorded on 200, 300, 400 or 600 MHz NMR spectrometers from Varian. Chemical shifts are reported in ppm relative to TMS ($\delta_H$ = 0.00, $\delta_C$ = 0.0), the deuterated solvent as the internal standard (CDCl$_3$: $\delta_H$ = 7.26, $\delta_C$ = 77.1; CD$_2$Cl$_2$: $\delta_H$ = 5.30, $\delta_C$ = 53.4; acetone-d$_6$: $\delta_H$ = 2.05, $\delta_C$ = 29.8, DMSO-d$_6$: $\delta_H$ = 2.50, $\delta_C$ = 39.5), or – in the case of $^{19}$F-NMR – fluorotrichloromethane ($\delta_F$ = 0). $^{13}$C-NMR spectra were recorded while $^1$H broadband decoupling. CH couplings have therefore been omitted in the descriptions of the $^{13}$C-NMR spectra. The following abbreviations were used to designate chemical shift multiplicities: br s = broad singlet, s = singlet, d = doublet, t = triplet, q = quartet, quint = quintet, sex = sextet, m = multiplet. Signal assignment was achieved by integration, HMBC and DEPT spectra. For structurally related carbon atoms, similar relaxation times were assumed, which allowed the derivation of the number of equivalent carbon atoms in $^{13}$C-NMR spectra. EI-MS was performed on a MAT 95 (Thermo Finnigan), while ESI-MS was performed on a LTQ FT (Thermo Finnigan). Reactions were followed by a 6890N GC-MS system (Agilent Technologies). Melting points were determined by a Büchi B-540 apparatus. An Elementar vario EL or an Elementar vario micro cube were used for elemental analyses.

*Kinetic instruments.* Reactions were either followed by conductimetry or photospectrometry. For conductimetry, reactions with half-times > 10 s were recorded by conventional methods using a Tacussel CD810 or a Radiometer MeterLab CDM230, both instruments equipped with Pt electrodes. Reactions with half-times τ < 10 s were followed by a Hi-Tech Scientific SF-61 DX2 stopped-flow device (cell volume 21 µL, Pt electrodes), controlled by the Hi-Tech KinetAsyst3 software. For photospectrometry, slow reactions (τ > 10 s) were followed by conventional UV-vis-spectrometry using a J&M TIDAS instrument equipped with an insertion quartz probe (Hellma) and a halogen or a deuterium light source. Faster reactions (τ < 10 s) were studied by using either a SX.18MV-R (Applied Photophysics) or a SF-61 DX2 (Hi-Tech Scientific) stopped-flow reactor. In all experiments, the temperature was controlled by water baths and water circuits.

*Laser-flash equipment.* The laser pulse (7 ns pulse width, 266 nm, 40-60 mJ/pulse) originated from a Nd-YAG laser (Innolas Spitlight 600). The UV-vis detection unit comprised a 150 W Xe-light source (Hamamatsu Photonics), a spectrograph (Acton), a photomultiplier, and a pulse generator. For the data acquisition a 350 MHz-oscilloscope was used. A shutter was used to prevent the sample from unnecessary exposure to the light from the Xe lamp. The timing was controlled by a BNC 565 delay generator (Berkeley Nucleonics Corporation).

*Computational chemistry.* For theoretical calculations Intel Xeon (2.67 GHz) and AMD Opteron (2.4 GHz) PCs with Debian GNU/Linux 4.0 (64-Bit-Version) were used. The conformational space has been searched using the MM3 force field and the systematic search routine in the TINKER program.[1] All quantum chemical calculations were performed with Gaussian 03, revision D.01.[2] Conformational minima on the energy hypersurface have been confirmed by frequency calculations. Single point energy calculations were performed with the option SCF=tight.

*Electrochemical equipment.* The half-wave reduction potentials of all compounds were examined by steady-state cyclic voltammetry in acetonitrile solution at ambient room temperature using the conventional three electrode configuration and a EG&G Princeton Applied Research potentiostat (model 273). A Powerlab/4sp AD converter controlled by Powerlab Chart 5 software (ADInstruments) allowed recording, analysis and storage of the data by a PC. The electrodes and an inlet for gaseous nitrogen were fixed in a four-necked glass vessel. The whole setup was encased in a faradaic cage. A network of 8 platinum microelectrodes (each Pt wire had a radius of 10 $\mu$m) in a glass tube served as working electrode. The counter electrode was a platinum spiral with a large and smooth surface. The reference electrode – a silver wire covered with a silver oxide layer – was pocketed in a glass capillary with a frit to avoid direct contact with the substrates. The frit was filled with acetonitrile and a small amount of tetraethylammonium perchlorate as conductive salt. No conductive salt was used in the bulk solution. The acetonitrile was freshly distilled from $CaH_2$ and stored over MS-4 Å for at least 24 h prior to use. The typical procedure was as follows: After the glass vessel was flushed with nitrogen, the first portion of acetonitrile (usually 5 mL) was added and degassed for about 30 seconds. The substrate was added, the voltammogram recorded with a scan rate of usually 20 mV/s. Additional solvent was injected in 5 mL steps, each time degassing the solution (30 s) and recording the voltammogram. The

Experimental Part 171

substrate concentrations ranged from $10^{-3}$ to $10^{-2}$ mol $L^{-1}$ At last, a small amount of the internal standard ferrocene was added.

### 1.2. Materials

*Solvents*. Dichloromethane was predried over $CaCl_2$ before it was distilled from $CaH_2$. DMSO, acetonitrile, and acetone (VWR, ≥ 99.9 %) were used as purchased. Water was purified by using a Millipore MilliQ (final specific resistance ≥ 18.2 MΩcm). Tetrahydrofuran and diethyl ether were dried over sodium/benzophenone and distilled prior to use.

*Chemicals*. The N-heterocyclic carbene boranes **1** and **2** were a gift of Prof. Dennis P. Curran (University of Pittsburgh, Pittsburgh, Pennsylvania). Benzhydrylium tetrafluoroborates and benzhydryl chlorides were taken from the stock of the work group. Triphenylsilane was used as purchased, liquid silanes, cycloheptatriene, tributylstannane, and tin tetrachloride were purchased and distilled prior to use. Gallium trichloride (anhydrous, 99.99 %) was used as purchased. Tetrabutylstannane, tris(trimethylsilyl)silane, 2-propyl-1,3-dioxolane, and 2-phenyl-1,3-dioxolane were prepared according to standard procedures. All reagents used in the synthetic procedures have been bought from commercial suppliers (Aldrich, Acros, ABCR, Apollo).

## 2. Synthetic Procedures

When it was necessary to avoid moisture, the glassware was heated and evacuated prior to use. A protecting gas atmosphere (nitrogen, argon) was applied during these reactions.

### 2.1. Preparation of triarylmethanols

General procedure 1 (GP1):
Grignard compounds derived from substituted bromobenzenes were prepared by slow addition of the bromoarene to a stirred suspension of magnesium turnings in dry THF and subsequent refluxing for 30 min. After cooling to r.t., the solution of a substituted benzophenone or alkyl benzoate in dry THF was added slowly. The mixture was refluxed for 1 h and then cooled with an ice bath. Water and 2 M HCl were added until the precipitate had

dissolved completely. The phases were separated and the aqueous phase twice extracted with diethyl ether. The combined organic layers were washed with saturated aqueous sodium bicarbonate, water and brine. After drying over $Na_2SO_4$ or $MgSO_4$, the solvent was evaporated under reduced pressure, and the product – usually a viscous oil – crystallized.

*Tris(3,5-difluorophenyl)methanol*
This compound was not isolated, but directly converted to the corresponding bromomethane.

*Bis(3,5-difluorophenyl)phenylmethanol*
According to GP1 from 9.30 g (48.2 mmol) 3,5-difluorobromobenzene, 1.19 g (49.0 mmol) magnesium and 3.62 g (24.1 mmol) ethyl benzoate in 30 mL tetrahydrofuran. 5.97 g (18.0 mmol, 75 %) of a colorless solid were obtained.

$^1$**H-NMR** (300 MHz, CDCl$_3$): $\delta$/ppm = 2.81 (s, 1 H, OH), 6.70-6.90 (m, 6 H, H$_{arom.}$), 7.18-7.26 (m, 2 H, H$_{arom.}$), 7.35-7.40 (m, 3 H, H$_{arom.}$).
$^{13}$**C-NMR** (75 MHz, CDCl$_3$): $\delta$/ppm = 81.2 (quint, 1 C, $^4J_{C,F}$ = 2.1 Hz, COH), 103.2 (t, 2 C, $^2J_{C,F}$ = 25.4 Hz, CH), 110.7-111.1 (m, 4 C, CH), 127.6 (s, 2 C, CH), 128.4 (s, 1 C, CH), 128.6 (s, 2 C, CH), 144.7 (s, 1 C, C$_{quat.}$), 149.6 (t, 2 C, $^3J_{C,F}$ = 8.1 Hz, C$_{quat.}$), 162.7 (dd, 4 C, $^1J_{C,F}$ = 249 Hz, $^3J_{C,F}$ = 12.6 Hz, C$_{quat.}$).
$^{19}$**F-NMR** (282 MHz, CDCl$_3$): $\delta$/ppm = –109.
**HR-MS** (EI, pos.): $m/z$ calculated for [C$_{19}$H$_{12}$F$_4$O]$^+$: 332.0819, found: 332.0813.
**Mp**: 71-72 °C.

*(3,5-Difluorophenyl)bis(3-fluorophenyl)methanol*
According to GP1 from 2.41 g (12.5 mmol) 3,5-difluorobromobenzene, 316 mg (13.0 mmol) magnesium, and 2.66 g (12.2 mmol) 3,3'-difluorobenzophenone. 1.21 g (3.64 mmol, 30 %) of a slightly yellow solid were obtained.

$^1$**H-NMR** (300 MHz, CDCl$_3$): $\delta$/ppm = 2.83 (s, 1 H, OH), 6.78 (tt, 1 H, $J$ = 8.7 Hz, $J$ = 2.3 Hz, H$_{arom.}$), 6.83-6.90 (m, 2 H, H$_{arom.}$), 7.00-7.10 (m, 6 H, H$_{arom.}$), 7.30-7.38 (m, 2 H, H$_{arom.}$).
$^{13}$**C-NMR** (75 MHz, CDCl$_3$): $\delta$/ppm = 81.0 (m, 1 C, COH), 103.3 (t, 1 C, $^2J_{C,F}$ = 25.3 Hz, CH), 110.7-111.1 (m, 2 C, CH), 114.9 (d, 2 C, $^2J_{C,F}$ = 23.1 Hz, CH), 115.0 (d, 2 C, $^2J_{C,F}$ = 21.1 Hz, CH), 123.3 (d, 2 C, $^4J_{C,F}$ = 3.0 Hz, CH), 129.9 (d, 2 C, $^3J_{C,F}$ = 8.2 Hz, CH), 147.7

Experimental Part                                                                 173

(d, 2 C, $^3J_{C,F}$ = 6.6 Hz, $C_{quat.}$), 149.7 (t, 1 C, $^3J_{C,F}$ = 8.1 Hz, $C_{quat.}$), 162.7 (d, 2 C, $^1J_{C,F}$ = 247 Hz, CF), 162.9 (d, 2 C, $^1J_{C,F}$ = 249 Hz, CF).
**$^{19}$F-NMR** (282 MHz, CDCl$_3$): $\delta$/ppm = –109, –112.
**HR-MS** (EI, pos.): $m/z$ calculated for [C$_{19}$H$_{12}$OF$_4$]$^+$: 332.0819, found: 332.0819.
**Mp**: 98-99 °C.


*Tris(3-fluorophenyl)methanol*
According to GP1 from 5.25 g (30.0 mmol) 3-fluorobromobenzene, 753 mg (31.0 mmol) magnesium, and 1.12 g (9.48 mmol) diethyl carbonate. 1.85 g (5.89 mmol, 62 %) of a colorless solid were obtained.

**$^1$H-NMR** (300 MHz, CDCl$_3$): $\delta$/ppm = 2.87 (br s, 1 H, OH), 7.02-7.08 (m, 9 H, H$_{arom.}$), 7.27-7.36 (m, 3 H, H$_{arom.}$).
**$^{13}$C-NMR** (75 MHz, CDCl$_3$): $\delta$/ppm = 81.1 (m, 1 C, COH), 114.8 (d, 3 C, $^2J_{C,F}$ = 21.2 Hz, CH), 115.1 (d, 3 C, $^2J_{C,F}$ = 23.0 Hz, CH), 123.5 (d, 3 C, $^4J_{C,F}$ = 2.9 Hz, CH), 129.8 (d, 3 C, $^3J_{C,F}$ = 8.1 Hz, CH), 148.4 (d, 3 C, $^3J_{C,F}$ = 6.5 Hz, $C_{quat.}$), 162.7 (d, 3 C, $^1J_{C,F}$ = 247 Hz, CF).
**$^{19}$F-NMR** (282 MHz, CDCl$_3$): $\delta$/ppm = –113.
**Mp**: 112.5-116 °C (Lit.:[3] 118.5-119 °C).


*(3,5-Difluorophenyl)diphenylmethanol*
According to GP1 from 6.50 g (33.7 mmol) 3,5-difluorobromobenzene, 830 mg (34.1 mmol) magnesium, and 6.14 g (33.7 mmol) benzophenone. 6.63 g (22.4 mmol, 66 %) of a colorless solid were obtained.

**$^1$H-NMR** (600 MHz, CDCl$_3$): $\delta$/ppm = 2.81 (s, 1 H, OH), 6.73 (tt, 1 H, $J$ = 8.73 Hz, $J$ = 2.33 Hz, H$_{arom.}$), 6.86-6.91 (m, 2 H, H$_{arom.}$), 7.24-7.27 (m, 4 H, H$_{arom.}$), 7.31-7.36 (m, 6 H, H$_{arom.}$).
**$^{13}$C-NMR** (150 MHz, CDCl$_3$): $\delta$/ppm = 81.6 (t, 1 C, $^4J_{C,F}$ = 2.2 Hz, COH), 102.6 (t, 1 C, $^2J_{C,F}$ = 25.4 Hz, CH), 110.9-111.2 (m, 2 C, CH), 127.7 (s, 4 C, CH), 127.8 (s, 2 C, CH), 128.2 (s, 4 C, CH), 145.8 (s, 2 C, $C_{quat.}$), 150.8 (t, 1 C, $^3J_{C,F}$ = 8.2 Hz, $C_{quat.}$), 162.6 (dd, 2 C, $^1J_{C,F}$ = 249 Hz, $^3J_{C,F}$ = 12.6 Hz, CF).
**HR-MS** (EI, pos.): $m/z$ calculated for [C$_{19}$H$_{14}$OF$_2$]$^+$: 296.107, found: 296.1014.
**Mp**: 87.5-88.5 °C.

*Bis(3-fluorophenyl)phenylmethanol*
According to GP1 from 10.5 g (60.0 mmol) 3-fluorobromobenzene, 1.46 g (60.1 mmol) magnesium, and 4.51 g (30.0 mmol) ethyl benzoate. 5.50 g (18.6 mmol, 62 %) of a colorless solid were obtained.

$^1$**H-NMR** (300 MHz, CDCl$_3$): $\delta$/ppm = 2.83 (s, 1 H, OH), 6.96-7.10 (m, 6 H, H$_{arom.}$), 7.25-7.40 (m, 7 H, H$_{arom.}$).
$^{13}$**C-NMR** (75 MHz, CDCl$_3$): $\delta$/ppm = 81.4 (s, 1 C, COH), 114.4 (d, 2 C, $^2J_{C-F}$ = 21.2 Hz, CH), 115.0 (d, 2 C, $^2J_{C-F}$ = 22.9 Hz, CH), 123.5 (d, 2 C, $^4J_{C-F}$ = 2.9 Hz, CH), 127.7 (s, 2 C, CH), 127.8 (s, 1 C, CH), 128.2 (s, 2 C, CH), 129.5 (d, 2 C, $^3J_{C-F}$ = 8.1 Hz, CH), 145.8 (s, 1 C, C$_{quat.}$), 148.9 (d, 2 C, $^3J_{C-F}$ = 6.5 Hz, C$_{quat.}$), 162.6 (d, 2 C, $^1J_{C-F}$ = 246 Hz, CF).
**HR-MS** (EI, pos.): *m/z* calculated for [C$_{19}$H$_{14}$OF$_2$]$^+$: 296.1007, found: 296.1012.
**Mp**: 113-114 °C (Lit.:[3] 114-114.5 °C).

*(3-Fluorophenyl)diphenylmethanol*
According to GP1 from 6.30 g (36.0 mmol) 3-fluorobromobenzene, 880 mg (36.2 mmol) magnesium, and 6.56 g (36.0 mmol) benzophenone. 7.47 g (26.8 mmol, 74 %) of a colorless solid were obtained.

$^1$**H-NMR** (300 MHz, CDCl$_3$): $\delta$/ppm = 2.85 (s, 1 H, OH), 6.96-7.04 (m, 1 H, H$_{arom.}$), 7.07-7.13 (m, 2 H, H$_{arom.}$), 7.26-7.40 (m, 11 H, H$_{arom.}$).
$^{13}$**C-NMR** (75 MHz, CDCl$_3$): $\delta$/ppm = 81.7 (s, 1 C, COH), 114.1 (d, 1C, $^2J_{C-F}$ = 21.2 Hz, CH), 115.1 (d, 1 C, $^2J_{C-F}$ = 22.8 Hz, CH), 123.6 (d, 1 C, $^4J_{C-F}$ = 2.8 Hz, CH), 127.5 (s, 2 C, CH), 127.8 (s, 4 C, CH), 128.1 (s, 4 C, CH), 129.3 (d, 1 C, $^3J_{C-F}$ = 8.1 Hz, CH), 146.3 (s, 2 C, C$_{quat.}$), 149.4 (d, 1 C, $^3J_{C-F}$ = 6.6 Hz, C$_{quat.}$), 162.5 (d, 1 C, $^1J_{C-F}$ = 246 Hz, CF).
$^{19}$**F-NMR** (282 MHz, CDCl$_3$): $\delta$/ppm = –114.
**HR-MS** (EI, pos.): *m/z* calculated for [C$_{19}$H$_{15}$OF]$^+$: 278.1101, found: 278.1089.
**Mp**: 114-115 °C (Lit.:[3] 117 °C).

*(4-Chlorophenyl)diphenylmethanol*
According to GP1 from 2.13 g (13.6 mmol) bromobenzene, 330 mg (13.6 mmol) magnesium and 2.94 g (13.6 mmol) 4-chlorobenzophenone in 40 mL tetrahydrofuran. After recrystallization from pentane 2.52 g (8.55 mmol, 63 %) of a colorless solid were obtained.

Experimental Part 175

**¹H-NMR** (600 MHz, CDCl₃): δ/ppm = 2.77 (s, 1 H, OH), 7.22-7.34 (m, 14 H, H$_{arom.}$).
**¹³C-NMR** (150 MHz, CDCl₃): δ/ppm = 81.7 (s, 1 C, COH), 127.5 (s, 2 C, CH), 127.8 (s, 4 C, CH), 128.0 (s, 2 C, CH), 128.1 (s, 4 C, CH), 129.3 (s, 2 C, CH), 133.1 (s, 1 C, CCl), 145.3 (s, 1 C, C$_{quat.}$), 146.4 (s, 2 C, C$_{quat.}$).
**HR-MS** (EI, pos.): m/z calculated for [C$_{19}$H$_{15}$ClO]⁺: 294.0806, found: 294.0786.
**Mp**: 80-81 °C (Lit.:[4] 85-86 °C).

*(4-Fluorophenyl)diphenylmethanol*
According to GP1 from 2.26 g (14.4 mmol) bromobenzene, 350 mg (14.4 mmol) magnesium, and 2.88 g (14.4 mmol) 4-fluorobenzophenone. 3.65 g (13.1 mmol, 91 %) of a colorless solid were obtained.

**¹H-NMR** (600 MHz, CDCl₃): δ/ppm = 2.79 (1 H, s, OH), 7.00 (2 H, t, $J_{H,H}$ = 8.7 Hz, H$_{arom.}$), 7.25-7.35 (12 H, m, H$_{arom.}$).
**¹³C-NMR** (150 MHz, CDCl₃): δ/ppm = 81.7 (s, 1 C, COH), 114.7 (d, 2 C, $^2J_{C,F}$ = 21.3 Hz, CH), 127.4 (s, 2 C, CH), 127.8 (s, 4 C, CH), 128.0 (s, 4 C, CH), 129.7 (d, 2 C, $^3J_{C,F}$ = 8.1 Hz, CH), 142.7 (d, 1 C, $^4J_{C,F}$ = 3.2 Hz, C$_{quat.}$), 146.7 (s, 2 C, C$_{quat.}$), 161.9 (d, 1 C, $^1J_{C,F}$ = 247 Hz, CF).
**HR-MS** (EI, pos.): m/z calculated for [C$_{19}$H$_{15}$OF]⁺: 278.1101, found: 278.1118.
**Mp**: 122-124 °C (Lit.:[5] 121-122 °C).

*Bis(4-fluorophenyl)phenylmethanol*
According to GP1 from 5.97 g (34.1 mmol) 4-fluorobromobenzene, 830 mg (34.1 mmol) magnesium, and 5.00 g (25.0 mmol) 4-fluorobenzophenone. After recrystallization from pentane, 4.16 g (14.0 mmol, 56 %) of a colorless solid were obtained.

**¹H-NMR** (300 MHz, CDCl₃): δ/ppm = 2.78 (s, 1 H, OH), 6.98-7.25 (m, 4 H, H$_{arom.}$), 7.22-7.34 (m, 9 H, H$_{arom.}$).
**¹³C-NMR** (75 MHz, CDCl₃): δ/ppm = 81.4 (s, 1 C, COH), 114.9 (d, 4 C, $^2J_{C,F}$ = 21.3 Hz, CH), 127.7 (s, 1 C, CH), 127.8 (s, 2 C, CH), 128.2 (s, 2 C, CH), 129.7 (d, 4 C, $^3J_{C,F}$ = 8.1 Hz, CH), 142.6 (d, 2 C, $^4J_{C,F}$ = 3.2 Hz, C$_{quat.}$), 146.6 (s, 1 C, C$_{quat.}$), 162.1 (d, 2 C, $^1J_{C,F}$ = 247 Hz, CF).
**¹⁹F-NMR** (282 MHz, CDCl₃): δ/ppm = –115.
**HR-MS** (EI, pos.): m/z calculated for [C$_{19}$H$_{12}$OF$_2$]⁺: 296.1007, found: 296.1016.

Mp: 96-97 °C (Lit.:[3] 100 °C).

*(4-Methylphenyl)diphenylmethanol*
According to GP 1 from 5.13 g (30.0 mmol) 4-bromotoluene, 750 mg (30.9 mmol) magnesium and 5.45 g (29.9 mmol) benzophenone. 7.11 g (25.9 mmol, 87 %) of a colorless solid were obtained after crystallization from pentane.

$^1$H-NMR (300 MHz, CDCl$_3$): $\delta$/ppm = 2.38 (s, 3 H, Me), 2.80 (s, 1 H, OH), 7.14-7.20 (m, 4 H, H$_{arom.}$), 7.30-7.36 (m, 10 H, H$_{arom.}$).
$^{13}$C-NMR (75 MHz, CDCl$_3$): $\delta$/ppm = 21.0 (s, 1 C, Me), 81.9 (s, 1 C, COH), 127.1 (s, 2 C, CH), 127.9 (s, 10 C, CH), 128.6 (s, 2 C, CH), 136.9 (s, 1 C, C$_{quat.}$), 144.0 (s, 1 C, C$_{quat.}$), 147.0 (s, 2 C, C$_{quat.}$).
Mp: 68-69 °C (Lit.:[6] 75.5-76.4 °C).

*Bis(4-methylphenyl)phenylmethanol*
According to GP 1 from 13.7 g (80.1 mmol) 4-bromotoluene, 1.95 g (80.2 mmol) magnesium and 6.00 g (40.0 mmol) ethyl benzoate. 7.57 g (26.3 mmol, 66 %) of a colorless solid were obtained after crystallization from pentane.

$^1$H-NMR (300 MHz, CDCl$_3$): $\delta$/ppm = 2.38 (s, 6 H, Me), 2.77 (s, 1 H, OH), 7.13-7.22 (m, 8 H, H$_{arom.}$), 7.32-7.34 (m, 5 H, H$_{arom.}$).
$^{13}$C-NMR (75 MHz, CDCl$_3$): $\delta$/ppm = 21.2 (s, 2 C, Me), 81.9 (s, 1 C, COH), 127.2 (s, 1 C, CH), 128.0 (s, 8 C, CH), 128.7 (s, 4 C, CH), 137.0 (s, 2 C, C$_{quat.}$), 144.3 (s, 2 C, C$_{quat.}$), 147.1 (s, 1 C, C$_{quat.}$).
Mp: 73-74 °C (Lit.:[6] 75.5-76.4 °C).

*Tris(4-methylphenyl)methanol*
According to GP 1 from 11.3 g (66.0 mmol) 4-bromotoluene, 1.62 g (66.7 mmol) magnesium and 5.39 g (32.8 mmol) ethyl 4-methylbenzoate. 7.22 g (23.9 mmol, 73 %) of a colorless solid were obtained after crystallization from diethyl ether/pentane.

$^1$H-NMR (300 MHz, CDCl$_3$): $\delta$/ppm = 2.35 (s, 9 H, Me), 2.71 (s, 1 H, OH), 7.10-7.19 (m, 12 H, H$_{arom.}$).

$^{13}$C-NMR (75 MHz, CDCl$_3$): $\delta$/ppm = 21.0 (s, 3 C, Me), 81.6 (s, 1 C, COH), 127.8 (s, 6 C, CH), 128.5 (s, 6 C, CH), 136.7 (s, 3 C, C$_{quat.}$), 144.3 (s, 3 C, C$_{quat.}$).
**Mp**: 93-94 °C (Lit.:[7] 94 °C).

*(4-Methoxyphenyl)diphenylmethanol*
According to GP 1 from 6.44 g (34.4 mmol) 4-methoxybromobenzene, 850 mg (35.0 mmol) magnesium and 6.38 g (35.0 mmol) benzophenone. 7.25 g (25.0 mmol, 73 %) of a colorless solid were obtained after crystallization from pentane.

$^1$**H-NMR** (600 MHz, CDCl$_3$): $\delta$/ppm = 2.77 (s, 1 H, OH), 3.80 (s, 3 H, Me), 6.83 (d, $^3J_{H,H}$ = 8.8 Hz, 2 H, H$_{arom.}$), 7.18 (d, $^3J_{H,H}$ = 8.8 Hz, 2 H, H$_{arom.}$), 7.25-7.33 (m, 10 H, H$_{arom.}$).
$^{13}$**C-NMR** (150 MHz, CDCl$_3$): $\delta$/ppm = 55.2 (s, 1 C, Me), 81.7 (s, 1 C, COH), 113.2 (s, 2 C, CH), 127.1 (s, 2 C, CH), 127.8 (s, 4 C, CH), 127.9 (s, 4 C, CH), 129.2 (2 C, CH), 139.2 (s, 1 C, C$_{quat.}$), 147.1 (s, 2 C, C$_{quat.}$), 158.7 (s, 1 C, COMe).
**Mp**: 77-78 °C (Lit.:[8] 58-61 °C).

*Bis(4-methoxyphenyl)phenylmethanol*
According to GP 1 from 2.25 g (12.0 mmol) 4-methoxybromobenzene, 300 mg (12.3 mmol) magnesium and 2.33 g (11.0 mmol) 4-methoxybenzophenone. 2.82 g (8.80 mmol, 80 %) of a colorless solid were obtained after crystallization from pentane.

$^1$**H-NMR** (600 MHz, CDCl$_3$): $\delta$/ppm = 2.65 (s, 1 H, OH), 3.71 (s, 6 H, Me), 6.75 (d, $^3J_{H,H}$ = 8.5 Hz, 4 H, H$_{arom.}$), 7.09 (d, $^3J_{H,H}$ = 8.5 Hz, 4 H, H$_{arom.}$), 7.15-7.25 (m, 5 H, H$_{arom.}$).
$^{13}$**C-NMR** (150 MHz, CDCl$_3$): $\delta$/ppm = 55.2 (s, 2 C, Me), 81.4 (s, 1 C, COH), 113.2 (s, 4 C, CH), 127.0 (s, 1 C, CH), 127.7 (s, 2 C, CH), 127.8 (s, 2 C, CH), 129.1 (s, 4 C, CH), 139.4 (s, 2 C, C$_{quat.}$), 147.3 (s, 1 C, C$_{quat.}$), 158.6 (s, 2 C, COMe).
**Mp**: 75-76 °C (Lit.:[9] 75-77 °C).

*Tris(4-methoxyphenyl)methanol*
According to GP 1 from 4.66 g (24.9 mmol) 4-methoxybromobenzene, 615 mg (25.3 mmol) magnesium and 2.00 g (12.0 mmol) methyl 4-methoxybenzoate. 2.26 g (6.45 mmol, 54 %) of a colorless solid were obtained after crystallization from diethyl ether/pentane.

**$^1$H-NMR** (600 MHz, CDCl$_3$): $\delta$/ppm = 2.73 (s, 1 H, OH), 3.80 (s, 9 H, Me), 6.83 (d, $^3J_{H,H}$ = 8.8 Hz, 6 H, H$_{arom.}$), 7.17 (d, $^3J_{H,H}$ = 8.8 Hz, 6 H, H$_{arom.}$).
**$^{13}$C-NMR** (150 MHz, CDCl$_3$): $\delta$/ppm = 55.2 (s, 3 C, Me), 81.1 (s, 1 C, COH), 113.1 (s, 6 C, CH), 129.0 (s, 6 C, CH), 139.7 (s, 3 C, C$_{quat.}$), 158.5 (s, 3 C, COMe).
**Mp**: 79-80 °C (Lit.:[10] 80 °C).

*(4-Dimethylaminophenyl)diphenylmethanol*
This compound was not isolated, but directly converted to the corresponding tetrafluoroborate salt.

*(4-Dimethylaminophenyl)(4-methoxyphenyl)phenylmethanol*
According to GP 1 from 2.28 g (12.2 mmol) 4-methoxybromobenzene, 300 mg (12.3 mmol) magnesium and 2.22 g (12.2 mmol) benzophenone. 2.50 g (7.50 mmol, 62 %) of a slightly red solid were obtained after crystallization from diethyl ether/pentane.

**$^1$H-NMR** (300 MHz, CDCl$_3$): $\delta$/ppm = 2.74 (s, 1 H, OH), 2.97 (s, 6 H, NMe$_2$), 3.82 (s, 3 H, OMe), 6.66-6.72 (m, 2 H, H$_{arom.}$), 6.84-6.87 (m, 2 H, H$_{arom.}$), 7.09-7.14 (m, 2 H, H$_{arom.}$), 7.21-7.35 (m, 7 H, H$_{arom.}$).
**$^{13}$C-NMR** (75 MHz, CDCl$_3$): $\delta$/ppm = 40.5 (s, 2 C, NMe$_2$), 55.2 (s, 1 C, OMe), 81.5 (s, 1 C, COH), 111.7 (s, 2 C, CH), 113.0 (s, 2 C, CH), 126.8 (s, 1 C, CH), 127.7 (s, 2 C, CH), 127.8 (s, 2 C, CH), 128.8 (s, 2 C, CH), 129.1 (s, 2 C, CH), 135.2 (1 C, C$_{quat.}$), 139.8 (1 C, C$_{quat.}$), 147.6 (1 C, C$_{quat.}$), 149.5 (1 C, CNMe$_2$), 158.5 (1 C, COMe).
**Mp**: 91-92 °C (Lit.:[11] 83-85 °C).

*Bis(4-dimethylaminophenyl)phenylmethanol*
According to GP 1 from 6.00 g (30.0 mmol) 4-(dimethylamino)bromobenzene, 730 mg (30.0 mmol) magnesium and 2.25 g (15.0 mmol) ethyl benzoate. 3.52 g (10.2 mmol, 68 %) of a slightly green solid were obtained after crystallization from diethyl ether/pentane.

**$^1$H-NMR** (300 MHz, CDCl$_3$): $\delta$/ppm = 2.68 (s, 1 H, OH), 2.96 (s, 12 H, Me), 6.65-6.71 (m, 4 H, H$_{arom.}$), 7.15-7.17 (m, 4 H, H$_{arom.}$), 7.24-7.37 (m, 5 H, H$_{arom.}$).
**$^{13}$C-NMR** (75 MHz, CDCl$_3$): $\delta$/ppm = 40.5 (s, 4 C, Me), 81.5 (s, 1 C, COH), 111.7 (s, 4 C, CH), 126.6 (s, 1 C, CH), 127.6 (s, 2 C, CH), 127.8 (s, 2 C, CH), 128.8 (s, 4 C, CH), 135.6 (s, 2 C, C$_{quat.}$), 147.9 (s, 1 C, C$_{quat.}$), 149.5 (s, 2 C, CNMe$_2$).

**Mp**: 108-109 °C (Lit.:[12] 109-110 °C).

*Preparation of 3,3'-difluorobenzophenone*
1.78 g (14 mmol) of oxalyl chloride were cooled to –70 °C in 30 mL dichloromethane. Gas started to evolve upun the addition of 2.19 g (28 mmol) of DMSO. 3.0 g (14 mmol) bis(3-fluorophenyl)methanol were added dropwise during 10 min, and the solution was stirred for 15 min. After addition of 2.19 g (28 mmol) triethylamine, the mixture was stirred for further 5 min before warming to room temperature. Water was added, the phases separated and the organic phase washed with water before it was dried. Evaporation of the solvent yielded 2.64 g (12.1 mmol, 89 %) of a slightly yellow solid.

**[1]H-NMR** (300 MHz, CDCl$_3$): $\delta$/ppm = 7.31 (tdd, 2 H, $J$ = 8.2 Hz, $J$ = 2.6 Hz, $J$ = 1.1 Hz, H$_{arom.}$), 7.43-7.58 (m, 6 H, H$_{arom.}$).
**[13]C-NMR** (75 MHz, CDCl$_3$): $\delta$/ppm = 116. 8 (d, 2 C, $^2J_{C,F}$ = 22.6 Hz, CH), 119.9 (d, 2 C, $^2J_{C,F}$ = 21.4 Hz, CH), 125.9 (d, 2 C, $^4J_{C,F}$ = 3.1 Hz, CH), 130.2 (d, 2 C, $^3J_{C,F}$ = 7.7 Hz, CH), 139.2 (d, 2 C, $^3J_{C,F}$ = 6.5 Hz, C$_{quat.}$), 162.6 (d, 2 C, $^1J_{C,F}$ = 249 Hz, CF), 193.9 (s, 1 C, C=O).
**HR-MS** (EI, pos.): *m/z* calculated for [C$_{13}$H$_8$F$_2$O]$^+$: 218.0538, found: 218.0538.
**[19]F-NMR** (282 MHz, CDCl$_3$): $\delta$/ppm = –112.
**Mp**: 58-60 °C (Lit.:[13] 58.5-59.0 °C).

## 2.2. Preparation of tritylium tetrafluoroborates

General procedure 2 (GP2):[14]
The corresponding triarylmethanol was dissolved in acetic anhydride or diethyl ether, and the solution stirred vigorously at 0 °C. A solution of tetrafluoroboric acid in water or diethyl ether (50-54 wt%) was added dropwise. After the mixture was stirred for 5 min, the solvent was removed under reduced pressure. The residue was washed with diethyl ether, and the colored product dried in vacuo.

*Bis(4-fluorophenyl)phenylmethylium tetrafluoroborate*
218 mg (0.736 mmol) of bis(4-fluorophenyl)phenylmethanol were dissolved in 5 mL diethyl ether and the solution cooled to 0 °C. 0.5 mL (3.1 mmol) of an ethereal solution (54 % w/v) of tetrafluoroboric acid were added dropwise. After completion, the yellow precipitate was

filtered off, washed with diethyl ether, and dried in vacuum. 100 mg (0.273 mmol, 37 %) of a yellow solid were obtained. The product was not stable and decomposed within several days.

**HR-MS** (ESI, pos.): $m/z$ calculated for $[C_{19}H_{13}F_2]^+$: 279.0980, found: 279.0979.
**Mp**: 120 °C.

*(4-Fluorophenyl)diphenylmethylium tetrafluoroborate*
According to GP 2 from 3.3 g (12 mmol) of (4-fluorophenyl)diphenylmethanol, 20 mL of diethyl ether and 3.0 mL (18 mmol) of an ethereal solution (54 % w/v) of tetrafluoroboric acid. 3.0 g (8.6 mmol, 72 %) of a yellow solid were obtained.

**$^1$H-NMR** (400 MHz, CD$_2$Cl$_2$): $\delta$/ppm = 7.60 (t, 2 H, $J$ = 8.5 Hz, H$_{arom.}$), 7.67 (d, 4 H, $J_{H,H}$ = 7.5 Hz, H$_{arom.}$), 7.78-7.82 (m, 2 H, H$_{arom.}$), 7.89 (t, 4 H, $J_{H,H}$ = 7.5 Hz, H$_{arom.}$), 8.25 (t, 2 H, $J$ = 7.5 Hz, H$_{arom.}$).
**$^{13}$C-NMR** (100 MHz, CD$_2$Cl$_2$): $\delta$/ppm = 118.9 (d, 2 C, $^2J_{C,F}$ = 22.5 Hz, CH), 130.5 (s, 4 C, CH), 136.4 (s, 1 C, C$_{quat.}$), 139.8 (s, 2 C, C$_{quat.}$), 142.3 (s, 4 C, CH), 143.1 (s, 2 C, CH), 146.4. (d, 2 C, $^3J_{C,F}$ = 12.7 Hz, CH), 172.9 (d, 1 C, $^1J_{C,F}$ = 227 Hz, CF), 208.2 (s, 1 C, C$^+$).
**$^{19}$F-NMR** (376 MHz, CD$_2$Cl$_2$): $\delta$/ppm = −83, 152.
**HR-MS** (ESI, pos.): $m/z$ calculated for $[C_{19}H_{14}F]^+$: 261.1074, found: 261.1074.

*(4-Methylphenyl)diphenylmethylium tetrafluoroborate*
According to GP 2 from 1.85 g (6.73 mmol) of (4-methyl-phenyl)diphenylmethanol, 12 mL of acetic anhydride and 1.5 mL of an aqueous solution of tetrafluoroboric acid (50 wt%, 12 mmol). 1.57 g (4.56 mmol, 68 %) of a green solid were obtained.

**$^1$H-NMR** (300 MHz, CDCl$_3$): $\delta$/ppm = 2.70 (s, 3 H, Me), 7.60-7.65 (m, 6 H, H$_{arom.}$), 7.72 (d, $^3J_{H,H}$ = 6.7 Hz, 2 H, H$_{arom.}$), 7.84 (t, $^3J_{H,H}$ = 7.8 Hz, 4 H, H$_{arom.}$), 8.16 (t, $^3J_{H,H}$ = 7.5 Hz, 2 H, H$_{arom.}$).
**$^{13}$C-NMR** (75 MHz, CDCl$_3$): $\delta$/ppm = 23.4 (s, 1 C, Me), 130.3 (s, 4 C, CH), 132.1 (s, 2 C, CH), 137.8 (s, 1 C, C$_{quat.}$), 139.6 (s, 2 C, C$_{quat.}$), 141.6 (s, 4 C, CH), 142.0 (s, 2 C, CH), 143.5 (s, 2 C, CH), 159.7 (s, 1 C, C$_{quat.}$), 208.0 (s, 1 C, C$^+$).
**HR-MS** (ESI, pos.): $m/z$ calculated for $[C_{20}H_{17}]^+$: 257.1325, found: 257.1325.
**Mp**: 163-168 °C (decomposition).

Experimental Part                                                                                          181

*Bis(4-methylphenyl)phenylmethylium tetrafluoroborate*
According to GP 2 from 2.50 g (8.67 mmol) of bis(4-methylphenyl)phenylmethanol, 10 mL of diethyl ether and 1.28 mL of an ethereal solution of tetrafluoroboric acid (50 wt%, 8.67 mmol). 2.26 g (6.31 mmol, 73 %) of a green solid were obtained.

**$^1$H-NMR** (600 MHz, CDCl$_3$): $\delta$/ppm = 2.68 (s, 6 H, Me), 7.57 (d, $^3J_{H,H}$ = 8.1 Hz, 4 H, H$_{arom.}$), 7.59 (dd, $^3J_{H,H}$ = 8.3 Hz, $^4J_{H,H}$ = 1.2 Hz, 2 H, H$_{arom.}$), 7.68 (d, $^3J_{H,H}$ = 8.1 Hz, 4 H, H$_{arom.}$), 7.81 (dd, , $^3J_{1, H,H}$ = 8.3 Hz, $^3J_{2, H,H}$ = 7.5 Hz, 2 H, H$_{arom.}$), 8.12 (tt, $^3J_{H,H}$ = 7.6 Hz, $^4J_{H,H}$ = 1.2 Hz, 1 H, H$_{arom.}$).
**$^{13}$C-NMR** (150 MHz, CDCl$_3$): $\delta$/ppm = 23.2 (s, 2 C, Me), 130.1 (s, 2 C, CH), 131.7 (s, 4 C, CH), 137.5 (s, 2 C, C$_{quat.}$), 139.5 (s, 1 C, C$_{quat.}$), 140.9 (s, 2 C, CH), 141.0 (s, 1 C, CH), 142.5 (s, 4 C, CH), 157.8 (s, 2 C, C$_{quat.}$), 205.9 (s, 1 C, C$^+$).
**MS** (EI): *m/z* (%) = 271 (28, [M–BF$_4$]$^+$), 211 (79), 197 (59), 119 (100).
**Mp**: 137-138 °C.


*Tris(4-methylphenyl)methylium tetrafluoroborate*
According to GP 2 from 2.50 g (8.27 mmol) of tris(4-methylphenyl)methanol, 10 mL diethyl ether and 1.29 mL of an ethereal solution of tetrafluoroboric acid (50 wt%, 8.74 mmol). 2.15 g (5.78 mmol, 70 %) of a green solid were obtained.

**$^1$H-NMR** (600 MHz, CDCl$_3$): $\delta$/ppm = 2.67 (s, 9 H, Me), 7.53 (d, $^3J_{H,H}$ = 8.1 Hz, 6 H, H$_{arom.}$), 7.65 (d, $^3J_{H,H}$ = 8.1 Hz, 6 H, H$_{arom.}$).
**$^{13}$C-NMR** (150 MHz, CDCl$_3$): $\delta$/ppm = 23.0 (s, 3 C, Me), 131.4 (s, 6 C, CH), 137.3 (s, 3 C, C$_{quat.}$), 141.7 (6 C, CH), 156.4 (3 C, C$_{quat.}$), 204.2 (1 C, C$^+$).
**MS** (EI): *m/z* (%) = 285 (81, [M–BF$_4$]$^+$), 225 (68), 211 (57), 119 (100).
**Mp**: 167-168 °C.


*Tris(3,5-di-*tert*-butylphenyl)methylium tetrafluoroborate*
According to GP 2 from 2.60 g (4.36 mmol) tris(3,5-di-*tert*-butylphenyl)methanol, 10 mL diethyl ether and 1.50 mL of an ethereal solution of tetrafluoroboric acid (50 wt%, 10.9 mmol). 2.51 g (2.76 mmol, 86 %) of an orange solid were obtained.

**$^1$H-NMR** (300 MHz, CDCl$_3$): $\delta$/ppm = 1.40 (s, 54 H, Me), 7.40 (d, 6 H, $^4J_{H,H}$ = 1.83 Hz, H$_{arom.}$), 8.27 (t, 3 H, $^4J_{H,H}$ = 1.83 Hz, H$_{arom.}$).

$^{13}$C-NMR (75 MHz, CDCl$_3$): $\delta$/ppm = 31.2 (s, 18 C, Me), 35.4 (s, 6 C, CMe$_3$), 137.4 (s, 6 C, CH), 138.0 (s, 3 C, CH), 140.9 (s, 3 C, C$_{quat.}$), 153.1 (s, 6 C, C$_{quat.}$).
**HR-MS** (EI, pos.): $m/z$ calculated for [C$_{43}$H$_{63}$]$^+$: 579.4924, found: 579.4926.
**Mp**: 247-248 °C.

*(4-Methoxyphenyl)diphenylmethylium tetrafluoroborate*
According to GP 2 from 7.25 g (25.0 mmol) (4-methoxyphenyl)diphenylmethanol, 20 mL acetic anhydride and 4.7 mL of an aqueous solution of tetrafluoroboric acid (50 wt%, 38 mmol). 7.74 g (21.5 mmol, 86 %) of a red solid were obtained.

$^1$**H-NMR** (600 MHz, CDCl$_3$): $\delta$/ppm = 4.31 (s, 3 H, Me), 7.51-7.54 (m, 6 H, H$_{arom.}$), 7.75 (t, $^3J_{H,H}$ = 7.9 Hz, 4 H, H$_{arom.}$), 7.85 (d, $^3J_{H,H}$ = 9.2 Hz, 2 H, H$_{arom.}$), 8.00 (t, $^3J_{H,H}$ = 7.5 Hz, 2 H, H$_{arom.}$).
$^{13}$**C-NMR** (150 MHz, CDCl$_3$): $\delta$/ppm = 58.9 (s, 1 C, Me), 119.3 (s, 2 C, CH), 129.7 (s, 4 C, CH), 133.4 (s, 1 C, C$_{quat.}$), 138.6 (s, 2 C, CH), 138.9 (s, 4 C, CH), 139.2 (s, 2 C, C$_{quat.}$), 147.9 (s, 2 C, CH), 177.0 (s, 1 C, COMe), 198.2 (s, 1 C, C$^+$).
**HR-MS** (ESI, pos.): $m/z$ calculated for [C$_{20}$H$_{17}$O]$^+$: 273.1274, found: 273.1268.
**Elemental analysis**: calculated (%): C 66.70, H 4.76, found (%): C 66.43, H 4.69.
**Mp**: 194-195 °C (Lit.:[15] 188-190 °C).

*Bis(4-methoxyphenyl)phenylmethylium tetrafluoroborate*
According to GP 2 from 800 mg (2.50 mmol) bis(4-methoxyphenyl)phenylmethanol, 6 mL of acetic anhydride and 0.43 mL of an aqueous solution of tetrafluoroboric acid (50 wt%, 3.5 mmol). 830 mg (2.13 mmol, 85 %) of a red solid were obtained.

$^1$**H-NMR** (600 MHz, CDCl$_3$): $\delta$/ppm = 4.15 (s, 6 H, Me), 7.36 (d, $^3J_{H,H}$ = 8.9 Hz, 4 H, H$_{arom.}$), 7.46 (d, $^3J_{H,H}$ = 8.4 Hz, 2 H, H$_{arom.}$), 7.65 (d, $^3J_{H,H}$ = 8.7 Hz, 4 H, H$_{arom.}$), 7.70 (t, $^3J_{H,H}$ = 7.9 Hz, 2 H, H$_{arom.}$), 7.93 (t, $^3J_{H,H}$ = 7.5 Hz, 1 H, H$_{arom.}$).
$^{13}$**C-NMR** (150 MHz, CDCl$_3$): $\delta$/ppm = 57.6 (s, 2 C, Me), 117.5 (s, 4 C, CH), 129.4 (s, 2 C, CH), 132.4 (s, 2 C, C$_{quat.}$), 137.3 (s, 1 C, CH), 137.8 (s, 2 C, CH), 139.1 (s, 1 C, C$_{quat.}$), 144.6 (s, 4 C, CH), 172.5 (s, 2 C, COMe), 194.4 (s, 1 C, C$^+$).
**HR-MS** (ESI, pos.): $m/z$ calculated for [C$_{21}$H$_{20}$O$_2$]$^+$: 303.1380, found: 303.1370.
**Mp**: 191-193 °C (Lit.:[16] 193-196 °C).

Experimental Part 183

*Tris(4-methoxyphenyl)methylium tetrafluoroborate*
According to GP 2 from 1.33 g (3.80 mmol) of tris(4-methoxyphenyl)methanol, 15 mL of diethyl ether and 0.70 mL of an aqueous solution of tetrafluoroboric acid (50 wt%, 4.74 mmol). 1.54 g (3.66 mmol, 96 %) of a red solid were obtained.

**$^1$H-NMR** (600 MHz, CDCl$_3$): $\delta$/ppm = 4.11 (s, 9 H, Me), 7.30 (d, $^3J_{H,H}$ = 8.4 Hz, 6 H, H$_{arom.}$), 7.57 (d, $^3J_{H,H}$ = 8.4 Hz, 6 H, H$_{arom.}$).
**$^{13}$C-NMR** (150 MHz, CDCl$_3$): $\delta$/ppm = 57.1 (s, 3 C, Me), 116.6 (s, 6 C, CH), 131.9 (s, 3 C, C$_{quat.}$), 142.8 (s, 6 C, CH), 170.4 (s, 3 C, COMe), 192.0 (s, 1 C, C$^+$).
**HR-MS** (ESI, pos.): *m/z* calculated for [C$_{22}$H$_{21}$O$_3$]$^+$: 333.1485, found: 333.1479.
**Mp**: 188-189 °C (Lit.:[16] 176-178 °C).


*(4-Dimethylaminophenyl)diphenylmethylium tetrafluoroborate*
According to GP 2 from (4-dimethylaminophenyl)di-phenylmethanol as a crude oil, 8 mL of acetic anhydride and 1.70 mL of an aqueous solution of tetrafluoroboric acid (50 wt%, 13.6 mmol). 2.07 g (5.55 mmol) of a violet solid were obtained.

**$^1$H-NMR** (600 MHz, CDCl$_3$): $\delta$/ppm = 3.63 (s, 6 H, Me), 7.21 (d, $^3J_{H,H}$ = 9.9 Hz, 2 H, H$_{arom.}$), 7.30 (d, $^3J_{H,H}$ = 7.1 Hz, 4 H, H$_{arom.}$), 7.53 (t, $^3J_{H,H}$ = 7.8 Hz, 4 H, H$_{arom.}$), 7.60 (d, $^3J_{H,H}$ = 9.9 Hz, 2 H, H$_{arom.}$), 7.66 (t, $^3J_{H,H}$ = 7.5 Hz, 2 H, H$_{arom.}$).
**$^{13}$C-NMR** (150 MHz, CDCl$_3$): $\delta$/ppm = 42.7 (s, 2 C, Me), 118.0 (s, 2 C, CH), 128.8 (s, 4 C, CH), 130.1 (s, 1 C, C$_{quat.}$), 133.2 (s, 2 C, CH), 134.3 (s, 4 C, CH), 139.1 (s, 2 C, C$_{quat.}$), 143.3 (s, 2 C, CH), 160.5 (s, 1 C, CNMe$_2$), 175.7 (s, 1 C, C$^+$).
**HR-MS** (ESI, pos.): *m/z* calculated for [C$_{21}$H$_{20}$N]$^+$: 286.1590, found: 286.1584.
**Mp**: 180-181 °C (decomposition).


*(4-Dimethylaminophenyl)(4-methoxyphenyl)phenylmethylium tetrafluoroborate*
According to GP 2 from 2.30 g (6.90 mmol) of (4-dimethylaminophenyl)(4-methoxyphenyl)-phenylmethanol, 5 mL of acetic anhydride and 1.00 mL of an aqueous solution of tetrafluoroboric acid (50 wt%, 8.03 mmol). 2.26 g (5.60 mmol, 81 %) of a red solid were obtained.

**$^1$H-NMR** (600 MHz, CDCl$_3$): $\delta$/ppm = 3.54 (br s, 6 H, NMe$_2$), 3.94 (s, 3 H, OMe), 7.06 (d, $^3J_{H,H}$ = 8.9 Hz, 2 H, H$_{arom.}$), 7.10 (dd, $^3J_{H,H}$ = 9.7 Hz, $^4J_{H,H}$ = 2.5 Hz, 1 H, H$_{arom.}$), 7.19 (dd, $^3J_{H,H}$ = 9.7 Hz, $^4J_{H,H}$ = 2.5 Hz, 1 H, H$_{arom.}$), 7.28-7.30 (m, 4 H, H$_{arom.}$), 7.45 (dd, $^3J_{H,H}$ = 9.7

Hz, $^4J_{H,H}$ = 2.1 Hz, 1 H, H$_{arom.}$), 7.53 (t, $^3J_{H,H}$ = 7.8 Hz, 2 H, H$_{arom.}$), 7.62 (dd, $^3J_{H,H}$ = 9.7 Hz, $^4J_{H,H}$ = 2.1 Hz, 1 H, H$_{arom.}$), 7.67 (t, $^3J_{H,H}$ = 7.5 Hz, 1 H, H$_{arom.}$).

$^{13}$C-NMR (75 MHz, CDCl$_3$): $\delta$/ppm = 42.1 (s, 2 C, NMe$_2$), 56.0 (s, 1 C, OMe), 114.8 (s, 2 C, CH), 116.7 (s, 1 C, CH), 117.1 (s, 1 C, CH), 128.7 (s, 2 C, CH), 129.2 (s, 1 C, C$_{quat.}$), 131.5 (s, 1 C, C$_{quat.}$), 133.4 (s, 1 C, CH), 134.6 (s, 2 C, CH), 137.8 (s, 2 C, CH), 139.1 (s, 1 C, C$_{quat.}$), 143.1 (s, 1 C, CH), 143.2 (s, 1 C, CH), 159.8 (s, 1 C, CNMe$_2$), 165.2 (s, 1 C, COMe), 176.9 (s, 1 C, C$^+$).

**HR-MS** (ESI, pos.): *m/z* calculated for [C$_{22}$H$_{22}$NO]$^+$: 316.1696, found: 316.1695.

**Mp**: 84-88 °C.

*Bis(4-dimethylaminophenyl)phenylmethylium tetrafluoroborate*
According to GP 2 from 1.70 g (4.91 mmol) of bis(4-dimethylaminophenyl)phenylmethanol, 10 mL of acetic anhydride and 0.8 mL of an aqueous solution of tetrafluoroboric acid (50 wt%, 6.42 mmol). 1.91 g (4.59 mmol, 94 %) of a deep red solid were obtained.

$^1$**H-NMR** (300 MHz, CDCl$_3$): $\delta$/ppm = 3.37 (s, 12 H, Me), 7.02 (d, $^3J_{H,H}$ = 9.2 Hz, 4 H, H$_{arom.}$), 7.31 (d, $^3J_{H,H}$ = 7.8 Hz, 2 H, H$_{arom.}$), 7.39 (d, $^3J_{H,H}$ = 9.1 Hz, 4 H, H$_{arom.}$), 7.53 (t, $^3J_{H,H}$ = 7.6 Hz, 2 H, H$_{arom.}$), 7.67 (t, $^3J_{H,H}$ = 7.4 Hz, 1 H, H$_{arom.}$).

$^{13}$**C-NMR** (75 MHz, CDCl$_3$): $\delta$/ppm = 41.5 (s, 4 C, Me), 114.6 (s, 4 C, CH), 128.5 (s, 2 C, C$_{quat.}$), 128.6 (s, 2 C, CH), 133.1 (s, 1 C, CH), 134.6 (s, 2 C, CH), 139.3 (s, 1 C, C$_{quat.}$), 140.7 (s, 4 C, CH), 156.4 (s, 2 C, CNMe$_2$), 176.7 (s, 1 C, C$^+$).

**HR-MS** (ESI, pos.): *m/z* calculated for [C$_{23}$H$_{25}$N$_2$]$^+$: 329.2012, found: 329.2011.

**Mp**: 117-120 °C (decomposition).

### 2.3. Preparation of triarylmethyl esters

General procedure 3 (GP3):

The corresponding triarylmethyl halide was dissolved in dry acetone, and an excess of the sodium carboxylate added. The mixture was refluxed for 6 hours and stirred for 6 hours at room temperature. After filtration through celite and washing with acetone, the solvent was evaporated.

*Bis(3-fluorophenyl)phenylmethyl acetate*
According to GP3 from 800 mg (2.23 mmol) bromobis(3-fluorophenyl)phenylmethane and 548 mg (6.68 mmol) sodium acetate. The resulting yellow oil crystallized upon the addition of pentane. 200 mg (0.591 mmol, 26 %) of a slightly brown solid were obtained.

$^1$**H-NMR** (300 MHz, CDCl$_3$): $\delta$/ppm = 2.21 (s, 3 H, CH$_3$), 6.99 (tdd, 2 H, $J$ = 8.2 Hz, $J$ = 2.51 Hz, $J$ = 1.0 Hz, H$_{arom.}$), 7.05-7.18 (m, 4 H, H$_{arom.}$), 7.26-7.40 (m, 7 H, H$_{arom.}$).
$^{13}$**C-NMR** (75 MHz, CDCl$_3$): $\delta$/ppm = 22.4 (s, 1 C, CH$_3$), 88.6 (t, 1 C, $^4J_{C,F}$ = 1.8 Hz, Ar$_3$C), 114.4 (d, 2 C, $^2J_{C,F}$ = 21.1 Hz, CH), 115.4 (d, 2 C, $^2J_{C,F}$ = 23.5 Hz, CH), 123.9 (d, 2 C, $^4J_{C,F}$ = 2.9 Hz, CH), 127.8 (s, 1 C, CH), 128.0 (s, 2 C, CH), 128.3 (s, 2 C, CH), 129.3 (d, 2 C, $^3J_{C,F}$ = 8.2 Hz, CH), 142.1 (s, 1 C, C$_{quat.}$), 145.5 (d, 2 C, $^3J_{C,F}$ = 7.0 Hz, C$_{quat.}$), 162.3 (d, 2 C, $^1J_{C,F}$ = 245 Hz, CF), 168.5 (s, 1 C, C=O).
$^{19}$**F-NMR** (282 MHz, CDCl$_3$): $\delta$/ppm = –113.
**HR-MS** (EI, pos.): m/z calculated for [C$_{21}$H$_{16}$F$_2$O$_2$]$^+$: 338.1113, found: 338.1119.
**Elemental analysis**: calculated (%): C 74.55, H 4.77; found (%): C 74.06, H 4.93.
**Mp**: 67-68 °C.


*(3-Fluorophenyl)diphenylmethyl acetate*
According to GP3 from 800 mg (2.70 mmol) chloro(3-fluorophenyl)diphenylmethane and 664 mg (8.09 mmol) sodium acetate. 849 mg (2.65 mmol, 98 %) of a colorless solid were obtained after crystallization at 4 °C.

$^1$**H-NMR** (300 MHz, CDCl$_3$): $\delta$/ppm = 2.22 (s, 3 H, CH$_3$), 6.99 (tdd, 1 H, $J$ = 8.2 Hz, $J$ = 2.5 Hz, $J$ = 1.1 Hz, H$_{arom.}$), 7.10-7.20 (m, 2 H, H$_{arom.}$), 7.28-7.45 (m, 11 H, H$_{arom.}$).
$^{13}$**C-NMR** (75 MHz, CDCl$_3$): $\delta$/ppm = 22.5 (s, 1 C, CH$_3$), 89.3 (d, 1 C, $^4J_{C,F}$ = 1.9 Hz, Ar$_3$C), 114.2 (d, 1 C, $^2J_{C,F}$ = 21.1 Hz, CH), 115.5 (d, 1 C, $^2J_{C,F}$ = 23.4 Hz, CH), 124.0 (d, 1 C, $^4J_{C,F}$ = 2.9 Hz, CH), 127.6 (s, 2 C, CH), 127.9 (s, 4 C, CH), 128.4 (s, 4 C, CH), 129.2 (d, 1 C, $^3J_{C,F}$ = 8.2 Hz, CH), 142.8 (s, 2 C, C$_{quat.}$), 146.2 (d, 1 C, $^3J_{C,F}$ = 6.9 Hz, C$_{quat.}$), 162.4 (d, 1 C, $^1J_{C,F}$ = 245 Hz, CF), 168.7 (s, 1 C, C=O).
$^{19}$**F-NMR** (282 MHz, CDCl$_3$): $\delta$/ppm = –113.
**Elemental analysis**: calculated (%): C 78.73, H 5.35; found (%): C 77.63, H 5.29.
**Mp**: 62-63 °C.

*Triphenylmethyl acetate*

According to GP3 from 1.58 g (5.67 mmol) triphenylmethyl chloride and 465 mg (5.67 mmol) sodium acetate in 20 mL acetone. After recrystallization from pentane 1.23 g (4.07 mmol, 72 %) of a colorless solid were obtained.

**$^1$H-NMR** (300 MHz, CDCl$_3$): $\delta$/ppm = 2.09 (s, 3 H, Me), 7.11-7.32 (m, 15 H, H$_{arom.}$).
**$^{13}$C-NMR** (75 MHz, CDCl$_3$): $\delta$/ppm = 22.5 (s, 1 C, Me), 89.8 (s, 1 C, Ph$_3$C), 127.2 (s, 3 C, CH), 127.7 (s, 6 C, CH), 128.3 (s, 6 C, CH), 143.3 (s, 3 C, C$_{quat.}$), 168.7 (s, 1 C, C=O).
**MS** (EI, pos.): m/z (%) = 302 (6, [M]$^+$), 260 (53), 259 (100), 243 (83).
**HR-MS** (ESI, pos.): m/z calculated for [C$_{21}$H$_{18}$O$_2$]$^+$: 302.1301, found: 302.1298.
**Elemental analysis**: calculated (%): C 83.42, H 6.00, found (%): C 83.43, H 5.90.
**Mp**: 82-83 °C (Lit.:[17] 83 °C).

*Triphenylmethyl benzoate*

According to GP3 from 1.00 g (3.59 mmol) triphenylmethyl chloride and 520 g (3.61 mmol) sodium benzoate in 20 mL acetone. 1.17 g (3.21 mmol, 89 %) of a colorless solid were obtained.

**$^1$H-NMR** (600 MHz, CDCl$_3$): $\delta$/ppm = 7.18 (t, $^3J_{H,H}$ = 7.3 Hz, 3 H, H$_{arom.}$), 7.24 (t, $^3J_{H,H}$ = 7.5 Hz, 6 H, H$_{arom.}$), 7.35-7.40 (m, 8 H, H$_{arom.}$), 7.49 (t, $^3J_{H,H}$ = 7.4 Hz, 1 H, H$_{arom.}$), 8.05 (d, $^3J_{H,H}$ = 7.2 Hz, 2 H, H$_{arom.}$).
**$^{13}$C-NMR** (150 MHz, CDCl$_3$): $\delta$/ppm = 90.5 (s, 1 C, Ph$_3$C), 127.3 (s, 3 C, CH), 127.8 (s, 6 C, CH), 128.4 (2 s, 8 C, CH), 129.8 (s, 2 C, CH), 131.3 (s, 1 C, C$_{quat.}$), 133.0 (s, 1 C, CH), 143.4 (s, 3 C, C$_{quart.}$), 164.4 (s, 1 C, C=O).
**HR-MS** (ESI, pos.): m/z calculated for [C$_{26}$H$_{20}$O$_2$]$^+$: 364.1458, found: 364.1478.
**Elemental analysis**: calculated (%): C 85.69, H 5.53, found (%): C 85.39, H 5.48.
**Mp**: 168-169 °C (Lit.:[18] 168-169 °C).

*Triphenylmethyl p-nitrobenzoate*

According to GP3 from 1.50 g (5.38 mmol) triphenylmethyl chloride and 1.02 g (5.39 mmol) sodium *p*-nitrobenzoate in 20 mL acetone. 1.21 g (2.96 mmol, 55 %) of a colorless solid were obtained.

**¹H-NMR** (400 MHz, acetone-d$_6$): $\delta$/ppm = 7.28-7.39 (m, 9 H, H$_{arom.}$), 7.48-7.56 (m, 6 H, H$_{arom.}$), 8.34-8.40 (m, 4 H, H$_{arom.}$).
**¹³C-NMR** (100 MHz, acetone-d$_6$): $\delta$/ppm = 91.5 (s, 1 C, Ph$_3$C), 124.0 (s, 2 C, CH), 127.7 (s, 3 C, CH), 128.1 (s, 6 C, CH), 128.5 (s, 6 C, CH), 131.1 (s, 2 C, CH), 136.8 (s, 1 C, C$_{quat.}$), 143.3 (s, 3 C, C$_{quat.}$), 151.0 (s, 1 C, CNO$_2$), 162.6 (s, 1 C, C=O).
**HR-MS** (ESI, pos.): m/z calculated for [C$_{19}$H$_{19}$]$^+$: 243.1168, found: 243.1163.
**Mp**: 177-180 °C.

*(4-Chlorophenyl)diphenylmethyl benzoate*
According to GP3 from 588 mg (1.88 mmol) chloro(4-chlorophenyl)diphenylmethane and 2.20 g (15.3 mmol) sodium benzoate in 30 mL acetone. After recrystallization from pentane 497 mg (1.25 mmol, 66 %) of a colorless solid were obtained.

**¹H-NMR** (300 MHz, CDCl$_3$): $\delta$/ppm = 7.25-7.55 (m, 16 H, H$_{arom.}$), 7.58-7.66 (m, 1 H, H$_{arom.}$), 8.12-8.17 (m, 2 H, H$_{arom.}$).
**¹³C-NMR** (75 MHz, CDCl$_3$): $\delta$/ppm = 90.1 (s, 1 C, Ar$_3$C), 127.7 (s, 2 C, CH), 128.1 (s, 4 C, CH), 128.2 (s, 2 C, CH), 128.3 (s, 4 C, CH), 128.6 (s, 2 C, CH), 129.9 (s, 2 C, CH), 130.2 (s, 2 C, CH), 131.2 (s, 1 C, CCl), 133.3 (s, 1 C, CH), 133.5 (s, 1 C, C$_{quat.}$), 142.1 (s, 1 C, C$_{quat.}$), 143.2 (s, 2 C, C$_{quat.}$), 164.6 (s, 1 C, C=O).
**MS** (EI, pos.): m/z (%) = 398 (1, [M]$^+$), 293 (34), 277 (84), 105 (100).
**HR-MS** (EI, pos.): m/z calculated for [C$_{26}$H$_{19}$$^{35}$ClO$_2$]$^+$: 398.1068, found: 398.1080.
**Elemental analysis**: calculated (%): C 78.29, H 4.80; found (%): C 79.17, H 4.81.
**Mp**: 102-103 °C.

*(4-Chlorophenyl)diphenylmethyl p-nitrobenzoate*
According to GP3 from 527 mg (1.68 mmol) chloro(4-chlorophenyl)diphenylmethane, 2.00 g (10.6 mmol) sodium p-nitrobenzoate and 20 mL dry acetone. 535 mg (1.21 mmol, 72 %) of a colorless solid were obtained.

**¹H-NMR** (400 MHz, acetone-d$_6$): $\delta$/ppm = 7.30-7.32 (m, 2 H, H$_{arom.}$), 7.36-7.42 (m, 6 H, H$_{arom.}$), 7.50-7.55 (m, 6 H, H$_{arom.}$), 8.35-8.40 (m, 4 H, H$_{arom.}$).
**¹³C-NMR** (100 MHz, acetone-d$_6$): $\delta$/ppm = 91.6 (s, 1 C, Ar$_3$C), 124.6 (s, 2 C, CH), 128.5 (s, 2 C, CH), 128.8 (s, 2 C, CH), 128.9 (s, 4 C, CH), 129.0 (s, 4 C, CH), 131.2 (s, 2 C, CH),

131.8 (s, 2 C, CH), 133.9 (s, 1 C, CCl), 137.2 (s, 1 C, $C_{quat.}$), 142.9 (s, 1 C, $C_{quat.}$), 143.6 (s, 2 C, $C_{quat.}$), 151.7 (s, 1 C, $CNO_2$), 163.3 (s, 1 C, C=O).
**MS** (EI, pos.): $m/z$ (%) = 443 (<1, [M]$^+$), 277 (63), 242 (53), 217 (67), 165 (89), 105 (100).
**HR-MS** (EI, pos.): $m/z$ calculated for [$C_{26}H_{18}{}^{35}ClNO_4$]$^+$: 443.0919, found: 443.0924.
**Mp**: 120-122 °C.

*Preparation of sodium p-nitrobenzoate*
463 mg (11.6 mmol) of NaOH were mixed with 1.94 g (11.6 mmol) $p$-nitrobenzoic acid in 20 mL acetone and the solution stirred overnight at room temperature. The precipitate was filtered off and washed with acetone. The solvent was evaporated and 2.12 g (11.2 mmol, 97 %) of a colorless solid were obtained.

**$^1$H-NMR** (200 MHz, $CD_3OD$): $\delta$/ppm = 8.05-8.25 (m, 4 H, $H_{arom.}$).
**Elemental analysis**: calculated (%): C 44.46, H 2.13, N 7.41, found (%): C 44.42, H 1.94, N 7.28

### 2.4. Preparation of triarylmethyl halides

General procedure 4 (GP4):
The triarylmethanol was mixed with an excess of acetyl bromide, thionyl chloride or acetyl chloride, respectively. The mixture was refluxed for 3 hours after which the liquid parts of the mixture were removed in vacuum. The remaining solid was washed with hexane and in some cases recrystallized to yield the desired product.

*Bromotris(3,5-difluorophenyl)methane*
The carbinol was prepared by a Grignard reaction from 6.29 g (32.6 mmol) 3,5-difluorobromobenzene, 809 mg (33.3 mmol) magnesium and 1.30 g (11.0 mmol) diethyl carbonate. The resulting crude product (yellow oil) was treated with 4.8 mL (65 mmol) acetyl bromide. After refluxing for 3 hours, and stirring overnight at room temperature, the acetyl bromide was removed in high vacuum, and the solid residue was washed with hexane. 835 mg (1.94 mmol, 18 %) of a slightly brown solid were obtained.

**$^1$H-NMR** (300 MHz, $CDCl_3$): $\delta$/ppm = 6.76-6.87 (m, 9 H, $H_{arom.}$).

Experimental Part                                                                                      189

$^{13}$C-NMR (75 MHz, CDCl$_3$): δ/ppm = 71.2-71.4 (m, 1 C, CBr), 104.4 (t, 3 C, $^2J_{C,F}$ = 25.3 Hz, CH), 113.3-113.8 (m, 6 C, CH), 147.2 (t, 3 C, $^3J_{C,F}$ = 8.7 Hz, C$_{quat.}$), 162.4 (dd, 6 C, $^1J_{C,F}$ = 250 Hz, $^3J_{C,F}$ = 12.8 Hz, CF).
$^{19}$F-NMR (282 MHz, CDCl$_3$): δ/ppm = –109.
**Elemental analysis**: calculated (%): C 52.93, H 2.10; found (%): C 52.49, H 1.99.
**Mp**: 104-105 °C.

*Bromo(3,5-difluorophenyl)bis(3-fluorophenyl)methane*
According to GP4 from 1.17 g (3.52 mmol) (3,5-difluorophenyl)bis(3-fluorophenyl)methanol and 1.8 mL (24 mmol) acetyl bromide. 344 mg (0.870 mmol, 25 %) of a slightly brown solid were obtained without recrystallization.

$^1$H-NMR (300 MHz, CDCl$_3$): δ/ppm = 6.80-6.90 (m, 3 H, H$_{arom.}$), 7.00-7.10 (m, 6 H, H$_{arom.}$), 7.28-7.36 (m, 2 H, H$_{arom.}$).
$^{13}$C-NMR (75 MHz, CDCl$_3$): δ/ppm = 73.3 (m, 1 C, CBr), 103.8 (t, 1 C, $^2J_{C,F}$ = 25.3 Hz, CH), 113.5-113.9 (m, 2 C, CH), 115.5 (d, 2 C, $^2J_{C,F}$ = 25.3 Hz, CH), 117.5 (d, 2 C, $^2J_{C,F}$ = 23.9 Hz, CH), 125.8 (d, 2 C, $^4J_{C,F}$ = 3.0 Hz, CH), 129.5 (d, 2 C, $^3J_{C,F}$ = 8.3 Hz, CH), 146.4 (d, 2 C, $^3J_{C,F}$ = 7.0 Hz, C$_{quat.}$), 148.4 (t, 1 C, $^3J_{C,F}$ = 8.7 Hz, C$_{quat.}$), 162.1 (d, 2 C, $^1J_{C,F}$ = 247 Hz, CF), 162.2 (dd, 2 C, $^1J_{C,F}$ = 249 Hz, $^3J_{C,F}$ = 12.8 Hz, CF).
**Elemental analysis**: calculated (%): C 57.75, H 2.81; found (%): C 57.33, H 2.39.
**Mp**: 70-72 °C.

*Chlorobis(3,5-difluorophenyl)phenylmethane*
The carbinol was prepared by a Grignard reaction from 5.48 g (28.4 mmol) 3,5-difluorobromobenzene, 700 mg (28.8 mmol) magnesium and 2.11 g (14.1 mmol) ethyl benzoate. The resulting crude product (yellow oil) was treated with 3.2 g (41 mmol) acetyl chloride. After refluxing for 2 hours, the acetyl chloride was removed in high vacuum, and the solid residue was washed with hexane. 1.07 g (3.05 mmol, 22 %) of a slightly brown solid were obtained.

$^1$H-NMR (300 MHz, CDCl$_3$): δ/ppm = 6.72-6.84 (m, 6 H, H$_{arom.}$), 7.18-7.24 (m, 2 H, H$_{arom.}$), 7.32-7.40 (m, 3 H, H$_{arom.}$).
$^{13}$C-NMR (75 MHz, CDCl$_3$): δ/ppm = 78.2 (m, 1 C, CCl), 103.9 (t, 2 C, $^2J_{C,F}$ = 25.3 Hz, CH), 112.6-113.1 (m, 4 C, CH), 128.3 (s, 2 C, CH), 128.7 (s, 1 C, CH), 129.1 (s, 2 C, CH), 142.9

(s, 1 C, C$_{quat.}$), 148.0 (t, 2 C, $^3J_{C,F}$ = 8.7 Hz, C$_{quat.}$), 162.4 (dd, 4 C, $^1J_{C,F}$ = 248 Hz, $^2J_{C,F}$ = 12.7 Hz, CF).

**HR-MS** (EI, pos.): m/z calculated for [C$_{19}$H$_{11}$$^{35}$ClF$_4$]$^+$: 350.0480, found: 350.0464.

**Elemental analysis**: calculated (%): C 65.06, H 3.16; found (%): C 65.04, H 3.11.

**Mp**: 64-65 °C.

*Bromobis(3,5-difluorophenyl)phenylmethane*

6.6 g (20 mmol) of the corresponding alcohol were refluxed for 3 hours with 10 mL (135 mmol) of acetyl bromide. After cooling to room temperature, all volatile compounds were removed in high vacuum. 4.9 g (12 mmol, 60 %) of a brown oil remained, which was not further purified.

**$^1$H-NMR** (300 MHz, CDCl$_3$): $\delta$/ppm = 6.80-6.90 (m, 6 H, H$_{arom.}$), 7.25-7.30 (m, 2 H, H$_{arom.}$), 7.33-7.40 (m, 3 H, H$_{arom.}$).

**$^{13}$C-NMR** (75 MHz, CDCl$_3$): $\delta$/ppm = 73.8 (quint, 1 C, $^4J_{C,F}$ = 2.15 Hz, CBr), 104.0 (t, 2 C, $^2J_{C,F}$ = 25.3 Hz, CH), 113.6-114.2 (m, 4 C, CH), 128.3 (s, 2 C, CH), 128.8 (s, 1 C, CH), 130.1 (s, 2 C, CH), 143.5 (s, 1 C, C$_{quat.}$), 148.6 (t, 2 C, $^3J_{C,F}$ = 8.73 Hz, C$_{quat.}$), 162.4 (dd, 4 C, $^1J_{C,F}$ = 249 Hz, $^3J_{C,F}$ = 12.8 Hz, CF).

**$^{19}$F-NMR** (282 MHz, CDCl$_3$): $\delta$/ppm = –109.

*Bromotris(3-fluorophenyl)methane*

According to GP4 from 1.85 g (5.89 mmol) tris(3-fluorophenyl)methanol and 3.1 mL (42 mmol) acetyl bromide. 405 mg (1.07 mmol, 18 %) of a slightly brown solid were obtained without recrystallization.

**$^1$H-NMR** (300 MHz, CDCl$_3$): $\delta$/ppm = 7.00-7.08 (m, 9 H, H$_{arom.}$), 7.28-7.35 (m, 3 H, H$_{arom.}$).

**$^{13}$C-NMR** (75 MHz, CDCl$_3$): $\delta$/ppm = 74.4 (q, 1 C, $^4J_{C,F}$ = 1.97 Hz, CBr), 115.3 (d, 3 C, $^2J_{C,F}$ = 21.2 Hz, CH), 117.7 (d, 3 C, $^2J_{C,F}$ = 23.9 Hz, CH), 126.0 (d, 3 C, $^4J_{C,F}$ = 2.95 Hz, CH), 129.3 (d, 3 C, $^3J_{C,F}$ = 8.25 Hz, CH), 147.1 (d, 3 C, $^3J_{C,F}$ = 7.00 Hz, C$_{quat.}$), 162.1 (d, 3 C, $^1J_{C,F}$ = 247 Hz, CF).

**$^{19}$F-NMR** (282 MHz, CDCl$_3$): $\delta$/ppm = –113.

**Elemental analysis**: calculated (%): C 60.50, H 3.21; found (%): C 60.40, H 2.87.

**Mp**: 75.6-80.1 °C.

Experimental Part 191

*Chlorotris(3-fluorophenyl)methane*
According to GP4 from 1.61 g (5.12 mmol) tris(3-fluorophenyl)methanol and 5.0 mL (70 mmol) acetyl chloride. After recrystallization from hexane, 1.08 g (3.25 mmol, 63 %) of a colorless solid were obtained.

$^1$**H-NMR** (300 MHz, CDCl$_3$): $\delta$/ppm = 6.94-7.36 (m, 12 H, H$_{arom.}$).
$^{13}$**C-NMR** (75 MHz, CDCl$_3$): $\delta$/ppm = 78.5 (s, 1 C, CCl), 115.3 (d, 3 C, $^2J_{C,F}$ = 21.2 Hz, CH), 116.8 (d, 3 C, $^2J_{C,F}$ = 23.8 Hz, CH), 125.1 (d, 3 C, $^4J_{C,F}$ = 3.0 Hz, CH), 129.4 (d, 3 C, $^3J_{C,F}$ = 8.2 Hz, CH), 146.7 (d, 3 C, $^3J_{C,F}$ = 7.0 Hz, C$_{quat.}$), 162.3 (d, 3 C, $^1J_{C,F}$ = 247 Hz, CF).
$^{19}$**F-NMR** (282 MHz, CDCl$_3$): $\delta$/ppm = –113.
**MS** (EI, pos.): m/z (%) = 297 (91, [M–Cl]$^+$), 218 (42), 201 (100), 123 (95).
**Elemental analysis**: calculated (%): C 68.58, H 3.63; found (%): C 68.65, H 3.54.
**Mp**: 78-80 °C (Lit.:[3] 92-93 °C).

*Chloro(3,5-difluorophenyl)diphenylmethane*
According to GP4 from 2.00 g (6.75 mmol) (3,5-difluorophenyl)diphenylmethanol and 2.0 mL (28 mmol) acetyl chloride. After recrystallization from hexane 1.05 g (3.34 mmol, 50 %) of a colorless solid were obtained.

$^1$**H-NMR** (300 MHz, CDCl$_3$): $\delta$/ppm = 6.72-6.90 (m, 3 H, H$_{arom.}$), 7.22-7.40 (m, 10 H, H$_{arom.}$).
$^{13}$**C-NMR** (75 MHz, CDCl$_3$): $\delta$/ppm = 79.8 (t, 1 C, $^4J_{C,F}$ = 2.2 Hz, CCl), 103.4 (t, 1 C, $^2J_{C,F}$ = 25.4 Hz, CH), 113.1 (m, 2 C, CH), 128.0 (s, 4 C, CH), 128.2 (s, 2 C, CH), 129.4 (s, 4 C, CH), 144.1 (s, 2 C, C$_{quat.}$), 149.3 (t, 1 C, $^3J_{C,F}$ = 8.7 Hz, C$_{quat.}$), 162.3 (dd, 2 C, $^1J_{C,F}$ = 249 Hz, $^3J_{C,F}$ = 12.8 Hz, CF).
$^{19}$**F-NMR** (282 MHz, CDCl$_3$): $\delta$/ppm = –110 (t, J = 8.4 Hz).
**HR-MS** (EI, pos.): m/z calculated for [C$_{19}$H$_{13}$F$_2$]$^+$: 279.0980, found: 279.0980.
**Mp**: 68.5-69.5 °C.

*Bromobis(3-fluorophenyl)phenylmethane*
According to GP4 from 2.87 g (9.68 mmol) bis(3-fluorophenyl)phenylmethanol and 5.0 mL (68 mmol) acetyl bromide. 1.88 g (5.23 mmol, 54 %) of a slightly brownish solid were obtained without recrystallization.

$^1$**H-NMR** (300 MHz, CDCl$_3$): $\delta$/ppm = 7.00-7.10 (m, 6 H, H$_{arom.}$), 7.25-7.38 (m, 7 H, H$_{arom.}$).

$^{13}$**C-NMR** (75 MHz, CDCl$_3$): δ/ppm = 75.8 (t, 1 C, $^4J_{C,F}$ = 2.0 Hz, CBr), 115.0 (d, 2 C, $^2J_{C,F}$ = 21.2 Hz, CH), 117.8 (d, 2 C, $^2J_{C,F}$ = 23.8 Hz, CH), 126.2 (d, 2 C, $^4J_{C,F}$ = 2.9 Hz, CH), 127.9 (s, 2 C, CH), 128.3 (s, 1 C, CH), 129.2 (d, 2 C, $^3J_{C,F}$ = 8.2 Hz, CH), 130.3 (s, 2 C, CH), 144.5 (s, 1 C, C$_{quat.}$), 147.7 (d, 2 C, $^3J_{C,F}$ = 7.0 Hz, C$_{quat.}$), 162.1 (d, 2 C, $^1J_{C,F}$ = 246 Hz, CF).
$^{19}$**F-NMR** (282 MHz, CDCl$_3$): δ/ppm = −113.
**Elemental analysis:** calculated (%): C 63.53, H 3.65; found (%): C 62.99, H 3.51.
**Mp**: 88.7-91.0 °C.

*Chlorobis(3-fluorophenyl)phenylmethane*
According to GP4 from 5.40 g (18.2 mmol) bis(3-fluorophenyl)phenylmethanol and 5.0 mL (69 mmol) thionyl chloride. After recrystallization from hexane, 3.17 g (10.1 mmol, 55 %) of a colorless solid were obtained.

$^1$**H-NMR** (300 MHz, CDCl$_3$): δ/ppm = 7.00-7.10 (m, 6 H, H$_{arom.}$), 7.24-7.40 (m, 7 H, H$_{arom.}$).
$^{13}$**C-NMR** (75 MHz, CDCl$_3$): δ/ppm = 79.5 (s, 1 C, CCl), 115.0 (d, 2 C, $^2J_{C,F}$ = 21.2 Hz, CH), 116.9 (d, 2 C, $^2J_{C,F}$ = 23.7 Hz, CH), 125.3 (d, 2 C, $^4J_{C,F}$ = 3.0 Hz, CH), 128.0 (s, 2 C, CH), 128.2 (s, 1 C, CH), 129.2 (d, 2 C, $^3J_{C,F}$ = 8.2 Hz, CH), 129.4 (s, 2 C, CH), 144.1 (s, 1 C, C$_{quat.}$), 147.3 (d, 2 C, $^3J_{C,F}$ = 7.0 Hz, C$_{quat.}$), 162.3 (d, 2 C, $^1J_{C,F}$ = 246 Hz, CF).
$^{19}$**F-NMR** (282 MHz, CDCl$_3$): δ/ppm = −113.
**HR-MS** (EI, pos.): m/z calculated for [C$_{19}$H$_{13}$$^{35}$ClF$_2$]$^+$: 314.0668, found: 314.0686.
**Mp**: 75-76 °C (Lit.:[3] 72.5-73 °C).

*Bromo(3-fluorophenyl)diphenylmethane*
According to GP4 from 1.79 g (6.43 mmol) (3-fluorophenyl)diphenylmethanol and 5.0 mL (68 mmol) acetyl bromide. 1.90 g (5.57 mmol, 87 %) of a colorless solid were obtained without recrystallization.

$^1$**H-NMR** (300 MHz, CDCl$_3$): δ/ppm = 7.00-7.35 (m, 14 H, H$_{arom.}$).
$^{13}$**C-NMR** (75 MHz, CDCl$_3$): δ/ppm = 77.3 (d, 1 C, $^4J_{C,F}$ = 2.0 Hz, CBr), 114.8 (d, 1 C, $^2J_{C,F}$ = 21.2 Hz, CH), 117.9 (d, 1 C, $^2J_{C,F}$ = 23.7 Hz, CH), 126.4 (d, 1 C, $^4J_{C,F}$ = 2.9 Hz, CH), 127.8 (s, 4 C, CH), 128.0 (s, 2 C, CH), 129.0 (d, 1 C, $^3J_{C,F}$ = 8.2 Hz, CH), 130.5 (s, 4 C, CH), 145.1 (s, 2 C, C$_{quat.}$), 148.2 (d, 1 C, $^3J_{C,F}$ = 7.1 Hz, C$_{quat.}$), 162.1 (d, 1 C, $^1J_{C,F}$ = 246 Hz, CF).
$^{19}$**F-NMR** (282 MHz, CDCl$_3$): δ/ppm = −114.
**MS** (EI, pos.): m/z (%) = 278 (20), 243 (44), 201 (90), 183 (75), 105 (100).

Experimental Part                                                                 193

**Mp**: 98-100 °C (Lit.:[19] 110-111 °C).


*Chloro(3-fluorophenyl)diphenylmethane*

5.0 mL (69 mmol) thionyl chloride were slowly added to an ice cooled flask containing 7.20 g (25.9 mmol) (3-fluorophenyl)diphenylmethanol. Gas evolved in a highly exothermic reaction. After 4 hours of refluxing, the liquid parts of the mixture were removed in vacuum. Two times a small volume of acetonitrile was added, and the liquid removed again. The yellow residue was recrystallized from hexane to give 4.54 g (15.3 mmol, 59 %) of a colorless solid.

**$^1$H-NMR** (300 MHz, CDCl$_3$): $\delta$/ppm = 7.00-7.10 (m, 3 H, H$_{arom.}$), 7.26-7.38 (m, 11 H, H$_{arom.}$).
**$^{13}$C-NMR** (75 MHz, CDCl$_3$): $\delta$/ppm = 80.4 (d, 1 C, $^4J_{C,F}$ = 1.95 Hz, CCl), 114.7 (d, 1 C, $^2J_{C,F}$ = 21.2 Hz, CH), 117.0 (d, 1 C, $^2J_{C,F}$ = 23.7 Hz, CH), 125.4 (d, 1 C, $^4J_{C,F}$ = 2.9 Hz, CH), 127.8 (s, 4 C, CH), 128.0 (s, 2 C, CH), 129.2 (d, 1 C, $^3J_{C,F}$ = 8.2 Hz, CH), 129.5 (s, 4 C, CH), 144.7 (s, 2 C, C$_{quat.}$), 147.9 (d, 1 C, $^3J_{C,F}$ = 7.0 Hz, C$_{quat.}$), 162.2 (d, 1 C, $^1J_{C,F}$ = 247 Hz, CF).
**$^{19}$F-NMR** (282 MHz, CDCl$_3$): $\delta$/ppm = –113.
**MS** (EI, pos.): *m/z* (%) = 261 (100, [M–Cl]$^+$), 183 (69), 165 (33).
**Mp**: 82.5-83.5 °C (Lit.:[9] 84-84.5 °C).


*Chloro(4-chlorophenyl)diphenylmethane*

According to GP4 from 2.49 g (8.45 mmol) (4-chlorophenyl)diphenylmethanol and 6.0 mL (83 mmol) thionyl chloride. After recrystallization from acetonitrile, 1.32 g (4.21 mmol, 50 %) of a colorless solid were obtained.

**$^1$H-NMR** (400 MHz, acetone-d$_6$): $\delta$/ppm = 7.20-7.25 (m, 6 H, H$_{arom.}$), 7.35-7.45 (m, 8 H, H$_{arom.}$).
**$^{13}$C-NMR** (100 MHz, acetone-d$_6$): $\delta$/ppm = 82.0 (s, 1 C, CCl), 129.0 (s, 2 C, CH), 129.1 (s, 4 C, CH), 129.2 (s, 2 C, CH), 130.5 (s, 4 C, CH), 132.3 (s, 2 C, CH), 134.6 (s, 1 C, CCl), 145.3 (s, 1 C, C$_{quat.}$), 144.7 (s, 2 C, C$_{quat.}$).
**HR-MS** (ESI, pos.): *m/z* calculated for [C$_{19}$H$_{14}$Cl]$^+$: 277.0779, found: 277.0775.
**Mp**: 88-89 °C (Lit.:[20] 88.8-89.7 °C).

*Bromotriphenylmethane*
According to GP4 from 3.22 g (12.4 mmol) triphenylmethanol and 5.0 mL (61 mmol) acetyl bromide. After washing with hexane 3.76 g (11.6 mmol, 94 %) of a colorless solid were obtained.

**$^1$H-NMR** (300 MHz, CDCl$_3$): $\delta$/ppm = 7.33-7.35 (m, 15 H, H$_{arom.}$).
**$^{13}$C-NMR** (75 MHz, CDCl$_3$): $\delta$/ppm = 78.9 (s, 1 C, CBr), 127.7 (s, 6 C, CH), 127.9 (s, 3 C, C$_{quat.}$), 130.7 (s, 6 C, CH), 145.7 (s, 3 C, C$_{quat.}$).
**MS** (EI, pos.): m/z (%) = 243 (17), 197 (11), 183 (100), 154 (19).
**Elemental analysis**: calculated (%): C 70.60, H 4.68; found (%): C 70.67, H 4.69.
**Mp**: 150-151 °C (Lit.:[21] 152 °C).

*Fluorotriphenylmethane*
1.53 g (4.73 mmol) bromotriphenylmethane were added to a suspension of 6.31 g of fluorinating agent[22] in 20 mL of acetonitrile. After the mixture was stirred for 60 min at 0 °C in a dark vessel, the solids were filtered off and the solvent evaporated. The residue was treated with diethyl ether and filtrated again. After evaporation of solvent, 1.07 g (4.08 mmol, 86 %) of a colorless solid were obtained. The NMR spectra showed the corresponding alcohol as impurity. The other peaks are in agreement with literature data.[23]

**$^1$H-NMR** (300 MHz, CDCl$_3$): $\delta$/ppm = 7.25-7.40 (m, 15 H, H$_{arom.}$).
**$^{13}$C-NMR** (75 MHz, CDCl$_3$): $\delta$/ppm = 101.2 (d, 1 C, $^1J_{C,F}$ = 174 Hz, CF), 127.8-127.9 (m, 12 C, CH), 128.1 (d, 3 C, $^5J_{C,F}$ = 2.2 Hz, CH), 143.3 (d, 3 C, $^2J_{C,F}$ = 24.4 Hz, C$_{quat.}$).
**$^{19}$F-NMR** (282 MHz, CDCl$_3$): $\delta$/ppm = –126.
**Mp**: 104 °C (Lit.:[24] 103.2-103.7 °C).

*Chloro(4-fluorophenyl)diphenylmethane*
According to GP4 from 3.64 g (13.1 mmol) (4-fluorophenyl)diphenylmethanol and 6.0 mL (83 mmol) thionyl chloride. After recrystallization from acetonitrile, 2.11 g (7.11 mmol, 54 %) of a slightly brown solid were obtained.

**$^1$H-NMR** (400 MHz, CD$_3$CN): $\delta$/ppm = 7.04-7.10 (m, 2 H, H$_{arom.}$), 7.20-7.26 (m, 6 H, H$_{arom.}$), 7.34-7.38 (m, 6 H, H$_{arom.}$).

Experimental Part 195

$^{13}$**C-NMR** (100 MHz, CD$_3$CN): $\delta$/ppm = 82.1 (s, 1 C, CCl), 115.5 (d, 2 C, $^2J_{C,F}$ = 21.8 Hz, CH), 129.0 (s, 4 C, CH), 129.1 (s, 2 C, CH), 130.3 (s, 4 C, CH), 132.5 (d, 2 C, $^3J_{C,F}$ = 8.3 Hz, CH), 142.3 (d, 1 C, $^4J_{C,F}$ = 3.3 Hz, C$_{quat.}$), 145.9 (s, 2 C, C$_{quat.}$), 163.1 (d, 1 C, $^1J_{C,F}$ = 247 Hz, CF).
**HR-MS** (EI, pos.): $m/z$ calculated for [C$_{19}$H$_{14}$F]$^+$: 261.1074, found: 261.1078.
**Mp**: 89-90 °C (Lit.:[5] 90-91 °C).

*Chlorobis(4-fluorophenyl)phenylmethane*
According to GP4 from 3.86 g (13.0 mmol) bis(4-fluorophenyl)phenylmethanol and 5.0 mL (69 mmol) thionyl chloride. After recrystallization from pentane, 780 mg (2.48 mmol, 19 %) of a slightly brown solid were obtained.

$^1$**H-NMR** (400 MHz, CDCl$_3$): $\delta$/ppm = 6.95-7.02 (m, 4 H, H$_{arom.}$), 7.18-7.35 (m, 9 H, H$_{arom.}$).
$^{13}$**C-NMR** (100 MHz, CDCl$_3$): $\delta$/ppm = 80.2 (s, 1 C, CCl), 114.6 (d, 4 C, $^2J_{C,F}$ = 21.6 Hz, CH), 127.9 (s, 2 C, CH), 128.1 (s, 1 C, CH), 129.5 (s, 2 C, CH), 131.4 (d, 4 C, $^3J_{C,F}$ = 8.2 Hz, CH), 141.1 (d, 2 C, $^4J_{C,F}$ = 3.4 Hz, C$_{quat.}$), 144.9 (s, 1 C, C$_{quat.}$), 162.1 (d, 2 C, $^1J_{C,F}$ = 249 Hz, CF).
$^{19}$**F-NMR** (376 MHz, CDCl$_3$): $\delta$/ppm = –114.
**HR-MS** (EI, pos.): $m/z$ calculated for [C$_{19}$H$_{13}$F$_2$]$^+$: 279.0980, found: 279.0985.
**Mp**: 43-45 °C (Lit.:[25] 56-57 °C).

*Chlorotris(4-fluorophenyl)phenylmethane*
The carbinol was prepared by a Grignard reaction from 8.4 g (48 mmol) 4-fluorobromobenzene, 1.2 g (49 mmol) magnesium and 1.9 g (16 mmol) diethyl carbonate. The resulting crude product (yellow oil) was treated with 7.0 g (59 mmol) thionyl chloride. After refluxing for 3 hours, the thionyl chloride was removed in high vacuum, and the solid residue was distilled. 3.5 g (11 mmol, 69 %) of an orange solid were obtained.

$^1$**H-NMR** (300 MHz, CDCl$_3$): $\delta$/ppm = 6.98-7.04 (m, 6 H, H$_{arom.}$), 7.18-7.26 (m, 6 H, H$_{arom.}$).
$^{13}$**C-NMR** (75 MHz, CDCl$_3$): $\delta$/ppm = 79.6 (s, 1 C, CCl), 114.8 (d, 6 C, $^2J_{C,F}$ = 21.6 Hz, CH), 131.3 (d, 6 C, $^3J_{C,F}$ = 8.3 Hz, CH), 140.9 (d, 3 C, $^4J_{C,F}$ = 3.4 Hz, C$_{quat.}$), 162.2 (d, 3 C, $^1J_{C,F}$ = 249 Hz, CF).
$^{19}$**F-NMR** (282 MHz, CDCl$_3$): $\delta$/ppm = –114.
**MS** (EI, pos.): $m/z$ (%) = 297 (34, [M–Cl]$^+$), 219 (50), 123 (100).

**Elemental analysis:** calculated (%): C 68.58, H 3.63; found (%): C 69.04, H 3.61.
**Mp:** 60-62 °C (Lit.:[25] 81-82 °C).

*Chloro(4-methylphenyl)diphenylmethane*
The carbinol was prepared by a Grignard reaction from 6.23 g (36.4 mmol) 4-bromotoluene, 894 mg (36.8 mmol) magnesium and 6.63 g (36.4 mmol) benzophenone. The resulting crude product (yellow oil) was treated with 8.0 g (67 mmol) thionyl chloride. After stirring for 3 hours, the thionyl chloride was removed in high vacuum, and the solid residue was washed with hexane. 7.35 g (25.1 mmol, 69 %) of a colorless solid were obtained.

**$^1$H-NMR** (600 MHz, CDCl$_3$): $\delta$/ppm = 2.40 (s, 3 H, Me), 7.12-7.20 (m, 4 H, H$_{arom.}$), 7.26-7.36 (m, 10 H, H$_{arom.}$).
**$^{13}$C-NMR** (150 MHz, CDCl$_3$): $\delta$/ppm = 21.1 (s, 1 C, Me), 81.5 (s, 1 C, CCl), 127.7 (s, 4 C, CH), 127.8 (s, 2 C, CH), 128.5 (s, 2 C, CH), 129.66 (s, 2 C, CH), 129.73 (s, 4 C, CH), 137.7 (s, 1 C, C$_{quat.}$), 142.5 (s, 1 C, C$_{quat.}$), 145.5 (s, 2 C, C$_{quat.}$).
**Elemental analysis:** calculated (%): C 82.04, H 5.85, Cl 12.11; found (%): C 81.85, H 5.87, Cl 12.27.
**Mp:** 99-100 °C (Lit.:[7] 97 °C).

*Bromo(4-methylphenyl)diphenylmethane*
According to GP4 from 10.6 g (38.6 mmol) (4-methylphenyl)diphenylmethanol and 10.0 mL (135 mmol) acetyl bromide. After recrystallization from hexane, 9.03 g (26.8 mmol, 69 %) of a colorless solid were obtained.

**$^1$H-NMR** (300 MHz, CDCl$_3$): $\delta$/ppm = 2.39 (s, 3 H, Me), 7.10-7.36 (m, 14 H, H$_{arom.}$).
**$^{13}$C-NMR** (75 MHz, CDCl$_3$): $\delta$/ppm = 21.1 (s, 1 C, Me), 79.2 (s, 1 C, CBr), 127.6 (s, 4 C, CH), 127.7 (s, 2 C, CH), 128.3 (s, 2 C, CH), 130.5 (s, 2 C, CH), 130.6 (s, 4 C, CH), 137.7 (s, 1 C, C$_{quat.}$), 142.8 (s, 1 C, C$_{quat.}$), 145.8 (s, 2 C, C$_{quat.}$).
**Mp:** 94-95 °C (Lit.:[26] 105-106 °C).

*Chlorobis(4-methylphenyl)phenylmethane*
The carbinol was prepared by a Grignard reaction from 11.9 g (69.6 mmol) 4-bromotoluene, 1.70 g (69.6 mmol) magnesium and 4.92 g (32.8 mmol) ethyl benzoate. The resulting crude product (yellow oil) was treated with 8.0 mL (0.11 mmol) acetyl chloride. After stirring

overnight, the acetyl chloride was removed in high vacuum, and the solid residue was washed with hexane. 4.27 g (13.9 mmol, 42 %) of a colorless solid were obtained.

$^1$**H-NMR** (300 MHz, CDCl$_3$): $\delta$/ppm = 2.40 (s, 6 H, Me), 7.12-7.20 (m, 8 H, H$_{arom.}$), 7.30-7.34 (m, 5 H, H$_{arom.}$).
$^{13}$**C-NMR** (75 MHz, CDCl$_3$): $\delta$/ppm = 21.1 (s, 2 C, Me), 81.6 (s, 1 C, CCl), 127.7 (s, 3 C, CH), 128.4 (s, 4 C, CH), 129.65 (s, 4 C, CH), 129.72 (s, 2 C, CH), 137.6 (s, 2 C, C$_{quat.}$), 142.6 (s, 2 C, C$_{quat.}$), 145.6 (s, 1 C, C$_{quat.}$).
**MS** (EI, pos.): $m/z$ (%) = 288 (21), 271 (50, [M–Cl]$^+$), 211 (100), 197 (77), 119 (98).
**Mp**: 106-107 °C (Lit.:[7] 105-106 °C).

*Chlorotris(4-methylphenyl)methane*
According to GP4 from 1.98 g (6.55 mmol) tris(4-methylphenyl)methanol and 5.0 mL (70 mmol) acetyl chloride. After recrystallization from pentane/diethyl ether 1.90 g (5.92 mmol, 90 %) of a colorless solid were obtained.

$^1$**H-NMR** (300 MHz, CDCl$_3$): $\delta$/ppm = 2.37 (s, 9 H, Me), 7.10-7.18 (m, 12 H, H$_{arom.}$).
$^{13}$**C-NMR** (75 MHz, CDCl$_3$): $\delta$/ppm = 21.2 (s, 3 C, Me), 81.8 (s, 1 C, CCl), 128.5 (s, 6 C, CH), 128.7 (s, 6 C, CH), 137.6 (s, 3 C, C$_{quat.}$), 142.9 (s, 3 C, C$_{quat.}$).
**MS** (EI, pos.): $m/z$ (%) = 285 (19, [M–Cl]$^+$), 211 (100), 119 (55).
**Mp**: 167-168 °C (Lit.:[27] 173 °C).

*Bromotris(4-methylphenyl)methane*
According to GP4 from 15 g (50 mmol) tris(4-methylphenyl)methanol and 15 mL (203 mmol) acetyl bromide. After recrystallization from pentane/diethyl ether 6.9 g (19 mmol, 38 %) of a colorless solid were obtained.

$^1$**H-NMR** (300 MHz, CDCl$_3$): $\delta$/ppm = 2.22 (s, 9 H, Me), 6.93-6.97 (m, 6 H, H$_{arom.}$), 7.05-7.10 (m, 6 H, H$_{arom.}$).
$^{13}$**C-NMR** (75 MHz, CDCl$_3$): $\delta$/ppm = 21.1 (s, 3 C, Me), 80.1 (s, 1 C, CBr), 128.3 (s, 6 C, CH), 130.6 (s, 6 C, CH), 137.6 (s, 3 C, C$_{quat.}$), 143.2 (s, 3 C, C$_{quat.}$).
**Mp**: 159-160 °C (Lit.:[28] 161-163 °C).

*Chlorotris(3,5-di*-tert-*butylphenyl)methane*
According to GP4 from 3.00 g (5.03 mmol) tris(3,5-di-*tert*-butylphenyl)methanol, 20 mL toluene and 430 µL (6.03 mmol) acetyl chloride. After recrystallization from hexane 1.12 g (1.82 mmol, 36 %) of a colorless solid were obtained.

**$^1$H-NMR** (600 MHz, CDCl$_3$): $\delta$/ppm = 1.23 (s, 54 H, Me), 7.01 (d, 6 H, $^4J_{H-H}$ = 1.8 Hz, H$_{arom.}$), 7.32 (t, 3 H, $^4J_{H-H}$ = 1.8 Hz, H$_{arom.}$).
**$^{13}$C-NMR** (150 MHz, CDCl$_3$): $\delta$/ppm = 31.4 (s, 18 C, Me), 34.8 (s, 6 C, C$_{quat.}$), 83.4 (s, 1 C, CCl), 120.8 (s, 3 C, CH), 124.5 (s, 6 C, CH), 145.1 (s, 3 C, C$_{quat.}$), 149.5 (s, 6 C, C$_{quat.}$).
**MS** (EI, pos.): *m/z* (%) = 581 (52), 580 (100), 566 (21), 565 (32), 564 (18), 407 (41).
**HR-MS** (EI, pos.): *m/z* calculated for [C$_{43}$H$_{63}$]$^+$: 579.4924, found: 579.4932.
**Mp**: 193-194 °C (Lit.:[29] 186-187 °C).

## 2.5. Preparation of nucleophiles

*Tetra-*n-*butyltin*[30]
A solution of 6.51 g (20.0 mmol) of tributyltin chloride in 20 mL THF was cooled to −78 °C. *n*-BuLi (1.55 M in hexane) was added dropwise, and the mixture stirred for 30 min, before it was warmed to room temperature. After the reaction was quenched with water, the phases were separated, and the organic phase dried over MgSO$_4$. Filtration, evaporation of the solvent and subsequent distillation (0.1 mbar, 190 °C) yielded 5.70 g (16.4 mmol, 82 %) of a colorless liquid.

**$^1$H-NMR** (300 MHz, CDCl$_3$): $\delta$/ppm = 0.78-0.82 (m, 8 H, CH$_2$), 0.90 (t, 12 H, $J_{H-H}$ = 7.2 Hz, Me), 1.24-1.38 (m, 8 H, CH$_2$), 1.42-1.52 (m, 8 H, CH$_2$).
**$^{13}$C-NMR** (75 MHz, CDCl$_3$): $\delta$/ppm = 8.9 (s, 4 C), 13.8 (s, 4 C), 27.5 (s, 4 C), 29.4 (s, 4 C).

*Tetrakis(trimethylsilyl)silane*[31]
2.52 g (363 mmol) of lithium turnings and 15.7 g (145 mmol) of chlorotrimethylsilane were suspended in 50 mL of THF. 3.40 mL (29.6 mmol) of tetrachlorosilane, dissolved in 20 mL of THF, were added over 1 h at −60 °C. After addition was complete, stirring was continued for 30 min while cooling, and 12 h at room temperature. The reaction was quenched by the addition of water and 2 M hydrochloric acid. Solids were filtered off, the THF evaporated.

The solution was extracted 2 times with diethylether, the ethereal phase dried over $MgSO_4$, and the ether evaporated. 5.10 g (15.9 mmol, 54 %) of a colorless solid were obtained.

**$^1$H NMR** (200 MHz, $CDCl_3$): $\delta$/ppm = 0.20 (s, 36 H, $CH_3$).
**MP:** 255 °C.

*Tris(trimethylsilyl)silane*[32]

10.2 g (31.8 mmol) of tetrakis(trimethylsilyl)silane and 3.40 g (30.3 mmol) of potassium tert-butoxide were dissolved in 30 mL of THF and stirred for 12 h. The reaction was quenched by addition of water and 2 M hydrochloric acid. Phase separation and evaporation of the ether was followed by distillation (2 mbar, 180 °C). 6.38 g (25.7 mmol, 81 %) of a colorless liquid were obtained.

**$^1$H NMR** (400 MHz, acetone-$d_6$): $\delta$/ppm = 0.23 (s, 27 H, $CH_3$), 2.24 (s, 1 H, SiH).
**$^{13}$C NMR** (100 MHz, acetone-$d_6$): $\delta$/ppm = 2.12 (s, 9 C, $CH_3$).
**$^{29}$Si NMR** (79 MHz, acetone-$d_6$): $\delta$/ppm = –11.82.

*3-Propylcyclopentene*[33]

13.9 g (210 mmol) Cyclopentadiene were cooled to –30 °C, and gaseous HCl was bubbled through until the weight of the liquid had increased by 6.5 g (178 mmol HCl). Distillation (rt, 3 mbar) yielded 12.2 g (119 mmol) 3-chlorocyclopentene, which was stored at –25 °C.[34]
A Grignard reagent was prepared from 19.7 g (160 mmol) of 1-bromopropane and 3.89 g (160 mmol) magnesium in 50 mL THF. 12 g (117 mmol) of 3-chlorocyclopentene was added over 1 h via a dropping funnel at 0 °C. After the mixture was stirred for further 2 h at room tempe-rature, the reaction was quenched with 2 M HCl. Extraction by diethyl ether, evaporation of solvent, and subsequent distillation (60 °C, 50 mbar) yielded 5.21 g (47.3 mmol, 40 %) of a colorless liquid.

**$^1$H NMR** (300 MHz, $CDCl_3$): $\delta$/ppm = 0.88-0.94 (m, 3 H, Me), 1.22-1.46 (m, 5 H), 1.99-2.10 (m, 1 H), 2.20-2.41 (m, 2 H), 2.51-2.52 (m, 1 H), 5.66-5.74 (m, 2 H).
**$^{13}$C-NMR** (75 MHz, $CDCl_3$): $\delta$/ppm = 14.5 (s, 1 C), 21.3 (s, 1 C), 30.1 (s, 1 C), 32.1 (s, 1 C), 38.7 (s, 1 C), 45.6 (s, 1 C), 130.1 (s, 1 C), 135.6 (s, 1 C).

*Tributyldeuteriosilane*[35]

217 mg (5.17 mmol) of LiAlD$_4$ were added to a solution of 1.95 g (8.30 mmol) of tributylchlorosilane in 10 mL of diethyl ether. After 3 hours of refluxing, the solution was further stirred for 12 hours at room temperature. Addition of water and 2 M hydrochloric acid was followed by phase separation and drying of the ethereal layer with Na$_2$SO$_4$. Filtration and evaporation of the solvent yielded 1.50 g (7.45 mmol, 90 %) of a colorless liquid.

$^1$H NMR (300 MHz, CDCl$_3$): $\delta$/ppm = 0.55-0.62 (m, 6 H, CH$_2$), 0.88-0.95 (m, 12 H, CH$_3$), 1.29-1.40 (m, 12 H, CH$_2$).
$^{13}$C NMR (75 MHz, CDCl$_3$): $\delta$/ppm = 11.1 (s, 3 C, CH$_2$), 13.9 (s, 3 C, CH$_3$), 26.5 (s, 3 C, CH$_2$), 27.0 (s, 3 C, CH$_2$).
$^{29}$Si NMR (53 MHz, CDCl$_3$): $\delta$/ppm = –5.90 (t, $^1J_{Si,D}$ = 27 Hz).

*2-Propyl-1,3-dioxolane*[36]

6.0 g (97 mmol) of glycol were mixed with 3.6 g (50 mmol) of butanal and 40 mg (0.21 mmol) of *p*-toluenesulfonic acid monohydrate. In the presence of 4 Å MS, the mixture was heated to 60 °C for 2 hours. After cooling to room temperature, diethyl ether and a sat. aqueous solution of NaHCO$_3$ were added, the phases separated, and the diethyl ether of the organic phase evaporated. Distillation delivered 1.5 g (13 mmol, 26 %) of a colorless liquid.

$^1$H NMR (300 MHz, CDCl$_3$): $\delta$/ppm = 0.90-1.00 (m, 3 H, CH$_3$), 1.36-1.50 (m, 2 H, CH$_2$), 1.57-1.66 (m, 2 H, CH$_2$), 3.80-3.97 (m, 4 H, 2 × CH$_2$), 4.81-4.86 (m, 1 H, CH).
$^{13}$C NMR (75 MHz, CDCl$_3$): $\delta$/ppm = 14.0 (s, 1 C, CH$_3$), 17.4 (s, 1 C, CH$_2$), 35.9 (s, 1 C, CH$_2$), 64.7 (s, 2 C, CH$_2$), 104.4 (s, 1 C, CH).

*2-Phenyl-1,3-dioxolane*[37]

10.6 g (100 mmol) of benzaldehyde, 6.2 g (100 mmol) of glycol and 30 mg (0.16 mmol) of *p*-toluenesulfonic acid monohydrate were refluxed in 50 mL benze with a Dean-Stark trap. After 2 hours, 1.8 mL (100 mmol) of water have been separated, and the reaction was stopped. The solvent was evaporated, and the remaining oil purified by column chromatography (alumina, Et$_2$O: pentane = 1:5). 12.3 (82 mmol, 82 %) of 2-phenyl-1,3-dioxolane as a colorless liquid were obtained.

¹H NMR (300 MHz, CDCl₃): δ/ppm = 4.01-4.16 (m, 4 H, CH₂), 5.83 (s, 1 H), 7.36-7.43 (m, 3 H, H$_{arom.}$), 7.45-7.52 (m, 2 H, H$_{arom.}$).
¹³C NMR (75 MHz, CDCl₃): δ/ppm = 65.4 (s, 2 C), 103.8 (s, 1 C), 126.5 (s, 2 C), 128.4 (s, 2 C), 129.2 (s, 1 C), 138.0 (s, 1 C).

*Diethyl 2,6-dimethyl-1,4-dihydropyridine-3,5-dicarboxylate*[38]
16.3 g (125 mmol) of ethyl acetoacetate, 7.25 g (94.1 mmol) of ammonium acetate and 1.88 g (62.4 mmol) of paraformaldehyde were mixed and stirred for 30 min at 70 °C. After coolong to room temperature, the reaction was quenched by addition of water. The yellow precipitate was filtered off and thoroughly washed with water. After drying, 6.01 g (23.7 mmol, 40 %) of a yellow solid remained.

¹H-NMR (200 MHz, CDCl₃): δ/ppm = 1.27 (t, 6 H, ³$J_{H,H}$ = 7.1 Hz, CH₂C*H*₃), 2.17 (s, 6 H, Me), 3.25 (s, 2 H), 4.14 (q, 4 H, ³$J_{H,H}$ = 7.1 Hz, C*H*₂CH₃), 5.38 (br s, 1 H, NH).
**HR-MS** (ESI, pos.): *m/z* calculated for [C₁₃H₁₉O₄N]⁺: 254.1387, found: 254.1343.
**Elemental analysis**: calculated (%): C 61.64, H 7.56, N 5.53; found (%): C 61.59, H 6.77, N 5.55.
**Mp**: 176-177 °C (Lit.:[39] 183-185 °C).

*Diethyl 2,4,6-trimethyl-1,4-dihydropyridine-3,5-dicarboxylate*
18.2 g (140 mmol) of ethyl acetoacetate, 6.17 g (80.0 mmol) of ammonium acetate and 3.53 g (80.0 mmol) of acetaldehyde were mixed and stirred for 24 h overnight. After the reaction was quenched with water, the yellow precipitate was filtered off and thoroughly washed with water. After drying, 9.62 g (36.0 mmol, 45 %) of a yellow solid remained.

¹H-NMR (300 MHz, CDCl₃): δ/ppm = 0.95 (d, 3 H, ³$J_{H,H}$ = 6.5 Hz, Me), 1.28 (t, 6 H, ³$J_{H,H}$ = 7.1 Hz, CH₂C*H*₃), 2.25 (s, 6 H, Me), 3.84 (q, 1 H, ³$J_{H,H}$ = 6.5 Hz), 4.10-4.25 (m, 4 H, C*H*₂CH₃), 5.76 (br s, 1 H, NH).
¹³**C-NMR** (75 MHz, CDCl₃): δ/ppm = 14.5 (s, 2 C), 19.5 (s, 2 C), 22.4 (s, 1 C), 28.6 (s, 1 C), 59.7 (s, 2 C), 104.7 (s, 2 C), 144.5 (s, 2 C), 168.0 (s, 2 C).
**Mp**: 129-130 °C (Lit.:[40] 130-131 °C).

## 3. Product Studies

*Reaction of $Me_2Tr^+BF_4^-$ with $HSi(SiMe_3)_3$*
205 mg (0.824 mmol) of $HSi(SiMe_3)_3$ were added to a solution of 293 mg (0.818 mmol) of $Me_2Tr^+BF_4^-$ in 10 mL of dichloromethane. During 30 s the green color turned into brown, while a gas developed. Evaporation of solvent and column chromatography (silica gel, pentane) yielded bis(4-methylphenyl)phenylmethane as a colorless oil (150 mg, 0.551 mmol, 67 %). Other products, stemming from the silane, were not identifiable.

**$^1$H-NMR** (300 MHz, $CDCl_3$): $\delta$/ppm = 2.48 (s, 6 H, $CH_3$), 5.65 (s, 1 H, $Ar_3CH$), 7.16-7.46 (m, 13 H, $H_{arom.}$).
**$^{13}$C-NMR** (75 MHz, $CDCl_3$): $\delta$/ppm = 21.1 (s, 2 C, $CH_3$), 56.2 (s, 1 C, $Ar_3C$), 126.3 (s, 1 C, CH), 128.4 (s, 2 C, CH), 129.1 (s, 4 C, CH), 129.4 (s, 4 C, CH), 129.5 (s, 2 C, CH), 135.8 (s, 2 C, $C_{quat.}$), 141.3 (s, 2 C, $C_{quat.}$), 144.5 (s, 1 C, $C_{quat.}$).

*Reaction of $(MeO)Tr^+BF_4^-$ with $HSiEt_3$*
376 mg (3.23 mmol) of $HSiEt_3$ were added to a solution of 1.00 g (2.78 mmol) of $(MeO)Tr^+BF_4^-$ in 10 mL dichloromethane. After fading of the red color with concomitant gas evolution, the solvent was evaporated. Column chromatography (silica gel, diethyl ether/pentane = 1/13) yielded 710 mg (2.59 mmol, 93 %) of (4-methoxyphenyl)diphenylmethane as a colorless solid.

**$^1$H-NMR** (300 MHz, $CDCl_3$): $\delta$/ppm = 3.83 (s, 3 H, Me), 5.57 (s, 1 H, $H_{arom.}$), 6.86-6.90 (m, 2 H, $H_{arom.}$), 7.08-7.38 (m, 12 H, $H_{arom.}$).
**$^{13}$C-NMR** (75 MHz, $CDCl_3$): $\delta$/ppm = 55.3 (s, 1 C), 56.1 (s, 1 C), 113.8 (s, 2 C, CH), 126.3 (s, 2 C, CH), 128.4 (s, 4 C, CH), 129.5 (s, 4 C, CH), 130.5 (s, 2 C, CH), 136.2 (s, 1 C, $C_{quat.}$), 144.4 (s, 2 C, $C_{quat.}$), 158.1 (s, 1 C, COMe).
**Mp**: 57-58 °C (Lit.:[41] 55-57 °C).

*Reaction of $MeTr^+BF_4^-$ with $HSiPh_3$*
290 mg (1.11 mmol) of $HSiPh_3$ were added to a solution of 393 mg (1.14 mmol) of $MeTr^+BF_4^-$ in 10 mL dichloromethane. Vigorous bubbling and a color-change from green to brown occurred within a few seconds. The black precipitate, which was formed during the reaction, was filtered off, and the solvent of the filtrate evaporated. 560 mg of a slightly

brown oil were obtained. GC/MS analysis revealed the presence of FSiPh$_3$, HSiPh$_3$, and (4-methylphenyl)diphenylmethane, which could not be further separated.

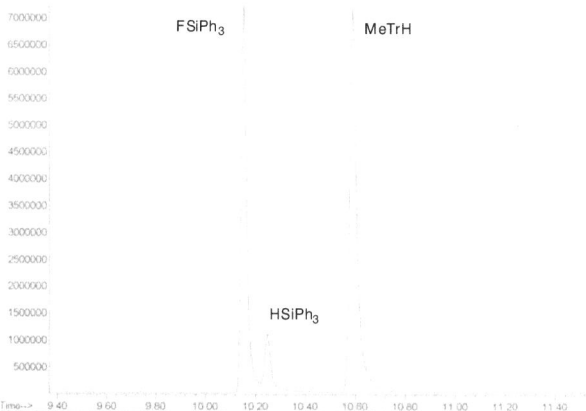

Chromatogram of the product mixture (GC/MS).

*Reaction of Tr$^+$BF$_4^-$ with DSiBu$_3$*

480 mg (2.38 mmol) of DSiBu$_3$ were added to a solution of 774 mg (2.34 mmol) of Tr$^+$BF$_4^-$ in 10 mL of dichloromethane. The yellow color of the solution disappeared immediately and a gas eluded. After stirring for 10 hours, the colorless precipitate was filtered off. 380 mg (1.55 mmol, 66 %) of triphenyldeuteriomethane as a colorless solid were obtained.

**$^1$H-NMR** (300 MHz, CDCl$_3$): $\delta$/ppm = 7.19-7.40 (m, 15 H, H$_{arom.}$).
**$^{13}$C-NMR** (75 MHz, CDCl$_3$): $\delta$/ppm = 56.5 (t, 1 C, $^1J_{C,D}$ = 18.8 Hz, CD), 126.4 (s, 3 C, CH), 128.5 (s, 6 C, CH), 129.6 (s, 6 C, CH), 144.0 (s, 3 C, C$_{quat.}$).
**HR-MS** (EI, pos.): *m/z* calculated for [C$_{19}$H$_{15}$D]$^+$: 245.1309; found: 245.1310.
**Mp:** 92 °C.

*Reaction of (MeO)Tr$^+$BF$_4^-$ with DSiBu$_3$*

580 mg (2.88 mmol) of DSiBu$_3$ were added to a solution of 600 mg (1.67 mmol) of (MeO)Tr$^+$BF$_4^-$ in 10 mL of dichloromethane. After fading of the red color with concomitant gas evolution, the solvent was evaporated. Distillation (1 bar, 225 °C) yielded 0.59 g of a colorless liquid (mixture of tributyldeuteriosilane and tributylfluorosilane). The solid and

colorless residue consisted of 370 mg (1.34 mmol, 80 %) of (4-methoxyphenyl)diphenyl-deuteriomethane.

**$^1$H-NMR** (300 MHz, CDCl$_3$,): $\delta$/ppm = 3.82 (s, 3 H, CH$_3$), 6.85-6.90 (m, 2 H, H$_{arom.}$), 7.05-7.10 (m, 2 H, H$_{arom.}$), 7.15-7.20 (m, 4 H, H$_{arom.}$), 7.22-7.36 (m, 6 H, H$_{arom.}$).
**$^{13}$C-NMR** (75 MHz, CDCl$_3$): $\delta$/ppm = 55.4 (s, 1 C, CH$_3$), 55.7 (t, 1 C, $^1J_{C,D}$ = 19.6 Hz, CD), 113.8 (s, 2 C, CH), 126.4 (s, 2 C, CH), 128.4 (s, 4 C, CH), 129.5 (s, 4 C, CH), 130.5 (s, 2 C, CH), 136.2 (s, 1 C, C$_{quat.}$), 144.3 (s, 2 C, C$_{quat.}$), 158.2 (s, 1 C, COMe).
**HR-MS** (EI, pos.): calculated for [C$_{20}$H$_{17}$DO]$^+$: 275.1415, found: 275.1416.
**Mp:** 63-64 °C.

*Reaction of Tr$^+$BF$_4^-$ with HSiEt$_3$*

200 mg (1.72 mmol) of HSiEt$_3$ were added to a solution of 500 mg (1.51 mmol) of Tr$^+$BF$_4^-$ in 10 mL of acetonitrile. The solution turned black immediately, and was stirred for 5 min. The acetonitrile was evaporated, and the black residue subjected to column chromatography (silica gel, pentane). 290 mg (1.19 mmol, 79 %) of triphenylmethane as a colorless solid were obtained.

**$^1$H-NMR** (300 MHz, CDCl$_3$,): $\delta$/ppm = 5.60 (s, 1 H, CH), 7.14-7.36 (m, 15 H, H$_{arom.}$).
**$^{13}$C-NMR** (75 MHz, CDCl$_3$,): $\delta$/ppm = 56.8 (s, 1 C, Ar$_3$C), 126.3 (s, 3 C, CH), 128.3 (s, 6 C, CH), 129.4 (s, 6 C, CH), 143.9 (s, 3 C, C$_{quat.}$).
**Mp:** 93 °C (Lit.:[42] 94 °C).

*Reaction of (mF)(mF)'Tr$^+$ with 2-propyl-1,3-dioxolane (isolation)*

330 mg (1.05 mmol) of chlorobis(3-fluorophenyl)phenylmethane were mixed with 120 mg (0.682 mmol) GaCl$_3$ in 20 mL CH$_2$Cl$_2$, before 220 mg (1.89 mmol) of 2-propyl-1,3-dioxolane were added. The mixture was stirred for 48 hours at room temperature, before the reaction was quenched with water. Separation of phases and evaporation of solvent yielded a brown oil, which was subjected to column chromatography (silica gel, ethyl acetate/pentane = 1/10). 200 mg (0.765 mmol, 73 %), of bis(3-fluorophenyl)phenylmethane as a slightly yellow oil were obtained.

Experimental Part 205

**¹H-NMR** (600 MHz, CDCl₃): $\delta$/ppm = 5.53 (s, 1 H, Ar₃CH), 6.79-6.82 (m, 2 H, H$_{arom.}$), 6.90-6.92 (m, 2 H, H$_{arom.}$), 6.92-6.95 (m, 2 H, H$_{arom.}$), 7.09-7.12 (m, 2 H, H$_{arom.}$), 7.24-7.27 (m, 3 H, H$_{arom.}$), 7.30-7.34 (m, 2 H, H$_{arom.}$).
**¹³C-NMR** (150 MHz, CDCl₃): $\delta$/ppm = 56.3 (t, 1 C, $^4J_{C,F}$ = 1.7 Hz, Ar₃CH), 113.6 (d, 2 C, $^2J_{C,F}$ = 21.2 Hz, CH), 116.4 (d, 2 C, $^2J_{C,F}$ = 21.8 Hz, CH), 125.1 (d, 2 C, $^4J_{C,F}$ = 2.8 Hz, CH), 126.9 (s, 1 C, CH), 128.7 (s, 2 C, CH), 129.4 (s, 2 C, CH), 129.9 (d, 2 C, $^3J_{C,F}$ = 8.3 Hz, CH), 142.6 (s, 1 C, C$_{quat.}$), 145.9 (d, 2 C, $^3J_{C,F}$ = 6.9 Hz, C$_{quat.}$), 163.0 (d, 2 C, $^1J_{C,F}$ = 246 Hz, CF).
**¹⁹F-NMR** (282 MHz, CDCl₃): $\delta$/ppm = –114.

*Reaction of (mF)(mF)'Tr⁺ with 2-propyl-1,3-dioxolane (NMR)*
30 mg (95 μmol) of chlorobis(3-fluorophenyl)phenylmethane were mixed with 34 mg (191 μmol) of GaCl₃ in 0.5 mL CD₂Cl₂, before 12 mg (103 μmol) of 2-propyl-1,3-dioxolane were added. The spectra showed the formation of bis(3-fluorophenyl)phenylmethane (a) and 2-propyl-1,3-dioxolenium ion (b).

**¹H-NMR** (600 MHz, CD₂Cl₂): $\delta$/ppm = 1.14 (t, 3 H, $^3J_{H,H}$ = 7.5 Hz, Me (b)), 1.98 (sex, 2 H, $^3J_{H,H}$ = 7.5 Hz, CH₂ (b)), 3.17 (t, 2 H, $^3J_{H,H}$ = 7.5 Hz, CH₂ (b)), 5.55 (s, 4 H, CH₂ (b)), 5.57 (s, 1 H, Ar₃CH (a)), 6.82-6.85 (m, 2 H, H$_{arom.}$ (a)), 6.94-6.96 (m, 3 H, H$_{arom.}$ (a)), 7.12-7.15 (m, 2 H, H$_{arom.}$ (a)), 7.26-7.34 (m, 5 H, H$_{arom.}$ (a)).
**¹³C-NMR** (150 MHz, CDCl₃): $\delta$/ppm = 13.2 (s, 1 C (b)), 17.6 (s, 1 C (b)), 30.9 (s, 1 C (b)), 56.1 (s, 1 C, Ar₃CH (a)), 75.8 (s, 2 C (b)), 113.3 (d, 2 C, $^2J_{C,F}$ = 21.1 Hz, CH (a)), 116.1 (d, 2 C, $^2J_{C,F}$ = 21.9 Hz, CH (a)), 125.1 (d, 2 C, $^4J_{C,F}$ = 2.8 Hz, CH (a)), 126.8 (s, 1 C, CH (a)), 128.5 (s, 2 C, CH (a)), 129.2 (s, 2 C, CH (a)), 129.9 (d, 2 C, $^3J_{C,F}$ = 8.3 Hz, CH (a)), 142.6 (s, 1 C, C$_{quat.}$ (a)), 146.0 (d, 2 C, $^3J_{C,F}$ = 6.9 Hz, C$_{quat.}$ (a)), 162.9 (d, 2 C, $^1J_{C,F}$ = 245 Hz, CF (a)), 194.2 (s, 1 C, C⁺ (b)).

*Reaction of (mF)(mF)'Tr⁺ with cycloheptatriene*
212 mg (0.674 mmol) of (mF)(mF)'TrCl were mixed with 810 mg (3.11 mmol) of SnCl₄ in 6 mL dichloromethane, resulting in a deeply red solution. 113 mg (1.23 mmol) of cycloheptatriene were added and the mixture stirred for 5 min at room temperature. After the reaction was quenched with water, the organic phase was treated consecutively with sat. aq. NaHCO₃, H₂O and brine. Filtration through MgSO₄ and evaporation of the solvent delivered 160 mg (0.571 mmol, 85 %) of bis(3-fluorophenyl)phenylmethane as a colorless oil.

**¹H-NMR** (300 MHz, CDCl₃): $\delta$/ppm = 5.63 (s, 1 H, Ar₃CH), 6.90-7.08 (m, 6 H, H$_{arom.}$), 7.18-7.44 (m, 7 H, H$_{arom.}$).
**¹³C-NMR** (75 MHz, CDCl₃): $\delta$/ppm = 56.2 (s, 1 C, Ar₃CH), 113.5 (d, 2 C, $^2J_{C,F}$ = 21.1 Hz, CH), 116.3 (d, 2 C, $^2J_{C,F}$ = 21.8 Hz, CH), 125.0 (d, 2 C, $^4J_{C,F}$ = 2.8 Hz, CH), 126.8 (s, 1 C, CH), 128.5 (s, 2 C, CH), 129.2 (s, 2 C, CH), 129.8 (d, 2 C, $^3J_{C,F}$ = 8.3 Hz, CH), 142.5 (s, 1 C, C$_{quat.}$), 145.8 (d, 2 C, $^3J_{C,F}$ = 6.9 Hz, C$_{quat.}$), 162.9 (d, 2 C, $^1J_{C,F}$ = 246 Hz, CF).
**¹⁹F-NMR** (282 MHz, CDCl₃): $\delta$/ppm = –113.

*Reaction of Tr⁺BF₄⁻ with diethyl 2,6-dimethyl-1,4-dihydropyridine-3,5-dicarboxylate (NMR)*
30 mg (91 µmol) of Tr⁺BF₄⁻ were mixed with 23 mg (91 µmol) of diethyl 2,6-dimethyl-1,4-dihydropyridine-3,5-dicarboxylate in 0.5 mL CDCl₃.

**¹H-NMR** (200 MHz, CDCl₃): $\delta$/ppm = 1.42 (t, 6 H, $^3J_{H,H}$ = 7.1 Hz, CH₂C*H*₃), 3.11 (s, 6 H, Me), 4.46 (q, 4 H, $^3J_{H,H}$ = 7.1 Hz, C*H*₂CH₃), 5.52 (s, 1 H, Ph₃CH), 7.08-7.30 (m, 15 H, H$_{arom.}$), 9.22 (s, 1 H, H$_{arom.}$), 11.52 (br s, 1 H, NH).

*Reaction of Me₃Tr⁺BF₄⁻ with diethyl 2,4,6-trimethyl-1,4-dihydropyridine-3,5-dicarboxylate (NMR)*
33 mg (87 µmol) of Me₃Tr⁺BF₄⁻ were mixed with 24 mg (90 µmol) of diethyl 2,4,6-trimethyl-1,4-dihydropyridine-3,5-dicarboxylate in 0.5 mL CDCl₃.

**¹H-NMR** (200 MHz, CDCl₃): $\delta$/ppm = 1.44 (t, 6 H, $^3J_{H,H}$ = 7.1 Hz, CH₂C*H*₃), 2.32 (s, 9 H, Me), 4.52 (q, 4 H, $^3J_{H,H}$ = 7.1 Hz, C*H*₂CH₃), 5.45 (s, 1 H, Ar₃CH), 6.96-7.15 (m, 12 H, H$_{arom.}$), 13.50 (br s, 1 H, NH).

*Reaction of (MeO)Tr⁺BF₄⁻ with diethyl 2,4,6-trimethyl-1,4-dihydropyridine-3,5-dicarboxylate (NMR)*
28 mg (78 µmol) of (MeO)Tr⁺BF₄⁻ were mixed with 19 mg (72 µmol) of diethyl 2,4,6-trimethyl-1,4-dihydropyridine-3,5-dicarboxylate in 0.5 mL CDCl₃.

**¹H-NMR** (200 MHz, CDCl₃): $\delta$/ppm = 1.43 (t, 6 H, $^3J_{H,H}$ = 7.1 Hz, CH₂C*H*₃), 2.55 (s, 3 H, Me), 2.84 (s, 6 H, Me), 3.78 (s, 3 H, OMe), 4.51 (q, 4 H, $^3J_{H,H}$ = 7.1 Hz, C*H*₂CH₃), 5.50 (s, 1 H, Ar₃CH), 6.57-7.30 (m, 14 H, H$_{arom.}$), 13.50 (br s, 1 H, NH).

## 4. Kinetic Data

### 4.1. Solvolyses of triarylmethyl esters in aqueous acetonitrile

All measurements have been performed at 25 °C.

Tr-OAc. NEt$_3$ as additive, conventional conductimetry.

| solvent | $k_{ion}$ (s$^{-1}$) | |
|---|---|---|
| 90AN10W | 1.48 × 10$^{-5}$ [a] | **1.47 × 10$^{-5}$** |
|  | 1.46 × 10$^{-5}$ [a] | |
| 80AN20W | 5.87 × 10$^{-5}$ [b] | **5.88 × 10$^{-5}$** |
|  | 5.88 × 10$^{-5}$ [c] | |
| 60AN40W | 2.64 × 10$^{-4}$ [d] | **2.70 × 10$^{-4}$** |
|  | 2.75 × 10$^{-4}$ [c,e] | |
| 50AN50W | 5.63 × 10$^{-4}$ [f] | **5.57 × 10$^{-4}$** |
|  | 5.51 × 10$^{-4}$ [g,e] | |

[a] Initial concentration: 8.80 × 10$^{-4}$ M; [b] initial concentration: 8.70 × 10$^{-4}$ M; [c] initial concentration: 7.90 × 10$^{-4}$ M; [d] initial concentration: 8.00 × 10$^{-4}$ M; [e] 1,8-bis(dimethylamino)naphthalene was used as additive; [f] initial concentration: 7.00 × 10$^{-4}$ M; [g] initial concentration: 8.33 × 10$^{-4}$ M.

MeTr-OAc (initial concentration: 1.06 × 10$^{-3}$ M). The substrate has been synthesized in solution by mixing equimolar amounts of ($n$-Bu)$_4$N$^+$AcO$^-$ and MeTr$^+$BF$_4^-$. NEt$_3$ as additive, conventional conductimetry.

| solvent | $k_{ion}$ (s$^{-1}$) | |
|---|---|---|
| 90AN10W | 1.07 × 10$^{-4}$ | **1.03 × 10$^{-4}$** |
|  | 9.98 × 10$^{-5}$ | |
| 80AN20W | 3.58 × 10$^{-4}$ [a] | **3.59 × 10$^{-4}$** |
|  | 3.60 × 10$^{-4}$ | |
| 60AN40W | 1.36 × 10$^{-3}$ [a] | **1.46 × 10$^{-3}$** |
|  | 1.56 × 10$^{-3}$ [a] | |
| 50AN50W | 3.04 × 10$^{-3}$ | **3.01 × 10$^{-3}$** |
|  | 2.98 × 10$^{-3}$ | |

[a] Initial concentration: 2.00 × 10$^{-4}$ M.

Me$_2$Tr-OAc (initial concentration: $1.00 \times 10^{-3}$ M). The substrate has been synthesized in solution by mixing equimolar amounts of $(n\text{-Bu})_4\text{N}^+\text{AcO}^-$ and Me$_2$Tr$^+$BF$_4^-$. NEt$_3$ as additive, conventional conductimetry.

| solvent | $k_{ion}$ (s$^{-1}$) | |
|---|---|---|
| 90AN10W | $3.33 \times 10^{-4}$ [a] | $\mathbf{3.23 \times 10^{-4}}$ |
|  | $3.13 \times 10^{-4}$ [a] | |
| 80AN20W | $1.20 \times 10^{-3}$ | $\mathbf{1.21 \times 10^{-3}}$ |
|  | $1.22 \times 10^{-3}$ | |
| 60AN40W | $5.58 \times 10^{-3}$ | $\mathbf{5.62 \times 10^{-3}}$ |
|  | $5.65 \times 10^{-3}$ | |
| 50AN50W | $9.55 \times 10^{-3}$ | $\mathbf{9.59 \times 10^{-3}}$ |
|  | $9.62 \times 10^{-3}$ | |

[a] Initial concentration: $2.00 \times 10^{-4}$ M.

Me$_3$Tr-OAc (initial concentration: $1.99 \times 10^{-4}$ M. The substrate has been synthesized in solution by mixing equimolar amounts of $(n\text{-Bu})_4\text{N}^+\text{AcO}^-$ and Me$_3$Tr$^+$BF$_4^-$. NEt$_3$ as additive, conventional conductimetry.

| solvent | $k_{ion}$ (s$^{-1}$) | |
|---|---|---|
| 90AN10W | $1.30 \times 10^{-3}$ | $\mathbf{1.30 \times 10^{-3}}$ |
|  | $1.29 \times 10^{-3}$ | |
| 80AN20W | $4.96 \times 10^{-3}$ | $\mathbf{4.98 \times 10^{-3}}$ |
|  | $4.99 \times 10^{-3}$ | |
| 60AN40W | $1.74 \times 10^{-2}$ | $\mathbf{1.77 \times 10^{-2}}$ |
|  | $1.80 \times 10^{-2}$ | |
| 50AN50W | $3.37 \times 10^{-2}$ [a] | $\mathbf{3.33 \times 10^{-2}}$ |
|  | $3.29 \times 10^{-2}$ [a] | |

[a] Initial concentration: $3.58 \times 10^{-4}$ M.

(MeO)Tr-OAc (initial concentration: $6.60 \times 10^{-4}$ M). The substrate has been synthesized in solution by mixing equimolar amounts of $(n\text{-Bu})_4\text{N}^+\text{AcO}^-$ and (MeO)Tr$^+$BF$_4^-$. NEt$_3$ as additive, conventional conductimetry.

| solvent | $k_{obs}$ (s$^{-1}$) | |
|---|---|---|
| 90AN10W | $1.19 \times 10^{-3}$ [a] | $\mathbf{1.20 \times 10^{-3}}$ |
|  | $1.21 \times 10^{-3}$ [a] | |
| 80AN20W | $4.54 \times 10^{-3}$ [b] | $\mathbf{4.53 \times 10^{-3}}$ |
|  | $4.51 \times 10^{-3}$ [b] | |
| 60AN40W | $1.57 \times 10^{-2}$ | $\mathbf{1.50 \times 10^{-2}}$ |
|  | $1.42 \times 10^{-2}$ | |
| 50AN50W | $2.42 \times 10^{-2}$ | $\mathbf{2.40 \times 10^{-2}}$ |
|  | $2.37 \times 10^{-2}$ | |

[a] Initial concentration: $3.20 \times 10^{-4}$ M; [b] initial concentration: $7.50 \times 10^{-4}$ M.

(MeO)$_2$Tr-OAc. The substrate has been synthesized in solution by mixing equimolar amounts of ($n$-Bu)$_4$N$^+$AcO$^-$ and (MeO)$_2$Tr$^+$BF$_4^-$. NEt$_3$ as additive.

| solvent | $k_{ion}$ (s$^{-1}$) | |
|---|---|---|
| 90AN10W | 3.98 × 10$^{-2}$ [a] | **4.04 × 10$^{-2}$** |
|  | 4.09 × 10$^{-2}$ [a] | |
| 80AN20W | 1.14 × 10$^{-1}$ [a] | **1.15 × 10$^{-1}$** |
|  | 1.16 × 10$^{-1}$ [a] | |
| 60AN40W [b] | - | **3.06 × 10$^{-1}$** |
| 50AN50W [b] | - | **4.41 × 10$^{-1}$** |

[a] Initial concentration: 2.00 × 10$^{-4}$ M, conventional conductimetry; [b] initial concentration: 1.00 × 10$^{-3}$ M, stopped-flow conductimetry.

(Me$_2$N)Tr-OAc (initial concentration: 6.00 × 10$^{-5}$ M). The substrate has been synthesized in solution by mixing (Me$_2$N)Tr$^+$BF$_4^-$ with different amounts of ($n$-Bu)$_4$N$^+$AcO$^-$; stopped-flow UV-vis spectroscopy, λ = 461 nm.

| [AcO$^-$] (mol L$^{-1}$) [a] | $k_{obs}$ (s$^{-1}$) | | | |
|---|---|---|---|---|
|  | 90AN10W | 80AN20W | 60AN40W | 50AN50W |
| 6.65 × 10$^{-5}$ | 1.07 | 1.97 | 4.35 | 7.50 |
| 1.32 × 10$^{-4}$ | 1.08 | 2.05 | 4.63 | 7.40 |
| 1.94 × 10$^{-4}$ | 1.10 | 2.00 | 4.62 | 7.67 |
| 2.58 × 10$^{-4}$ | 1.07 | 2.01 | 4.59 | 7.02 |
| 3.21 × 10$^{-4}$ | 1.07 | 1.96 | 4.34 | 7.41 |
| 1.87 × 10$^{-3}$ | 1.02 | - | - | 7.29 |
| 3.73 × 10$^{-3}$ | 1.04 | - | - | 7.34 |
| 6.78 × 10$^{-3}$ | 1.03 | - | - | 7.32 |
| 1.01 × 10$^{-2}$ | 1.05 | - | - | 7.22 |
| 1.35 × 10$^{-2}$ | 1.04 | - | - | 7.22 |
|  | **1.08** | **2.00** | **4.51** | **7.40** |

[a] free AcO$^-$: [($n$-Bu)$_4$N$^+$AcO$^-$]$_0$ – [(Me$_2$N)Tr$^+$BF$_4^-$]$_0$

(Me$_2$N)(MeO)Tr-OAc (initial concentration: 5.03 × 10$^{-5}$ M). The substrate has been synthesized in solution by mixing equimolar amounts of ($n$-Bu)$_4$N$^+$AcO$^-$ and (Me$_2$N)(MeO)Tr$^+$BF$_4^-$; stopped-flow UV-vis spectroscopy, λ = 506 nm.

| solvent | 90AN10W | 80AN20W | 60AN40W | 50AN50W |
|---|---|---|---|---|
| $k_{ion}$ (s$^{-1}$) | **6.23** | **1.22 × 10$^1$** | **2.49 × 10$^1$** | **3.93 × 10$^1$** |

(Me$_2$N)(MeO)Tr-OAc (initial concentration: 5.03 × 10$^{-5}$ M) in 90AN10W with different amounts of additional (n-Bu)$_4$N$^+$AcO$^-$. The substrate has been synthesized in solution by mixing equimolar amounts of (n-Bu)$_4$N$^+$AcO$^-$ and (Me$_2$N)(MeO)Tr$^+$BF$_4^-$; stopped-flow UV-vis spectroscopy, λ = 506 nm.

| [(n-Bu$_4$)N$^+$BF$_4^-$] (mol L$^{-1}$) [a] | 6.07 × 10$^{-4}$ | 1.75 × 10$^{-3}$ | 3.17 × 10$^{-3}$ | 4.53 × 10$^{-3}$ |
|---|---|---|---|---|
| $k_{ion}$ (s$^{-1}$) | 6.37 | 6.13 | 6.27 | 6.19 |

[a] additional salt.

(Me$_2$N)$_2$Tr-OAc (initial concentration: 1.92 × 10$^{-5}$ M). The substrate has been synthesized in solution by mixing (Me$_2$N)$_2$Tr$^+$BF$_4^-$ with 140 equiv. of (n-Bu)$_4$N$^+$AcO$^-$; stopped-flow UV-vis spectroscopy, λ = 620 nm.

| solvent | 90AN10W | 80AN20W |
|---|---|---|
| $k_{ion}$ (s$^{-1}$) | **1.28 × 10$^2$** | **2.15 × 10$^2$** |

Tr-OBz. NEt$_3$ as additive, conventional conductimetry.

| solvent | $k_{ion}$ (s$^{-1}$) | |
|---|---|---|
| 90AN10W | 5.37 × 10$^{-5}$ [a] | **5.34 × 10$^{-5}$** |
|  | 5.30 × 10$^{-5}$ [a] |  |
| 80AN20W | 1.69 × 10$^{-4}$ [b] | **1.67 × 10$^{-4}$** |
|  | 1.64 × 10$^{-4}$ [b] |  |
| 60AN40W | 5.26 × 10$^{-4}$ [c] | **5.14 × 10$^{-4}$** |
|  | 5.02 × 10$^{-4}$ [c] |  |
| 50AN50W | 9.97 × 10$^{-4}$ [d] | **9.99 × 10$^{-4}$** |
|  | 1.00 × 10$^{-3}$ [d] |  |

[a] Initial concentration: 8.20 × 10$^{-4}$ M; [b] initial concentration: 3.48 × 10$^{-4}$ M; [c] initial concentration: 1.46 × 10$^{-4}$ M; [d] initial concentration: 8.10 × 10$^{-4}$ M.

MeTr-OBz (initial concentration: 1.00 × 10$^{-3}$ M). The substrate has been synthesized in solution by mixing equimolar amounts of (n-Bu)$_4$N$^+$BzO$^-$ and MeTr$^+$BF$_4^-$. NEt$_3$ as additive, conventional conductimetry.

| solvent | $k_{ion}$ (s$^{-1}$) | |
|---|---|---|
| 90AN10W | 2.43 × 10$^{-4}$ | **2.56 × 10$^{-4}$** |
|  | 2.68 × 10$^{-4}$ |  |
| 80AN20W | 8.10 × 10$^{-4}$ | **8.08 × 10$^{-4}$** |
|  | 8.04 × 10$^{-4}$ |  |
| 60AN40W | 2.76 × 10$^{-3}$ | **2.78 × 10$^{-3}$** |
|  | 2.79 × 10$^{-3}$ |  |
| 50AN50W | 5.00 × 10$^{-3}$ | **5.01 × 10$^{-3}$** |
|  | 5.02 × 10$^{-3}$ |  |

# Experimental Part 211

Me$_2$Tr-OBz. The substrate has been synthesized in solution by mixing equimolar amounts of (n-Bu)$_4$N$^+$BzO$^-$ and Me$_2$Tr$^+$BF$_4^-$. NEt$_3$ as additive, conventional conductimetry.

| solvent | $k_{ion}$ (s$^{-1}$) | |
|---------|---------------------|---|
| 90AN10W | 1.26 × 10$^{-3}$ [a] | **1.26 × 10$^{-3}$** |
|         | 1.25 × 10$^{-3}$ [a] | |
| 80AN20W | 3.62 × 10$^{-3}$ [b] | **3.55 × 10$^{-3}$** |
|         | 3.48 × 10$^{-3}$ [b] | |
| 60AN40W | 1.02 × 10$^{-2}$ [c] | **1.05 × 10$^{-2}$** |
|         | 1.08 × 10$^{-2}$ [c] | |
| 50AN50W | 1.69 × 10$^{-2}$ [a] | **1.71 × 10$^{-2}$** |
|         | 1.72 × 10$^{-2}$ [c] | |

[a] Initial concentration: 2.00 × 10$^{-4}$ M; [b] initial concentration: 2.70 × 10$^{-4}$ M; [c] initial concentration: 1.72 × 10$^{-4}$ M.

Me$_3$Tr-OBz (initial concentration: 2.50 × 10$^{-4}$ M. The substrate has been synthesized in solution by mixing equimolar amounts of (n-Bu)$_4$N$^+$BzO$^-$ and Me$_3$Tr$^+$BF$_4^-$. NEt$_3$ as additive, conventional conductimetry.

| solvent | $k_{ion}$ (s$^{-1}$) | |
|---------|---------------------|---|
| 90AN10W | 5.44 × 10$^{-3}$ | **5.43 × 10$^{-3}$** |
|         | 5.41 × 10$^{-3}$ | |
| 80AN20W | 1.50 × 10$^{-2}$ | **1.51 × 10$^{-2}$** |
|         | 1.51 × 10$^{-2}$ | |
| 60AN40W | 4.55 × 10$^{-2}$ | **4.55 × 10$^{-2}$** |
|         | 4.54 × 10$^{-2}$ | |
| 50AN50W | 6.86 × 10$^{-2}$ | **6.97 × 10$^{-2}$** |
|         | 7.08 × 10$^{-2}$ | |

(MeO)Tr-OBz. The substrate has been synthesized in solution by mixing equimolar amounts of (n-Bu)$_4$N$^+$BzO$^-$ and (MeO)Tr$^+$BF$_4^-$. NEt$_3$ as additive, conventional conductimetry.

| solvent | $k_{ion}$ (s$^{-1}$) | |
|---------|---------------------|---|
| 90AN10W | 4.51 × 10$^{-3}$ [a] | **4.45 × 10$^{-3}$** |
|         | 4.38 × 10$^{-3}$ [a] | |
| 80AN20W | 1.26 × 10$^{-2}$ [a] | **1.30 × 10$^{-2}$** |
|         | 1.33 × 10$^{-2}$ [a] | |
| 60AN40W | 3.81 × 10$^{-2}$ [b] | **3.86 × 10$^{-2}$** |
|         | 3.90 × 10$^{-2}$ [b] | |
| 50AN50W | 5.41 × 10$^{-2}$ [c] | **5.56 × 10$^{-2}$** |
|         | 5.70 × 10$^{-2}$ [c] | |

[a] Initial concentration: 9.18 × 10$^{-4}$ M; [b] initial concentration: 1.67 × 10$^{-3}$ M; [c] initial concentration: 2.00 × 10$^{-4}$ M.

(MeO)$_2$Tr-OBz. The substrate has been synthesized in solution by mixing equimolar amounts of ($n$-Bu)$_4$N$^+$BzO$^-$ and (MeO)$_2$Tr$^+$BF$_4^-$. NEt$_3$ as additive.

| solvent | $k_{ion}$ (s$^{-1}$) | |
|---|---|---|
| 90AN10W | 1.67 × 10$^{-1}$ [a] | **1.61 × 10$^{-1}$** |
|  | 1.55 × 10$^{-1}$ [a] | |
| 80AN20W [b] | - | **3.34 × 10$^{-1}$** |
| 60AN40W [b] | - | **6.67 × 10$^{-1}$** |
| 50AN50W [b] | - | **9.30 × 10$^{-1}$** |

[a] Initial concentration: 2.00 × 10$^{-4}$ M, conventional conductimetry; [b] initial concentration: 8.20 × 10$^{-4}$ M, stopped-flow conductimetry.

(Me$_2$N)Tr-OBz (initial concentration: 5.89 × 10$^{-5}$ M). The substrate has been synthesized in solution by mixing (Me$_2$N)Tr$^+$BF$_4^-$ with different amounts of ($n$-Bu)$_4$N$^+$BzO$^-$; stopped-flow UV-vis spectroscopy, λ = 461 nm.

| [BzO$^-$] | $k_{ion}$ (s$^{-1}$) | | | |
|---|---|---|---|---|
| (mol L$^{-1}$) [a] | 90AN10W | 80AN20W | 60AN40W | 50AN50W |
| 1.26 × 10$^{-5}$ | 5.37 | 8.35 | **1.40 × 10$^1$** | **2.04 × 10$^1$** |
| 1.15 × 10$^{-3}$ | 5.38 | 8.16 | 1.41 × 10$^1$ | 2.04 × 10$^1$ |

[a] Free BzO$^-$: [($n$-Bu)$_4$N$^+$BzO$^-$]$_0$ − [(Me$_2$N)Tr-OBz]$_0$

(Me$_2$N)(MeO)Tr-OBz (initial concentration: 4.96 × 10$^{-5}$ M). The substrate has been synthesized in solution by mixing (Me$_2$N)(MeO)Tr$^+$BF$_4^-$ and 2.56 equivalents of ($n$-Bu)$_4$N$^+$BzO$^-$; stopped-flow UV-vis spectroscopy, λ = 506 nm.

| solvent | 90AN10W | 80AN20W | 60AN40W | 50AN50W |
|---|---|---|---|---|
| $k_{ion}$ (s$^{-1}$) | **3.36 × 10$^1$** | **4.70 × 10$^1$** | **6.95 × 10$^1$** | **1.02 × 10$^2$** |

Tr-PNB (initial concentration: 7.30 × 10$^{-4}$ M). NEt$_3$ as additive, conventional conductimetry.

| solvent | $k_{ion}$ (s$^{-1}$) | |
|---|---|---|
| 90AN10W | 1.54 × 10$^{-3}$ | **1.57 × 10$^{-3}$** |
|  | 1.59 × 10$^{-3}$ | |
| 80AN20W | 4.14 × 10$^{-3}$ | **4.19 × 10$^{-3}$** |
|  | 4.25 × 10$^{-3}$ | |
| 60AN40W | 9.66 × 10$^{-3}$ | **9.68 × 10$^{-3}$** |
|  | 9.70 × 10$^{-3}$ | |
| 50AN50W | 1.82 × 10$^{-2}$ | **1.82 × 10$^{-2}$** |
|  | 1.82 × 10$^{-2}$ | |

Experimental Part                                                                 213

## 4.2. Solvolyses of triphenylmethyl esters in aqueous acetone

All measurements have been performed at 25 °C.

Tr-OAc (initial concentration: $7.10 \times 10^{-4}$ M). NEt$_3$ as additive, conventional conductimetry.

| solvent | $k_{ion}$ (s$^{-1}$) | |
|---|---|---|
| 80A20W | **1.38 × 10$^{-5}$** [a] | |
| 60A40W | $1.96 \times 10^{-4}$ | **1.99 × 10$^{-4}$** |
|  | $2.01 \times 10^{-4}$ | |
| 50A50W | $6.32 \times 10^{-4}$ | **6.40 × 10$^{-2}$** |
|  | $6.52 \times 10^{-4}$ | |

[a] Initial concentration: $8.60 \times 10^{-4}$ M, 1,8-bis(dimethylamino)naphthalene was used as additive.

Tr-OBz (initial concentration: $7.10 \times 10^{-4}$ M). NEt$_3$ as additive, conventional conductimetry.

| solvent | $k_{ion}$ (s$^{-1}$) | |
|---|---|---|
| 90A10W | $9.27 \times 10^{-6}$ | **9.31 × 10$^{-6}$** |
|  | $9.34 \times 10^{-6}$ | |
| 80A20W | $3.47 \times 10^{-5}$ | **3.50 × 10$^{-5}$** |
|  | $3.52 \times 10^{-5}$ | |
| 60A40W | $2.85 \times 10^{-4}$ | **2.87 × 10$^{-4}$** |
|  | $2.89 \times 10^{-4}$ | |
| 50A50W | $6.94 \times 10^{-4}$ | **6.95 × 10$^{-4}$** |
|  | $6.96 \times 10^{-4}$ | |

Tr-PNB (initial concentration: $7.30 \times 10^{-4}$ M). NEt$_3$ as additive, conventional conductimetry.

| solvent | $k_{ion}$ (s$^{-1}$) | |
|---|---|---|
| 90A10W | $3.66 \times 10^{-4}$ | **3.63 × 10$^{-4}$** |
|  | $3.59 \times 10^{-4}$ | |
| 80A20W | $1.49 \times 10^{-3}$ | **1.49 × 10$^{-3}$** |
|  | $1.48 \times 10^{-3}$ | |
| 60A40W | $1.11 \times 10^{-2}$ | **1.08 × 10$^{-2}$** |
|  | $1.05 \times 10^{-2}$ | |
| 50A50W | $3.21 \times 10^{-2}$ | **3.25 × 10$^{-2}$** |
|  | $3.30 \times 10^{-2}$ | |

### 4.3. Ionizations of trianisylmethyl esters in AN/W in the presence of piperidine

The substrate has been synthesized in solution by mixing equimolar amounts of $(MeO)_3Tr^+BF_4^-$ and $(n\text{-}Bu)_4N^+AcO^-$ or $(n\text{-}Bu)_4N^+BzO^-$, respectivly. Piperidine as additive; stopped-flow conductimetry, 25 °C.

| nucleofuge | [substrate] (mol L$^{-1}$) | solvent | [piperidine] (mol L$^{-1}$) | $k_{ion}$ (s$^{-1}$) | |
|---|---|---|---|---|---|
| AcO$^-$ | 8.33 × 10$^{-4}$ | 90AN10W | 5.18 × 10$^{-3}$ | 6.74 × 10$^{-1}$ | **6.80 × 10$^{-1}$** |
| | | | 1.04 × 10$^{-2}$ | 6.79 × 10$^{-1}$ | |
| | | | 1.55 × 10$^{-2}$ | 6.65 × 10$^{-1}$ | |
| | | | 2.07 × 10$^{-2}$ | 6.80 × 10$^{-1}$ | |
| | 8.22 × 10$^{-4}$ | 80AN20W | 1.32 × 10$^{-2}$ | **1.58** | |
| | | 60AN40W | | **3.86** | |
| | | 50AN50W | | **5.56** | |
| BzO$^-$ | 1.01 × 10$^{-3}$ | 90AN10W | 1.04 × 10$^{-2}$ | 3.71 | **3.79** |
| | | | 1.55 × 10$^{-2}$ | 3.80 | |
| | | | 2.07 × 10$^{-2}$ | 3.76 | |
| | 7.18 × 10$^{-4}$ | 80AN20W | 1.32 × 10$^{-2}$ | **6.45** | |
| | | 60AN40W | | **1.12 × 10$^1$** | |
| | | 50AN50W | | **1.38 × 10$^1$** | |

### 4.4. Reactions of tritylium ions with water in aqueous acetonitrile

All measurements have been performed at 20 °C.

The carbenium ions were generated by laser-flash photolysis of the corresponding acetates, which have been synthesized by mixing equimolar amounts of $Ar_3Tr^+BF_4^-$ and $(n\text{-}Bu)_4N^+AcO^-$; UV-vis spectroscopy.

| system | [Ar$_3$C$^+$AcO$^-$]$_0$ (M) | solvent | $k_w$ (s$^{-1}$) | |
|---|---|---|---|---|
| Tr$^+$ | 4.10 × 10$^{-4}$ | 90AN10W | 1.18 × 10$^5$ | **1.19 × 10$^5$** |
| | | | 1.17 × 10$^5$ | |
| | | | 1.22 × 10$^{5}$ [a] | |
| | | 80AN20W | **1.58 × 10$^5$** [a] | |
| | | 60AN40W | 1.67 × 10$^5$ | **1.69 × 10$^5$** |
| | | | 1.70 × 10$^5$ | |
| | | 50AN50W | 1.64 × 10$^5$ | **1.62 × 10$^5$** |
| | | | 1.62 × 10$^5$ | |
| | | | 1.62 × 10$^5$ | |
| MeTr$^+$ | 6.65 × 10$^{-4}$ | 90AN10W | 2.24 × 10$^4$ | **2.44 × 10$^4$** |

Experimental Part

Table continued.

| system | [Ar$_3$C$^+$AcO$^-$]$_0$ (M) | solvent | $k_w$ (s$^{-1}$) | |
|---|---|---|---|---|
| MeTr$^+$ | 6.65 × 10$^{-4}$ | 90AN10W | 2.51 × 10$^4$ | |
| | | | 2.56 × 10$^4$ | |
| | | 80AN20W | 3.54 × 10$^4$ | **3.60 × 10$^4$** |
| | | | 3.77 × 10$^4$ | |
| | | | 3.48 × 10$^4$ | |
| | | 60AN40W | 4.30 × 10$^4$ | **4.29 × 10$^4$** |
| | | | 4.25 × 10$^4$ | |
| | | | 4.32 × 10$^4$ | |
| | | 50AN50W | 4.07 × 10$^4$ | **4.08 × 10$^4$** |
| | | | 4.08 × 10$^4$ | |
| | | | 4.10 × 10$^4$ | |
| Me$_2$Tr$^+$ | 1.63 × 10$^{-3}$ | 90AN10W | 7.73 × 10$^3$ | **7.85 × 10$^3$** |
| | | | 7.88 × 10$^3$ | |
| | | | 7.95 × 10$^3$ | |
| | | 80AN20W | 9.47 × 10$^3$ | **9.35 × 10$^3$** |
| | | | 9.21 × 10$^3$ | |
| | | | 9.38 × 10$^3$ | |
| | | 60AN40W | 9.85 × 10$^3$ | **9.84 × 10$^3$** |
| | | | 9.96 × 10$^3$ | |
| | | | 9.72 × 10$^3$ | |
| Me$_2$Tr$^+$ | 1.63 × 10$^{-3}$ | 50AN50W | 9.94 × 10$^3$ | **9.89 × 10$^3$** |
| | | | 9.81 × 10$^3$ | |
| | | | 9.91 × 10$^3$ | |
| Me$_3$Tr$^+$ | 9.58 × 10$^{-3}$ | 90AN10W | 2.74 × 10$^3$ | **2.77 × 10$^3$** |
| | | | 2.79 × 10$^3$ | |
| | | 80AN20W | 2.98 × 10$^3$ | **3.01 × 10$^3$** |
| | | | 3.03 × 10$^3$ | |
| | | 60AN40W | 3.19 × 10$^3$ | **3.17 × 10$^3$** |
| | | | 3.15 × 10$^3$ | |
| | | 50AN50W | 2.80 × 10$^3$ | **2.83 × 10$^3$** |
| | | | 2.86 × 10$^3$ | |
| (MeO)Tr$^+$ | 1.00 × 10$^{-4}$ | 90AN10W | 1.14 × 10$^3$ | **1.17 × 10$^3$** |
| | | | 1.19 × 10$^3$ | |
| | | | 1.17 × 10$^3$ | |
| | | 80AN20W | 1.43 × 10$^3$ | **1.43 × 10$^3$** |
| | | | 1.43 × 10$^3$ | |
| | | | 1.45 × 10$^3$ | |
| | | 60AN40W | 1.78 × 10$^3$ | **1.75 × 10$^3$** |

Experimental Part

Table continued.

| system | $[Ar_3C^+AcO^-]_0$ (M) | solvent | $k_w$ (s$^{-1}$) | |
|---|---|---|---|---|
| (MeO)Tr$^+$ | $1.00 \times 10^{-4}$ | 60AN40W | $1.72 \times 10^3$ | |
| | | | $1.74 \times 10^3$ | |
| | | 50AN50W | $1.72 \times 10^3$ | $\mathbf{1.73 \times 10^3}$ |
| | | | $1.73 \times 10^3$ | |
| | | | $1.75 \times 10^3$ | |
| (MeO)$_2$Tr$^+$ [b] | $8.70 \times 10^{-6}$ | 90AN10W | $\mathbf{4.16 \times 10^1}$ | |
| | | 80AN20W | $\mathbf{5.61 \times 10^1}$ | |
| | | 60AN40W | $\mathbf{5.47 \times 10^1}$ | |
| | | 50AN50W | $\mathbf{5.81 \times 10^1}$ | |
| (MeO)$_3$Tr$^+$ [b] | $1.56 \times 10^{-5}$ | 90AN10W | **3.73** | |
| | | 80AN20W | **4.78** | |
| | | 60AN40W | **4.93** | |
| | | 50AN50W | **4.88** | |

[a] This value represents an average of 6 runs; [b] stopped-flow UV-vis spectroscopy.

(Me$_2$N)Tr$^+$BF$_4^-$ (initial concentration: $3.22 \times 10^{-5}$ M). ($n$-Bu)$_4$N$^+$AcO$^-$ or ($n$-Bu)$_4$N$^+$BzO$^-$ have been added; conventional UV-vis spectroscopy.

| | $k_w$ (s$^{-1}$) | | |
|---|---|---|---|
| solvent | 6.7 equiv. ($n$-Bu)$_4$N$^+$AcO$^-$ | 4.6 equiv. ($n$-Bu)$_4$N$^+$BzO$^-$ | |
| 90AN10W | $2.57 \times 10^{-3}$ | $2.57 \times 10^{-3}$ | $\mathbf{2.57 \times 10^{-3}}$ |
| 80AN20W | $3.43 \times 10^{-3}$ | $3.43 \times 10^{-3}$ | $\mathbf{3.43 \times 10^{-3}}$ |
| 60AN40W | $3.78 \times 10^{-3}$ | $3.76 \times 10^{-3}$ | $\mathbf{3.77 \times 10^{-3}}$ |
| 50AN50W | $3.76 \times 10^{-3}$ | $3.77 \times 10^{-3}$ | $\mathbf{3.77 \times 10^{-3}}$ |

(Me$_2$N)(MeO)Tr$^+$BF$_4^-$ (initial concentration: $5.67 \times 10^{-5}$ M). ($n$-Bu)$_4$N$^+$AcO$^-$ ($4.95 \times 10^{-4}$ M) has been added; conventional UV-vis spectroscopy.

| solvent | $k_w$ (s$^{-1}$) | |
|---|---|---|
| 90AN10W | $1.53 \times 10^{-3}$ | $\mathbf{1.53 \times 10^{-3}}$ |
| | $1.52 \times 10^{-3}$ | |
| 80AN20W | $1.97 \times 10^{-3}$ | $\mathbf{1.97 \times 10^{-3}}$ |
| | $1.96 \times 10^{-3}$ | |
| 60AN40W | $2.16 \times 10^{-3}$ | $\mathbf{2.16 \times 10^{-3}}$ |
| | $2.16 \times 10^{-3}$ | |
| 50AN50W | $2.14 \times 10^{-3}$ | $\mathbf{2.14 \times 10^{-3}}$ |
| | $2.13 \times 10^{-3}$ | |

# Experimental Part

## 4.5. Reactions of tritylium ions in aqueous acetonitrile in the presence of additives

First-order rate constants for the reactions of $(MeO)_3Tr^+BF_4^-$ with variable amounts of $(n\text{-}Bu)_4N^+AcO^-$ in 90AN10W, stopped-flow UV-vis photospectrometry, $\lambda = 484$ nm, 25 °C.

| $[(MeO)_3Tr^+]$ (mol L$^{-1}$) | $[AcO^-]$ (mol L$^{-1}$) | equiv. | $k_{obs}$ (s$^{-1}$) |
|---|---|---|---|
| $2.62 \times 10^{-5}$ | 0 | 0 | 4.31 |
| $3.31 \times 10^{-5}$ | $5.97 \times 10^{-4}$ | 18 | $1.00 \times 10^1$ |
| $3.31 \times 10^{-5}$ | $1.39 \times 10^{-3}$ | 42 | $1.64 \times 10^1$ |
| $3.31 \times 10^{-5}$ | $1.86 \times 10^{-3}$ | 56 | $2.03 \times 10^1$ |
| $3.31 \times 10^{-5}$ | $2.59 \times 10^{-3}$ | 78 | $2.63 \times 10^1$ |
| $3.31 \times 10^{-5}$ | $3.18 \times 10^{-3}$ | 96 | $3.02 \times 10^1$ |

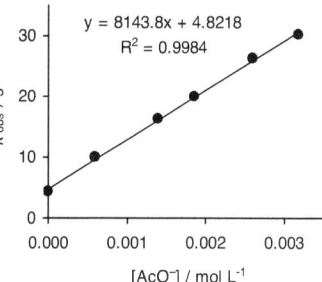

First-order rate constants for the reactions of $(MeO)_3Tr^+BF_4^-$ with variable amounts of $(n\text{-}Bu)_4N^+AcO^-$ in 50AN50W, stopped-flow UV-vis photospectrometry, $\lambda = 484$ nm, 25 °C.

| $[(MeO)_3Tr^+]$ (mol L$^{-1}$) | $[AcO^-]$ (mol L$^{-1}$) | equiv. | $k_{obs}$ (s$^{-1}$) |
|---|---|---|---|
| $2.71 \times 10^{-5}$ | $2.16 \times 10^{-3}$ | 80 | 7.23 |
| $2.71 \times 10^{-5}$ | $4.39 \times 10^{-3}$ | 162 | 7.32 |
| $2.71 \times 10^{-5}$ | $7.79 \times 10^{-3}$ | 288 | 7.85 |
| $2.71 \times 10^{-5}$ | $9.55 \times 10^{-3}$ | 352 | 8.06 |
| $2.71 \times 10^{-5}$ | $1.21 \times 10^{-2}$ | 447 | 8.50 |

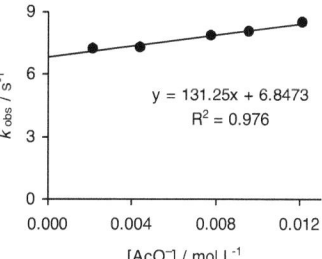

First-order rate constants for the reactions of $(MeO)_3Tr^+BF_4^-$ with variable amounts of $(n\text{-}Bu)_4N^+BzO^-$ in 90AN10W, stopped-flow UV-vis photospectrometry, $\lambda = 484$ nm, 25 °C.

| $[(MeO)_3Tr^+]$ (mol L$^{-1}$) | $[BzO^-]$ (mol L$^{-1}$) | equiv. | $k_{obs}$ (s$^{-1}$) |
|---|---|---|---|
| $2.62 \times 10^{-5}$ | 0 | 0 | 4.31 |
| $2.62 \times 10^{-5}$ | $6.60 \times 10^{-4}$ | 25 | $2.14 \times 10^1$ |
| $2.62 \times 10^{-5}$ | $1.32 \times 10^{-3}$ | 50 | $3.45 \times 10^1$ |
| $2.62 \times 10^{-5}$ | $1.82 \times 10^{-3}$ | 70 | $4.85 \times 10^1$ |
| $2.62 \times 10^{-5}$ | $2.31 \times 10^{-3}$ | 88 | $5.77 \times 10^1$ |

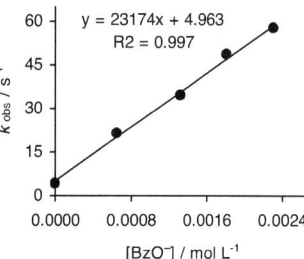

First-order rate constants for the reactions of $(MeO)_3Tr^+BF_4^-$ with variable amounts of DABCO in 90AN10W, stopped-flow UV-vis photospectrometry, $\lambda = 484$ nm, 25 °C.

| $[(MeO)_3Tr^+]$ (mol $L^{-1}$) | [DABCO] (mol $L^{-1}$) | equiv. | $k_{obs}$ (s$^{-1}$) |
|---|---|---|---|
| $2.71 \times 10^{-5}$ | $6.15 \times 10^{-3}$ | 227 | 7.22 |
| $2.71 \times 10^{-5}$ | $1.03 \times 10^{-2}$ | 380 | 8.61 |
| $2.71 \times 10^{-5}$ | $2.02 \times 10^{-2}$ | 745 | $1.50 \times 10^1$ |
| $2.71 \times 10^{-5}$ | $2.87 \times 10^{-2}$ | $1.06 \times 10^3$ | $1.81 \times 10^1$ |
| $2.71 \times 10^{-5}$ | $4.43 \times 10^{-2}$ | $1.64 \times 10^3$ | $2.52 \times 10^1$ |

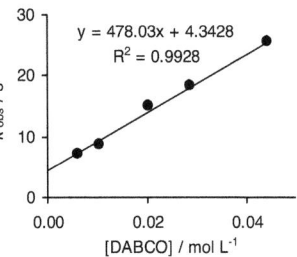

First-order rate constants for the reactions of $(MeO)_3Tr^+BF_4^-$ with variable amounts of DABCO in 50AN50W, stopped-flow UV-vis photospectrometry, $\lambda = 484$ nm, 25 °C.

| $[(MeO)_3Tr^+]$ (mol $L^{-1}$) | [DABCO] (mol $L^{-1}$) | equiv. | $k_{obs}$ (s$^{-1}$) |
|---|---|---|---|
| $2.71 \times 10^{-5}$ | $5.88 \times 10^{-3}$ | 217 | 7.89 |
| $2.71 \times 10^{-5}$ | $8.96 \times 10^{-3}$ | 331 | 8.26 |
| $2.71 \times 10^{-5}$ | $1.81 \times 10^{-2}$ | 668 | 9.10 |
| $2.71 \times 10^{-5}$ | $3.07 \times 10^{-2}$ | $1.13 \times 10^3$ | $1.02 \times 10^1$ |
| $2.71 \times 10^{-5}$ | $4.27 \times 10^{-2}$ | $1.59 \times 10^3$ | $1.16 \times 10^1$ |

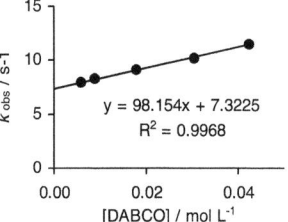

First-order rate constants for the reactions of $(MeO)_3Tr^+BF_4^-$ with variable amounts of ($n$-Bu)$_4$N$^+$OH$^-$ in 90AN10W, stopped-flow UV-vis photospectrometry, $\lambda = 484$ nm, 25 °C.

| $[(MeO)_3Tr^+]$ (mol $L^{-1}$) | [OH$^-$] (mol $L^{-1}$) | equiv. | $k_{obs}$ (s$^{-1}$) |
|---|---|---|---|
| $1.33 \times 10^{-5}$ | $2.01 \times 10^{-4}$ | 15 | $4.81 \times 10^1$ |
| $1.33 \times 10^{-5}$ | $4.03 \times 10^{-4}$ | 30 | $1.04 \times 10^2$ |
| $1.33 \times 10^{-5}$ | $6.04 \times 10^{-4}$ | 45 | $1.58 \times 10^2$ |
| $1.33 \times 10^{-5}$ | $8.05 \times 10^{-4}$ | 61 | $2.19 \times 10^2$ |
| $1.33 \times 10^{-5}$ | $1.21 \times 10^{-3}$ | 91 | $3.10 \times 10^2$ |

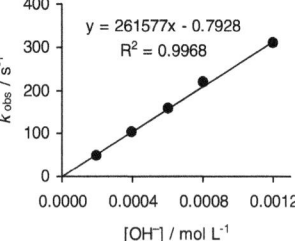

Experimental Part 219

First-order rate constants for the reactions of $(Me_2N)Tr^+BF_4^-$ with variable amounts of $(n\text{-}Bu)_4N^+AcO^-$ in 90AN10W, conventional UV-vis photospectrometry, $\lambda = 461$ nm, 25 °C.

| $[(Me_2N)Tr^+]$ (mol L$^{-1}$) | $[AcO^-]$ (mol L$^{-1}$) | equiv. | $k_{obs}$ (s$^{-1}$) |
|---|---|---|---|
| $5.06 \times 10^{-5}$ | $7.38 \times 10^{-4}$ | 15 | $3.69 \times 10^{-3}$ |
| $5.06 \times 10^{-5}$ | $1.48 \times 10^{-3}$ | 29 | $3.85 \times 10^{-3}$ |
| $5.06 \times 10^{-5}$ | $2.21 \times 10^{-3}$ | 44 | $3.99 \times 10^{-3}$ |
| $5.06 \times 10^{-5}$ | $2.95 \times 10^{-3}$ | 58 | $4.12 \times 10^{-3}$ |

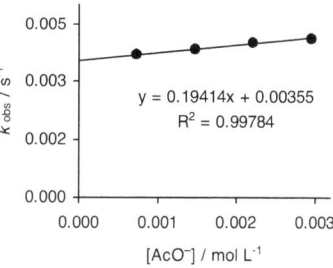

First-order rate constants for the reactions of $(Me_2N)Tr^+BF_4^-$ with variable amounts of $(n\text{-}Bu)_4N^+AcO^-$ in 50AN50W, conventional UV-vis photospectrometry, $\lambda = 461$ nm, 20 °C.

| $[(Me_2N)Tr^+]$ (mol L$^{-1}$) | $[AcO^-]$ (mol L$^{-1}$) | equiv. | $k_{obs}$ (s$^{-1}$) |
|---|---|---|---|
| $4.64 \times 10^{-5}$ | $3.54 \times 10^{-4}$ | 8 | $3.67 \times 10^{-3}$ |
| $4.64 \times 10^{-5}$ | $7.06 \times 10^{-4}$ | 15 | $3.67 \times 10^{-3}$ |
| $4.64 \times 10^{-5}$ | $1.42 \times 10^{-3}$ | 31 | $3.69 \times 10^{-3}$ |
| $4.64 \times 10^{-5}$ | $2.12 \times 10^{-3}$ | 46 | $3.70 \times 10^{-3}$ |
| $4.64 \times 10^{-5}$ | $2.83 \times 10^{-3}$ | 61 | $3.68 \times 10^{-3}$ |
| $4.64 \times 10^{-5}$ | $3.54 \times 10^{-3}$ | 76 | $3.75 \times 10^{-3}$ |

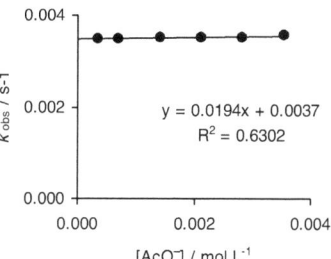

First-order rate constants for the reactions of $(Me_2N)Tr^+BF_4^-$ with variable amounts of $(n\text{-}Bu)_4N^+BzO^-$ in 90AN10W, conventional UV-vis photospectrometry, $\lambda = 461$ nm, 25 °C.

| $[(Me_2N)Tr^+]$ (mol L$^{-1}$) | $[AcO^-]$ (mol L$^{-1}$) | equiv. | $k_{obs}$ (s$^{-1}$) |
|---|---|---|---|
| $5.06 \times 10^{-5}$ | $4.18 \times 10^{-4}$ | 8 | $3.64 \times 10^{-3}$ |
| $5.06 \times 10^{-5}$ | $8.36 \times 10^{-4}$ | 17 | $3.79 \times 10^{-3}$ |
| $5.06 \times 10^{-5}$ | $1.25 \times 10^{-3}$ | 25 | $3.88 \times 10^{-3}$ |
| $5.06 \times 10^{-5}$ | $1.67 \times 10^{-3}$ | 33 | $4.04 \times 10^{-3}$ |
| $5.06 \times 10^{-5}$ | $2.09 \times 10^{-3}$ | 41 | $4.25 \times 10^{-3}$ |

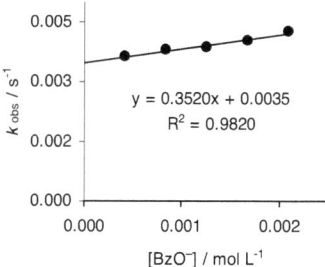

First-order rate constants for the reaction of $(Me_2N)Tr^+BF_4^-$ with variable amounts of DABCO in 90AN10W, conventional UV-vis photospectrometry, $\lambda = 461$ nm, 25 °C.

| $[(Me_2N)Tr^+]$ (mol L$^{-1}$) | [DABCO] (mol L$^{-1}$) | equiv. | $k_{obs}$ (s$^{-1}$) |
|---|---|---|---|
| 5.54 × 10$^{-5}$ | 6.27 × 10$^{-3}$ | 96 | 4.48 × 10$^{-3}$ |
| 5.54 × 10$^{-5}$ | 1.25 × 10$^{-2}$ | 191 | 5.21 × 10$^{-3}$ |
| 5.54 × 10$^{-5}$ | 1.88 × 10$^{-2}$ | 288 | 6.02 × 10$^{-3}$ |
| 5.54 × 10$^{-5}$ | 3.67 × 10$^{-2}$ | 561 | 8.44 × 10$^{-3}$ |
| 5.54 × 10$^{-5}$ | 4.89 × 10$^{-2}$ | 748 | 1.02 × 10$^{-2}$ |

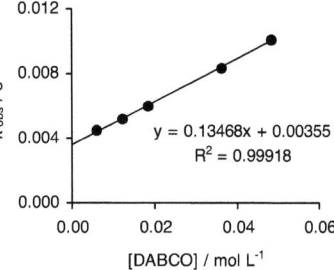

### 4.6. Reactions of tritylium ions with hydride donors

Reaction of $(mF)_2(mF)'_2Tr^+$ with 3-propylcyclopentene (conventional UV-vis spectrometry, 422 nm, CH$_2$Cl$_2$, 20 °C). The carbocation was generated in solution by mixing $(mF)_2(mF)'_2$TrCl with GaCl$_3$.

| [E] / mol L$^{-1}$ | [GaCl$_3$]/[E] | [Nu] / mol L$^{-1}$ | [Nu]/[E] | $k_{obs}$ / s$^{-1}$ |
|---|---|---|---|---|
| 9.20 × 10$^{-5}$ | 49 | 4.09 × 10$^{-3}$ | 45 | 2.38 × 10$^{-2}$ |
| 6.59 × 10$^{-5}$ | 81 | 4.54 × 10$^{-3}$ | 69 | 2.65 × 10$^{-2}$ |
| 6.77 × 10$^{-5}$ | 81 | 5.33 × 10$^{-3}$ | 79 | 3.15 × 10$^{-2}$ |
| 6.58 × 10$^{-5}$ | 81 | 5.84 × 10$^{-3}$ | 89 | 3.31 × 10$^{-2}$ |
| 6.59 × 10$^{-5}$ | 81 | 6.31 × 10$^{-3}$ | 96 | 3.54 × 10$^{-2}$ |

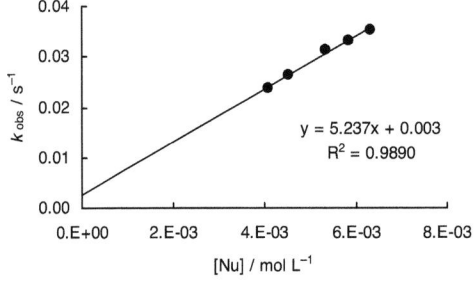

$k = 5.24 \text{ L mol}^{-1} \text{ s}^{-1}$

Experimental Part 221

Reaction of $(mF)_2(mF)'(mF)''Tr^+$ with $HSiPh_3$ (stopped-flow UV-vis spectrometry, 418 nm, $CH_2Cl_2$, 20 °C). The carbocation was generated in solution by mixing $(mF)_2(mF)'(mF)''TrBr$ with $GaCl_3$.

| [E] / mol L$^{-1}$ | [GaCl$_3$]/[E] | [Nu] / mol L$^{-1}$ | [Nu]/[E] | $k_{obs}$ / s$^{-1}$ |
|---|---|---|---|---|
| 6.07 × 10$^{-5}$ | 97 | 2.86 × 10$^{-3}$ | 47 | 1.71 × 10$^1$ |
| 6.07 × 10$^{-5}$ | 97 | 3.81 × 10$^{-3}$ | 63 | 2.32 × 10$^1$ |
| 6.07 × 10$^{-5}$ | 97 | 4.76 × 10$^{-3}$ | 78 | 2.87 × 10$^1$ |

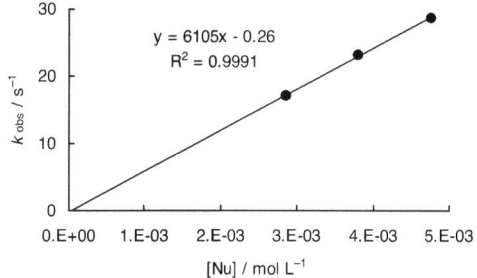

$k = 6.11 \times 10^3$ L mol$^{-1}$ s$^{-1}$

Reaction of $(mF)_2(mF)'(mF)''Tr^+$ with $HSiEt_3$ (stopped-flow UV-vis spectrometry, 418 nm, $CH_2Cl_2$, 20 °C). The carbocation was generated in solution by mixing $(mF)_2(mF)'(mF)''TrBr$ with $GaCl_3$.

| [E] / mol L$^{-1}$ | [GaCl$_3$]/[E] | [Nu] / mol L$^{-1}$ | [Nu]/[E] | $k_{obs}$ / s$^{-1}$ |
|---|---|---|---|---|
| 6.07 × 10$^{-5}$ | 97 | 5.16 × 10$^{-4}$ | 9 | 1.97 |
| 6.07 × 10$^{-5}$ | 97 | 1.03 × 10$^{-3}$ | 17 | 4.18 |
| 6.07 × 10$^{-5}$ | 97 | 1.55 × 10$^{-3}$ | 26 | 6.74 |
| 6.07 × 10$^{-5}$ | 97 | 2.06 × 10$^{-3}$ | 34 | 8.97 |
| 6.07 × 10$^{-5}$ | 97 | 2.58 × 10$^{-3}$ | 43 | 1.15 × 10$^1$ |

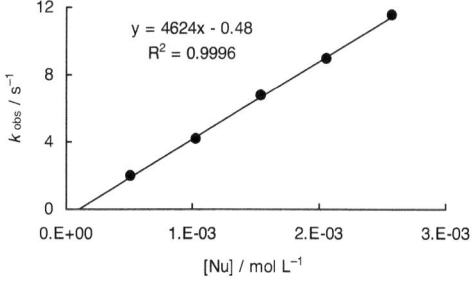

$k = 4.62 \times 10^3$ L mol$^{-1}$ s$^{-1}$

# Experimental Part

Reaction of $(mF)_2(mF)'(mF)''Tr^+$ with cycloheptatriene (stopped-flow UV-vis spectrometry, 418 nm, $CH_2Cl_2$, 20 °C). The carbocation was generated in solution by mixing $(mF)_2(mF)'(mF)''TrBr$ with $GaCl_3$.

| [E] / mol L$^{-1}$ | [GaCl$_3$]/[E] | [Nu] / mol L$^{-1}$ | [Nu]/[E] | $k_{obs}$ / s$^{-1}$ |
|---|---|---|---|---|
| 5.67 × 10$^{-5}$ | 118 | 1.30 × 10$^{-2}$ | 229 | 1.88 × 10$^{1}$ |
| 5.67 × 10$^{-5}$ | 118 | 2.60 × 10$^{-2}$ | 459 | 3.87 × 10$^{1}$ |
| 5.67 × 10$^{-5}$ | 118 | 3.91 × 10$^{-2}$ | 690 | 5.48 × 10$^{1}$ |
| 5.67 × 10$^{-5}$ | 118 | 5.21 × 10$^{-2}$ | 919 | 7.74 × 10$^{1}$ |
| 5.67 × 10$^{-5}$ | 118 | 6.51 × 10$^{-2}$ | 1148 | 9.09 × 10$^{1}$ |

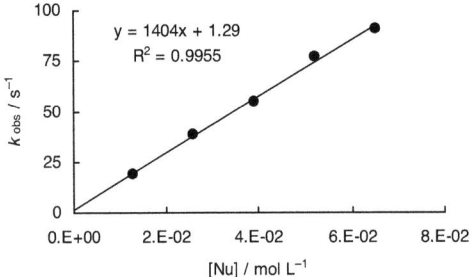

$k = 1.40 \times 10^3$ L mol$^{-1}$ s$^{-1}$

Reaction of $(mF)(mF)'(mF)''Tr^+$ with $HSiPh_3$ (stopped-flow UV-vis spectrometry, 420 nm, $CH_2Cl_2$, 20 °C). The carbocation was generated in solution by mixing $(mF)(mF)'(mF)''TrBr$ with $GaCl_3$.

| [E] / mol L$^{-1}$ | [GaCl$_3$]/[E] | [Nu] / mol L$^{-1}$ | [Nu]/[E] | $k_{obs}$ / s$^{-1}$ |
|---|---|---|---|---|
| 5.57 × 10$^{-5}$ | 53 | 1.63 × 10$^{-3}$ | 29 | 3.80 |
| 5.57 × 10$^{-5}$ | 53 | 5.03 × 10$^{-3}$ | 90 | 1.09 × 10$^{1}$ |
| 5.57 × 10$^{-5}$ | 53 | 9.39 × 10$^{-3}$ | 169 | 2.06 × 10$^{1}$ |
| 5.57 × 10$^{-5}$ | 53 | 1.33 × 10$^{-2}$ | 239 | 2.86 × 10$^{1}$ |
| 5.57 × 10$^{-5}$ | 53 | 1.80 × 10$^{-2}$ | 323 | 3.83 × 10$^{1}$ |

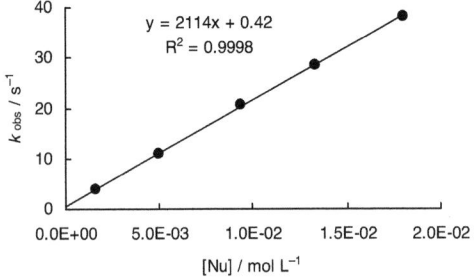

$k = 2.11 \times 10^3$ L mol$^{-1}$ s$^{-1}$

Experimental Part 223

Reaction of $(mF)(mF)'(mF)''Tr^+$ with HSiEt$_3$ (stopped-flow UV-vis spectrometry, 420 nm, CH$_2$Cl$_2$, 20 °C). The carbocation was generated in solution by mixing $(mF)(mF)'(mF)''$TrBr with GaCl$_3$.

| [E] / mol L$^{-1}$ | [GaCl$_3$]/[E] | [Nu] / mol L$^{-1}$ | [Nu]/[E] | $k_{obs}$ / s$^{-1}$ |
| --- | --- | --- | --- | --- |
| 5.57 × 10$^{-5}$ | 53 | 7.27 × 10$^{-3}$ | 130 | 1.51 × 10$^{1}$ |
| 5.57 × 10$^{-5}$ | 53 | 1.13 × 10$^{-2}$ | 202 | 2.32 × 10$^{1}$ |
| 5.57 × 10$^{-5}$ | 53 | 1.58 × 10$^{-2}$ | 284 | 3.38 × 10$^{1}$ |
| 5.57 × 10$^{-5}$ | 53 | 2.34 × 10$^{-2}$ | 421 | 4.81 × 10$^{1}$ |
| 5.57 × 10$^{-5}$ | 53 | 3.14 × 10$^{-2}$ | 564 | 6.48 × 10$^{1}$ |

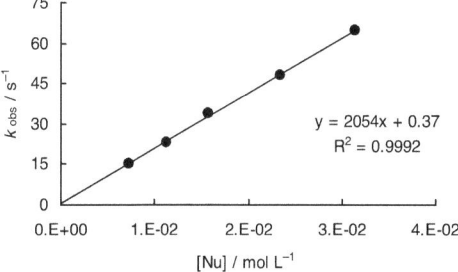

$k = 2.05 \times 10^3$ L mol$^{-1}$ s$^{-1}$

Reaction of $(mF)(mF)'(mF)''Tr^+$ with cycloheptatriene (stopped-flow UV-vis spectrometry, 420 nm, CH$_2$Cl$_2$, 20 °C). The carbocation was generated in solution by mixing $(mF)(mF)'(mF)''$TrBr with GaCl$_3$.

| [E] / mol L$^{-1}$ | [GaCl$_3$]/[E] | [Nu] / mol L$^{-1}$ | [Nu]/[E] | $k_{obs}$ / s$^{-1}$ |
| --- | --- | --- | --- | --- |
| 5.57 × 10$^{-5}$ | 53 | 1.01 × 10$^{-2}$ | 182 | 3.06 |
| 5.57 × 10$^{-5}$ | 53 | 1.61 × 10$^{-2}$ | 289 | 4.96 |
| 5.57 × 10$^{-5}$ | 53 | 2.47 × 10$^{-2}$ | 443 | 7.49 |
| 5.57 × 10$^{-5}$ | 53 | 3.37 × 10$^{-2}$ | 605 | 1.05 × 10$^{1}$ |
| 5.57 × 10$^{-5}$ | 53 | 5.67 × 10$^{-2}$ | 1.02 × 10$^{3}$ | 1.79 × 10$^{1}$ |

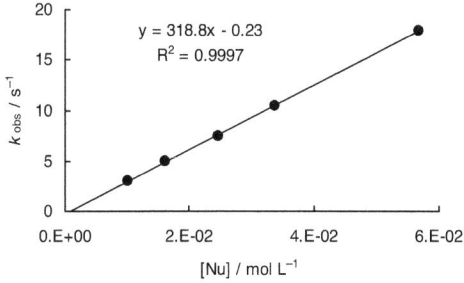

$k = 3.19 \times 10^2$ L mol$^{-1}$ s$^{-1}$

Reaction of (mF)(mF)′Tr⁺ with HSiPh₃ (stopped-flow UV-vis spectrometry, 415 nm, CH₂Cl₂, 20 °C). The carbocation was generated in solution by mixing (mF)(mF)′TrBr with GaCl₃.

| [E] / mol L$^{-1}$ | [GaCl$_3$]/[E] | [Nu] / mol L$^{-1}$ | [Nu]/[E] | $k_{obs}$ / s$^{-1}$ |
|---|---|---|---|---|
| 4.76 × 10$^{-5}$ | 3 | 2.56 × 10$^{-3}$ | 54 | 1.71 |
| 4.76 × 10$^{-5}$ | 3 | 2.98 × 10$^{-3}$ | 63 | 2.12 |
| 4.76 × 10$^{-5}$ | 3 | 3.41 × 10$^{-3}$ | 72 | 2.33 |
| 4.76 × 10$^{-5}$ | 3 | 3.84 × 10$^{-3}$ | 81 | 2.56 |
| 4.76 × 10$^{-5}$ | 3 | 4.26 × 10$^{-3}$ | 89 | 2.89 |

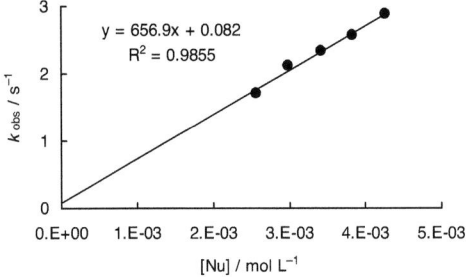

$k = 6.57 \times 10^2$ L mol$^{-1}$ s$^{-1}$

Reaction of (mF)(mF)′Tr⁺ with HSiPh₃ (stopped-flow UV-vis spectrometry, 415 nm, CH₂Cl₂, 20 °C). The carbocation was generated in solution by mixing (mF)(mF)′TrBr with GaCl₃.

| [E] / mol L$^{-1}$ | [GaCl$_3$]/[E] | [Nu] / mol L$^{-1}$ | [Nu]/[E] | $k_{obs}$ / s$^{-1}$ |
|---|---|---|---|---|
| 7.24 × 10$^{-5}$ | 110 | 1.67 × 10$^{-3}$ | 23 | 1.03 |
| 7.24 × 10$^{-5}$ | 110 | 3.35 × 10$^{-3}$ | 46 | 2.05 |
| 7.24 × 10$^{-5}$ | 110 | 5.02 × 10$^{-3}$ | 69 | 3.14 |
| 7.24 × 10$^{-5}$ | 110 | 6.70 × 10$^{-3}$ | 93 | 4.13 |
| 7.24 × 10$^{-5}$ | 110 | 1.13 × 10$^{-2}$ | 156 | 7.16 |

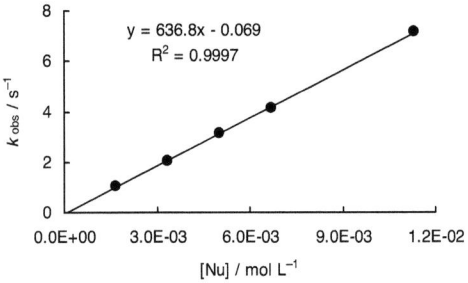

$k = 6.37 \times 10^2$ L mol$^{-1}$ s$^{-1}$

Experimental Part 225

Reaction of $(m\text{F})(m\text{F})'\text{Tr}^+$ with $\text{HSiEt}_3$ (stopped-flow UV-vis spectrometry, 415 nm, $\text{CH}_2\text{Cl}_2$, 20 °C). The carbocation was generated in solution by mixing $(m\text{F})(m\text{F})'\text{TrBr}$ with $\text{GaCl}_3$.

| [E] / mol L$^{-1}$ | [GaCl$_3$]/[E] | [Nu] / mol L$^{-1}$ | [Nu]/[E] | $k_{obs}$ / s$^{-1}$ |
|---|---|---|---|---|
| 7.24 × 10$^{-5}$ | 110 | 4.08 × 10$^{-3}$ | 56 | 2.84 |
| 7.24 × 10$^{-5}$ | 110 | 9.20 × 10$^{-3}$ | 127 | 7.41 |
| 7.24 × 10$^{-5}$ | 110 | 1.20 × 10$^{-2}$ | 166 | 9.61 |
| 7.24 × 10$^{-5}$ | 110 | 1.91 × 10$^{-2}$ | 264 | 1.48 × 10$^{1}$ |
| 7.24 × 10$^{-5}$ | 110 | 2.66 × 10$^{-2}$ | 367 | 2.09 × 10$^{1}$ |

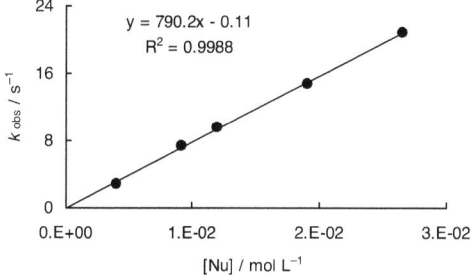

$k = 7.90 \times 10^2 \text{ L mol}^{-1} \text{ s}^{-1}$

Reaction of $(m\text{F})(m\text{F})'\text{Tr}^+$ with $\text{HSiBu}_3$ (stopped-flow UV-vis spectrometry, 415 nm, $\text{CH}_2\text{Cl}_2$, 20 °C). The carbocation was generated in solution by mixing $(m\text{F})(m\text{F})'\text{TrBr}$ with $\text{GaCl}_3$.

| [E] / mol L$^{-1}$ | [GaCl$_3$]/[E] | [Nu] / mol L$^{-1}$ | [Nu]/[E] | $k_{obs}$ / s$^{-1}$ |
|---|---|---|---|---|
| 8.69 × 10$^{-5}$ | 95 | 1.32 × 10$^{-3}$ | 15 | 1.86 |
| 8.69 × 10$^{-5}$ | 95 | 2.65 × 10$^{-3}$ | 30 | 3.94 |
| 8.69 × 10$^{-5}$ | 95 | 3.97 × 10$^{-3}$ | 46 | 5.73 |
| 8.69 × 10$^{-5}$ | 95 | 5.29 × 10$^{-3}$ | 61 | 7.79 |
| 8.69 × 10$^{-5}$ | 95 | 6.91 × 10$^{-3}$ | 80 | 1.01 × 10$^{1}$ |

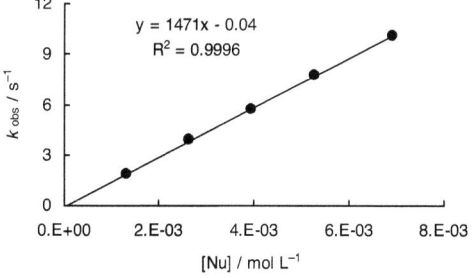

$k = 1.47 \times 10^3 \text{ L mol}^{-1} \text{ s}^{-1}$

Reaction of (mF)(mF)'Tr⁺ with cycloheptatriene (conventional UV-vis spectrometry, 415 nm, $CH_2Cl_2$, 20 °C). The carbocation was generated in solution by mixing (mF)(mF)'TrBr with $SnCl_4$.

| [E] / mol L⁻¹ | [GaCl₃]/[E] | [Nu] / mol L⁻¹ | [Nu]/[E] | $k_{obs}$ / s⁻¹ |
|---|---|---|---|---|
| $6.97 \times 10^{-5}$ | 488 | $5.86 \times 10^{-4}$ | 8 | $2.50 \times 10^{-2}$ |
| $6.65 \times 10^{-5}$ | 488 | $1.12 \times 10^{-3}$ | 17 | $4.38 \times 10^{-2}$ |
| $6.90 \times 10^{-5}$ | 488 | $1.74 \times 10^{-3}$ | 25 | $6.81 \times 10^{-2}$ |
| $6.85 \times 10^{-5}$ | 488 | $2.30 \times 10^{-3}$ | 34 | $9.07 \times 10^{-2}$ |
| $7.05 \times 10^{-5}$ | 488 | $3.56 \times 10^{-3}$ | 50 | $1.44 \times 10^{-1}$ |

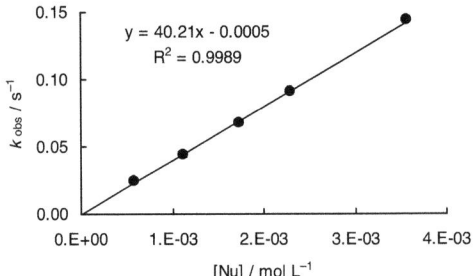

$k = 4.02 \times 10^1 \text{ L mol}^{-1} \text{ s}^{-1}$

Reaction of (mF)(mF)'Tr⁺ with Bu₄Sn (conventional UV-vis spectrometry, 415 nm, $CH_2Cl_2$, 20 °C). The carbocation was generated in solution by mixing (mF)(mF)'TrBr with $GaCl_3$.

| [E] / mol L⁻¹ | [GaCl₃]/[E] | [Nu] / mol L⁻¹ | [Nu]/[E] | $k_{obs}$ / s⁻¹ |
|---|---|---|---|---|
| $7.62 \times 10^{-5}$ | 16 | $2.42 \times 10^{-3}$ | 32 | $6.09 \times 10^{-3}$ |
| $7.48 \times 10^{-5}$ | 16 | $4.76 \times 10^{-3}$ | 64 | $1.50 \times 10^{-2}$ |
| $7.47 \times 10^{-5}$ | 16 | $7.12 \times 10^{-3}$ | 95 | $2.55 \times 10^{-2}$ |
| $7.54 \times 10^{-5}$ | 16 | $9.59 \times 10^{-3}$ | 127 | $3.38 \times 10^{-2}$ |
| $7.51 \times 10^{-5}$ | 16 | $1.19 \times 10^{-2}$ | 159 | $4.19 \times 10^{-2}$ |

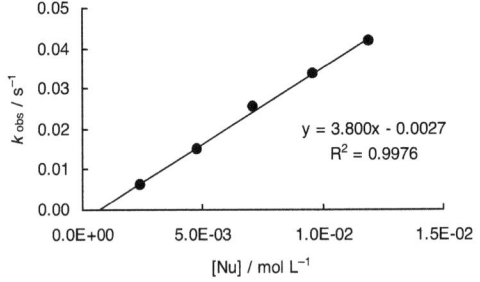

$k = 3.80 \text{ L mol}^{-1} \text{ s}^{-1}$

Reaction of $(mF)_2Tr^+$ with $Bu_4Sn$ (conventional UV-vis spectrometry, 412 nm, $CH_2Cl_2$, 20 °C). The carbocation was generated in solution by mixing $(mF)_2TrCl$ with $GaCl_3$.

| [E] / mol L$^{-1}$ | [GaCl$_3$]/[E] | [Nu] / mol L$^{-1}$ | [Nu]/[E] | $k_{obs}$ / s$^{-1}$ |
|---|---|---|---|---|
| 3.72 × 10$^{-5}$ | 34 | 1.49 × 10$^{-3}$ | 40 | 1.83 × 10$^{-3}$ |
| 3.79 × 10$^{-5}$ | 34 | 3.04 × 10$^{-3}$ | 80 | 6.49 × 10$^{-3}$ |
| 3.64 × 10$^{-5}$ | 34 | 5.85 × 10$^{-3}$ | 161 | 1.59 × 10$^{-2}$ |
| 3.65 × 10$^{-5}$ | 34 | 8.80 × 10$^{-3}$ | 241 | 2.40 × 10$^{-2}$ |
| 3.56 × 10$^{-5}$ | 34 | 1.14 × 10$^{-2}$ | 321 | 3.26 × 10$^{-2}$ |

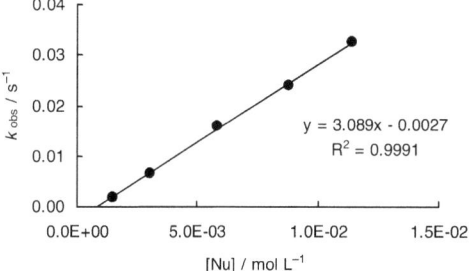

$k = 3.09$ L mol$^{-1}$ s$^{-1}$

Reaction of $(mF)Tr^+$ with $HSiPh_3$ (stopped-flow UV-vis spectrometry, 415 nm, $CH_2Cl_2$, 20 °C). The carbocation was generated in solution by mixing $(mF)TrBr$ with $GaCl_3$.

| [E] / mol L$^{-1}$ | [GaCl$_3$]/[E] | [Nu] / mol L$^{-1}$ | [Nu]/[E] | $k_{obs}$ / s$^{-1}$ |
|---|---|---|---|---|
| 3.52 × 10$^{-5}$ | 4 | 8.53 × 10$^{-4}$ | 24 | 1.95 × 10$^{-1}$ |
| 3.52 × 10$^{-5}$ | 4 | 1.28 × 10$^{-3}$ | 36 | 2.65 × 10$^{-1}$ |
| 3.52 × 10$^{-5}$ | 4 | 1.71 × 10$^{-3}$ | 49 | 3.62 × 10$^{-1}$ |
| 3.52 × 10$^{-5}$ | 4 | 2.13 × 10$^{-3}$ | 61 | 4.35 × 10$^{-1}$ |
| 3.52 × 10$^{-5}$ | 4 | 2.56 × 10$^{-3}$ | 73 | 5.34 × 10$^{-1}$ |

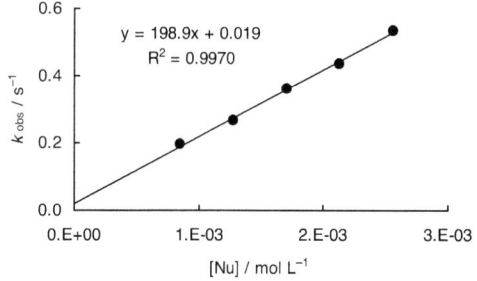

$k = 1.99 \times 10^2$ L mol$^{-1}$ s$^{-1}$

Reaction of $(mF)Tr^+$ with $HSiPh_3$ (stopped-flow UV-vis spectrometry, 415 nm, $CH_2Cl_2$, 20 °C). The carbocation was generated in solution by mixing $(mF)TrBr$ with $GaCl_3$.

| [E] / mol L$^{-1}$ | [GaCl$_3$]/[E] | [Nu] / mol L$^{-1}$ | [Nu]/[E] | $k_{obs}$ / s$^{-1}$ |
|---|---|---|---|---|
| 8.85 × 10$^{-5}$ | 99 | 1.67 × 10$^{-3}$ | 19 | 3.11 × 10$^{-1}$ |
| 8.85 × 10$^{-5}$ | 99 | 3.35 × 10$^{-3}$ | 38 | 6.24 × 10$^{-1}$ |
| 8.85 × 10$^{-5}$ | 99 | 5.02 × 10$^{-3}$ | 57 | 9.20 × 10$^{-1}$ |
| 8.85 × 10$^{-5}$ | 99 | 6.70 × 10$^{-3}$ | 76 | 1.24 |
| 8.85 × 10$^{-5}$ | 99 | 8.37 × 10$^{-3}$ | 95 | 1.59 |

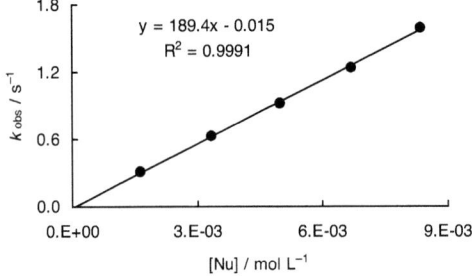

$k = 1.89 \times 10^2$ L mol$^{-1}$ s$^{-1}$

Reaction of $(mF)Tr^+$ with $HSiEt_3$ (stopped-flow UV-vis spectrometry, 415 nm, $CH_2Cl_2$, 20 °C). The carbocation was generated in solution by mixing $(mF)TrBr$ with $GaCl_3$.

| [E] / mol L$^{-1}$ | [GaCl$_3$]/[E] | [Nu] / mol L$^{-1}$ | [Nu]/[E] | $k_{obs}$ / s$^{-1}$ |
|---|---|---|---|---|
| 3.52 × 10$^{-5}$ | 4 | 1.51 × 10$^{-3}$ | 43 | 4.85 × 10$^{-1}$ |
| 3.52 × 10$^{-5}$ | 4 | 2.01 × 10$^{-3}$ | 57 | 6.54 × 10$^{-1}$ |
| 3.52 × 10$^{-5}$ | 4 | 2.52 × 10$^{-3}$ | 72 | 8.16 × 10$^{-1}$ |
| 3.52 × 10$^{-5}$ | 4 | 3.02 × 10$^{-3}$ | 86 | 1.00 |
| 3.52 × 10$^{-5}$ | 4 | 3.52 × 10$^{-3}$ | 100 | 1.14 |

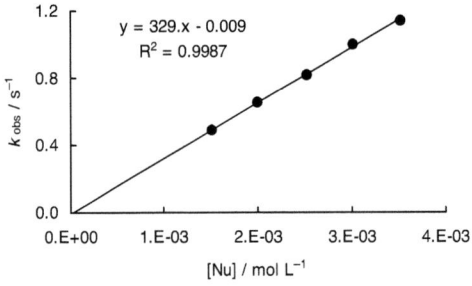

$k = 3.29 \times 10^2$ L mol$^{-1}$ s$^{-1}$

# Experimental Part

Reaction of $(mF)Tr^+$ with $HSiEt_3$ (stopped-flow UV-vis spectrometry, 415 nm, $CH_2Cl_2$, 20 °C). The carbocation was generated in solution by mixing $(mF)TrCl$ with $GaCl_3$.

| [E] / mol L$^{-1}$ | [GaCl$_3$]/[E] | [Nu] / mol L$^{-1}$ | [Nu]/[E] | $k_{obs}$ / s$^{-1}$ |
|---|---|---|---|---|
| 7.28 × 10$^{-5}$ | 189 | 2.52 × 10$^{-3}$ | 35 | 8.26 × 10$^{-1}$ |
| 7.28 × 10$^{-5}$ | 189 | 5.04 × 10$^{-3}$ | 69 | 1.66 |
| 7.28 × 10$^{-5}$ | 189 | 7.55 × 10$^{-3}$ | 104 | 2.48 |
| 7.28 × 10$^{-5}$ | 189 | 1.01 × 10$^{-2}$ | 139 | 3.39 |
| 7.28 × 10$^{-5}$ | 189 | 2.47 × 10$^{-2}$ | 339 | 8.51 |

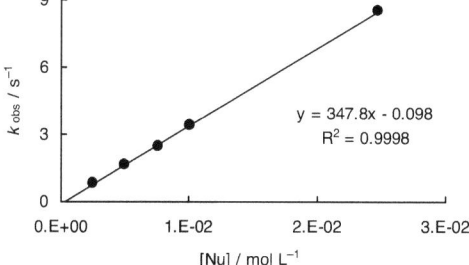

$k = 3.48 \times 10^2$ L mol$^{-1}$ s$^{-1}$

Reaction of $(mF)Tr^+$ with $HSiBu_3$ (stopped-flow UV-vis spectrometry, 415 nm, $CH_2Cl_2$, 20 °C). The carbocation was generated in solution by mixing $(mF)TrBr$ with $GaCl_3$.

| [E] / mol L$^{-1}$ | [GaCl$_3$]/[E] | [Nu] / mol L$^{-1}$ | [Nu]/[E] | $k_{obs}$ / s$^{-1}$ |
|---|---|---|---|---|
| 8.97 × 10$^{-5}$ | 67 | 1.32 × 10$^{-3}$ | 15 | 7.88 × 10$^{-1}$ |
| 8.97 × 10$^{-5}$ | 67 | 2.65 × 10$^{-3}$ | 30 | 1.65 |
| 8.97 × 10$^{-5}$ | 67 | 3.97 × 10$^{-3}$ | 44 | 2.49 |
| 8.97 × 10$^{-5}$ | 67 | 5.29 × 10$^{-3}$ | 59 | 3.39 |
| 8.97 × 10$^{-5}$ | 67 | 6.91 × 10$^{-3}$ | 77 | 4.35 |

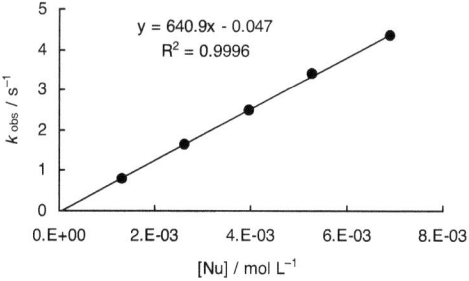

$k = 6.41 \times 10^2$ L mol$^{-1}$ s$^{-1}$

Reaction of (mF)Tr⁺ with cycloheptatriene (conventional UV-vis spectrometry, 415 nm, CH₂Cl₂, 20 °C). The carbocation was generated in solution by mixing (mF)TrCl with GaCl₃.

| [E] / mol L⁻¹ | [GaCl₃]/[E] | [Nu] / mol L⁻¹ | [Nu]/[E] | $k_{obs}$ / s⁻¹ |
|---|---|---|---|---|
| $5.90 \times 10^{-5}$ | 72 | $8.82 \times 10^{-4}$ | 15 | $1.02 \times 10^{-2}$ |
| $5.95 \times 10^{-5}$ | 54 | $1.78 \times 10^{-3}$ | 30 | $1.89 \times 10^{-2}$ |
| $5.73 \times 10^{-5}$ | 54 | $2.57 \times 10^{-3}$ | 45 | $2.68 \times 10^{-2}$ |
| $5.93 \times 10^{-5}$ | 54 | $3.54 \times 10^{-3}$ | 60 | $3.62 \times 10^{-2}$ |
| $5.91 \times 10^{-5}$ | 54 | $4.42 \times 10^{-3}$ | 75 | $4.36 \times 10^{-2}$ |

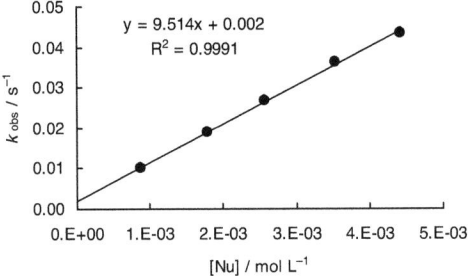

$k = 9.51$ L mol⁻¹ s⁻¹

Reaction of (mF)Tr⁺ with cycloheptatriene (conventional UV-vis spectrometry, 415 nm, CH₂Cl₂, 20 °C). The carbocation was generated in solution by mixing (mF)TrCl with SnCl₄.

| [E] / mol L⁻¹ | [GaCl₃]/[E] | [Nu] / mol L⁻¹ | [Nu]/[E] | $k_{obs}$ / s⁻¹ |
|---|---|---|---|---|
| $7.56 \times 10^{-5}$ | 63 | $8.79 \times 10^{-4}$ | 12 | $1.01 \times 10^{-2}$ |
| $6.38 \times 10^{-5}$ | 30 | $1.24 \times 10^{-3}$ | 19 | $1.12 \times 10^{-2}$ |
| $6.54 \times 10^{-5}$ | 39 | $1.77 \times 10^{-3}$ | 27 | $1.74 \times 10^{-2}$ |
| $6.54 \times 10^{-5}$ | 63 | $2.54 \times 10^{-3}$ | 39 | $2.65 \times 10^{-2}$ |
| $6.82 \times 10^{-5}$ | 63 | $6.31 \times 10^{-3}$ | 93 | $5.88 \times 10^{-2}$ |
| $6.65 \times 10^{-5}$ | 63 | $7.69 \times 10^{-3}$ | 116 | $7.11 \times 10^{-2}$ |

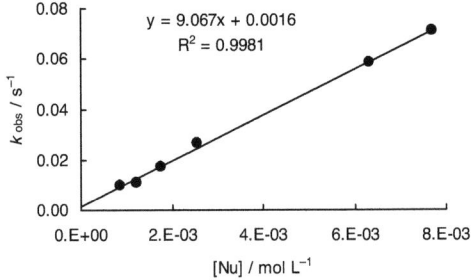

$k = 9.07$ L mol⁻¹ s⁻¹

Experimental Part                                                              231

Reaction of $(m\text{F})\text{Tr}^+$ with $\text{Bu}_4\text{Sn}$ (conventional UV-vis spectrometry, 415 nm, $\text{CH}_2\text{Cl}_2$, 20 °C). The carbocation was generated in solution by mixing $(m\text{F})\text{TrBr}$ with $\text{GaCl}_3$.

| [E] / mol L$^{-1}$ | [GaCl$_3$]/[E] | [Nu] / mol L$^{-1}$ | [Nu]/[E] | $k_{obs}$ / s$^{-1}$ |
|---|---|---|---|---|
| $6.60 \times 10^{-5}$ | 19 | $4.18 \times 10^{-3}$ | 63 | $2.10 \times 10^{-3}$ |
| $6.68 \times 10^{-5}$ | 19 | $8.47 \times 10^{-3}$ | 127 | $4.25 \times 10^{-3}$ |
| $6.78 \times 10^{-5}$ | 19 | $1.29 \times 10^{-2}$ | 190 | $6.83 \times 10^{-3}$ |
| $6.60 \times 10^{-5}$ | 19 | $1.67 \times 10^{-2}$ | 253 | $8.92 \times 10^{-3}$ |
| $6.60 \times 10^{-5}$ | 19 | $2.09 \times 10^{-2}$ | 317 | $1.11 \times 10^{-2}$ |

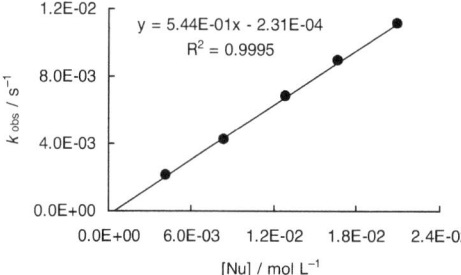

$k = 5.44 \times 10^{-1}$ L mol$^{-1}$ s$^{-1}$

Reaction of $(p\text{F})\text{Tr}^+\text{BF}_4^-$ with $\text{HSiPh}_3$ (stopped-flow UV-vis spectrometry, 436 nm, $\text{CH}_2\text{Cl}_2$, 20 °C).

| [E] / mol L$^{-1}$ | [Nu] / mol L$^{-1}$ | [Nu]/[E] | $k_{obs}$ / s$^{-1}$ |
|---|---|---|---|
| $4.66 \times 10^{-5}$ | $3.98 \times 10^{-4}$ | 9 | $1.94 \times 10^{-2}$ |
| $4.69 \times 10^{-5}$ | $8.00 \times 10^{-4}$ | 17 | $3.56 \times 10^{-2}$ |
| $4.64 \times 10^{-5}$ | $1.19 \times 10^{-3}$ | 26 | $5.31 \times 10^{-2}$ |
| $4.80 \times 10^{-5}$ | $1.64 \times 10^{-3}$ | 34 | $7.43 \times 10^{-2}$ |
| $4.55 \times 10^{-5}$ | $1.94 \times 10^{-3}$ | 43 | $9.01 \times 10^{-2}$ |

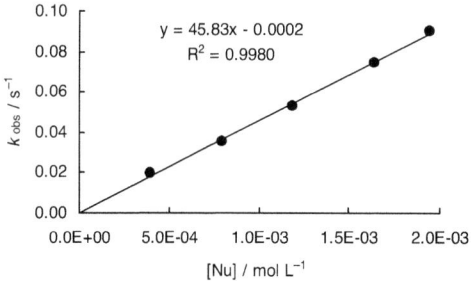

$k = 4.58 \times 10^{1}$ L mol$^{-1}$ s$^{-1}$

Reaction of $(p\text{F})\text{Tr}^+\text{BF}_4^-$ with $\text{HSiEt}_3$ (stopped-flow UV-vis spectrometry, 436 nm, $\text{CH}_2\text{Cl}_2$, 20 °C).

| [E] / mol L$^{-1}$ | [Nu] / mol L$^{-1}$ | [Nu]/[E] | $k_{obs}$ / s$^{-1}$ |
|---|---|---|---|
| $1.38 \times 10^{-4}$ | $3.22 \times 10^{-2}$ | 233 | 2.69 |
| $1.38 \times 10^{-4}$ | $4.21 \times 10^{-2}$ | 305 | 3.47 |
| $1.38 \times 10^{-4}$ | $5.41 \times 10^{-2}$ | 392 | 4.38 |
| $1.38 \times 10^{-4}$ | $6.58 \times 10^{-2}$ | 477 | 5.33 |
| $1.38 \times 10^{-4}$ | $8.43 \times 10^{-2}$ | 611 | 6.72 |

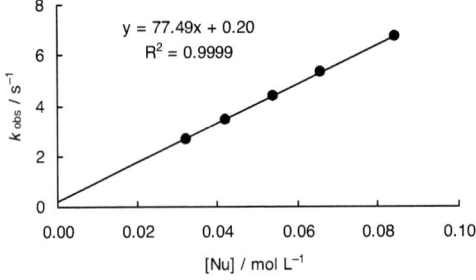

$k = 7.75 \times 10^1 \text{ L mol}^{-1} \text{ s}^{-1}$

Reaction of $(p\text{F})\text{Tr}^+\text{BF}_4^-$ with cycloheptatriene (conventional UV-vis spectrometry, 436 nm, $\text{CH}_2\text{Cl}_2$, 20 °C).

| [E] / mol L$^{-1}$ | [Nu] / mol L$^{-1}$ | [Nu]/[E] | $k_{obs}$ / s$^{-1}$ |
|---|---|---|---|
| $7.51 \times 10^{-5}$ | $3.81 \times 10^{-3}$ | 51 | $5.19 \times 10^{-3}$ |
| $7.67 \times 10^{-5}$ | $7.77 \times 10^{-3}$ | 101 | $1.02 \times 10^{-2}$ |
| $7.43 \times 10^{-5}$ | $1.13 \times 10^{-2}$ | 152 | $1.43 \times 10^{-2}$ |
| $7.35 \times 10^{-5}$ | $1.49 \times 10^{-2}$ | 203 | $1.97 \times 10^{-2}$ |
| $7.38 \times 10^{-5}$ | $1.87 \times 10^{-2}$ | 253 | $2.38 \times 10^{-2}$ |

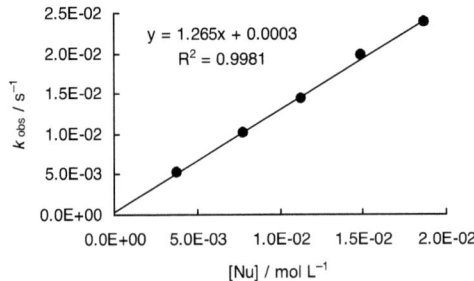

$k = 1.27 \text{ L mol}^{-1} \text{ s}^{-1}$

Experimental Part                                                                                  233

Reaction of $(p\text{F})_2\text{Tr}^+\text{BF}_4^-$ with $\text{HSiPh}_3$ (stopped-flow UV-vis spectrometry, 437 nm, $\text{CH}_2\text{Cl}_2$, 20 °C).

| [E] / mol L$^{-1}$ | [Nu] / mol L$^{-1}$ | [Nu]/[E] | $k_{obs}$ / s$^{-1}$ |
|---|---|---|---|
| $1.09 \times 10^{-4}$ | $7.28 \times 10^{-3}$ | 67 | $3.39 \times 10^{-1}$ |
| $1.09 \times 10^{-4}$ | $9.10 \times 10^{-3}$ | 83 | $4.23 \times 10^{-1}$ |
| $1.09 \times 10^{-4}$ | $1.09 \times 10^{-2}$ | 100 | $4.81 \times 10^{-1}$ |
| $1.09 \times 10^{-4}$ | $1.82 \times 10^{-2}$ | 167 | $6.87 \times 10^{-1}$ |

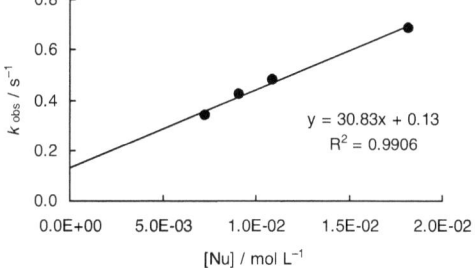

$k = 3.08 \times 10^1$ L mol$^{-1}$ s$^{-1}$

Reaction of $(p\text{F})_2\text{Tr}^+\text{BF}_4^-$ with $\text{HSiEt}_3$ (stopped-flow UV-vis spectrometry, 437 nm, $\text{CH}_2\text{Cl}_2$, 20 °C).

| [E] / mol L$^{-1}$ | [Nu] / mol L$^{-1}$ | [Nu]/[E] | $k_{obs}$ / s$^{-1}$ |
|---|---|---|---|
| $1.09 \times 10^{-4}$ | $1.18 \times 10^{-2}$ | 108 | $6.93 \times 10^{-1}$ |
| $1.09 \times 10^{-4}$ | $1.76 \times 10^{-2}$ | 161 | $9.72 \times 10^{-1}$ |
| $1.09 \times 10^{-4}$ | $2.35 \times 10^{-2}$ | 216 | 1.27 |
| $1.09 \times 10^{-4}$ | $2.94 \times 10^{-2}$ | 270 | 1.51 |

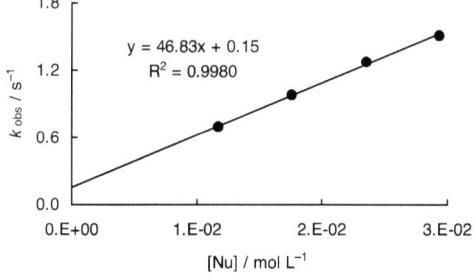

$k = 4.68 \times 10^1$ L mol$^{-1}$ s$^{-1}$

Reaction of $(p\text{F})_2\text{Tr}^+\text{BF}_4^-$ with $\text{HSiBu}_3$ (stopped-flow UV-vis spectrometry, 437 nm, $\text{CH}_2\text{Cl}_2$, 20 °C).

| [E] / mol L$^{-1}$ | [Nu] / mol L$^{-1}$ | [Nu]/[E] | $k_{obs}$ / s$^{-1}$ |
|---|---|---|---|
| $1.09 \times 10^{-4}$ | $8.86 \times 10^{-3}$ | 81 | $8.40 \times 10^{-1}$ |
| $1.09 \times 10^{-4}$ | $1.33 \times 10^{-2}$ | 122 | 1.18 |
| $1.09 \times 10^{-4}$ | $1.77 \times 10^{-2}$ | 162 | 1.61 |
| $1.09 \times 10^{-4}$ | $2.22 \times 10^{-2}$ | 204 | 1.89 |

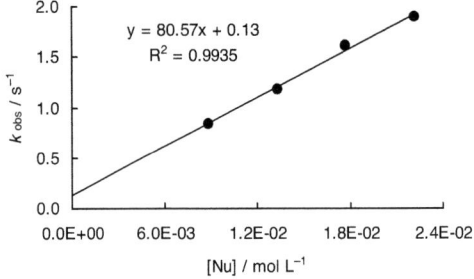

$k = 8.06 \times 10^1 \text{ L mol}^{-1} \text{ s}^{-1}$

Reaction of $(p\text{F})_2\text{Tr}^+\text{BF}_4^-$ with cycloheptatriene (conventional UV-vis spectrometry, 437 nm, $\text{CH}_2\text{Cl}_2$, 20 °C).

| [E] / mol L$^{-1}$ | [Nu] / mol L$^{-1}$ | [Nu]/[E] | $k_{obs}$ / s$^{-1}$ |
|---|---|---|---|
| $1.05 \times 10^{-4}$ | $3.79 \times 10^{-3}$ | 36 | $3.51 \times 10^{-3}$ |
| $6.31 \times 10^{-5}$ | $7.57 \times 10^{-3}$ | 120 | $7.14 \times 10^{-3}$ |
| $6.29 \times 10^{-5}$ | $1.13 \times 10^{-2}$ | 180 | $1.10 \times 10^{-3}$ |
| $6.34 \times 10^{-5}$ | $1.52 \times 10^{-2}$ | 240 | $1.39 \times 10^{-2}$ |
| $6.07 \times 10^{-5}$ | $1.82 \times 10^{-2}$ | 300 | $1.60 \times 10^{-2}$ |

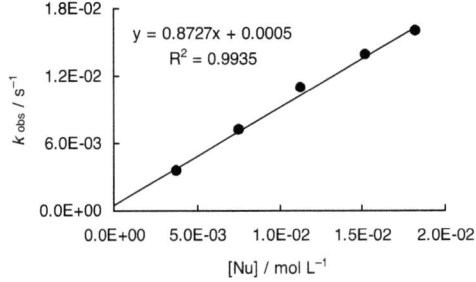

$k = 8.73 \times 10^{-1} \text{ L mol}^{-1} \text{ s}^{-1}$

Experimental Part 235

Reaction of $(pF)_3Tr^+$ with $HSiPh_3$ (stopped-flow UV-vis spectrometry, 437 nm, $CH_2Cl_2$, 20 °C). The carbocation was generated in solution by mixing $(pF)_3TrCl$ with $GaCl_3$.

| [E] / mol L$^{-1}$ | [GaCl$_3$]/[E] | [Nu] / mol L$^{-1}$ | [Nu]/[E] | $k_{obs}$ / s$^{-1}$ |
|---|---|---|---|---|
| 3.51 × 10$^{-5}$ | 3 | 2.81 × 10$^{-4}$ | 8 | 6.56 × 10$^{-3}$ |
| 3.55 × 10$^{-5}$ | 3 | 5.68 × 10$^{-4}$ | 16 | 1.36 × 10$^{-2}$ |
| 3.57 × 10$^{-5}$ | 3 | 8.58 × 10$^{-4}$ | 24 | 2.02 × 10$^{-2}$ |
| 3.48 × 10$^{-5}$ | 3 | 1.12 × 10$^{-3}$ | 32 | 2.64 × 10$^{-2}$ |

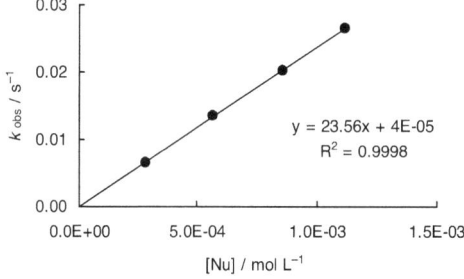

$k = 2.36 \times 10^1$ L mol$^{-1}$ s$^{-1}$

Reaction of $(pF)_3Tr^+$ with $HSiEt_3$ (stopped-flow UV-vis spectrometry, 437 nm, $CH_2Cl_2$, 20 °C). The carbocation was generated in solution by mixing $(pF)_3TrCl$ with $GaCl_3$.

| [E] / mol L$^{-1}$ | [GaCl$_3$]/[E] | [Nu] / mol L$^{-1}$ | [Nu]/[E] | $k_{obs}$ / s$^{-1}$ |
|---|---|---|---|---|
| 3.60 × 10$^{-5}$ | 3 | 2.54 × 10$^{-4}$ | 7 | 6.29 × 10$^{-3}$ |
| 3.42 × 10$^{-5}$ | 5 | 4.82 × 10$^{-4}$ | 14 | 1.36 × 10$^{-2}$ |
| 3.43 × 10$^{-5}$ | 3 | 7.26 × 10$^{-4}$ | 21 | 1.86 × 10$^{-2}$ |
| 3.44 × 10$^{-5}$ | 3 | 9.71 × 10$^{-4}$ | 28 | 2.58 × 10$^{-2}$ |

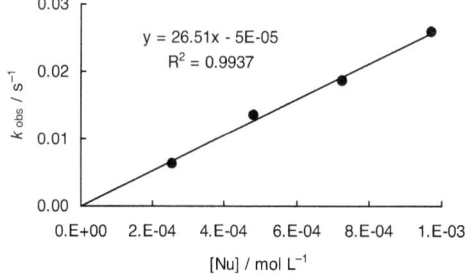

$k = 2.65 \times 10^1$ L mol$^{-1}$ s$^{-1}$

Reaction of Tr⁺BF₄⁻ with HSiPh₃ (stopped-flow UV-vis spectrometry, 430 nm, CH₂Cl₂, 20 °C).

| [E] / mol L$^{-1}$ | [Nu] / mol L$^{-1}$ | [Nu]/[E] | $k_{obs}$ / s$^{-1}$ |
|---|---|---|---|
| $1.03 \times 10^{-4}$ | $3.69 \times 10^{-3}$ | 36 | $3.02 \times 10^{-1}$ |
| $1.03 \times 10^{-4}$ | $5.53 \times 10^{-3}$ | 54 | $4.49 \times 10^{-1}$ |
| $1.03 \times 10^{-4}$ | $7.37 \times 10^{-3}$ | 72 | $5.77 \times 10^{-1}$ |
| $1.03 \times 10^{-4}$ | $9.22 \times 10^{-3}$ | 90 | $7.23 \times 10^{-1}$ |

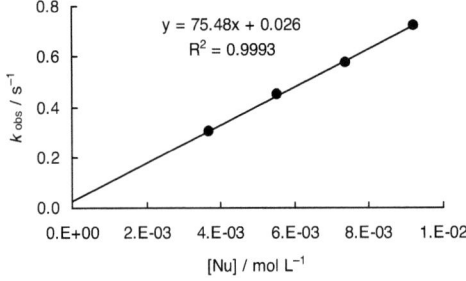

$k = 7.55 \times 10^1$ L mol$^{-1}$ s$^{-1}$

Reaction of Tr⁺SbF₆⁻ with HSiPh₃ (conventional UV-vis spectrometry, 432 nm, CH₂Cl₂, 20 °C).

| [E] / mol L$^{-1}$ | [Nu] / mol L$^{-1}$ | [Nu]/[E] | $k_{obs}$ / s$^{-1}$ |
|---|---|---|---|
| $5.99 \times 10^{-5}$ | $6.74 \times 10^{-4}$ | 11 | $4.56 \times 10^{-2}$ |
| $6.01 \times 10^{-5}$ | $1.35 \times 10^{-3}$ | 23 | $9.29 \times 10^{-2}$ |
| $5.90 \times 10^{-5}$ | $1.99 \times 10^{-3}$ | 34 | $1.53 \times 10^{-1}$ |
| $5.84 \times 10^{-5}$ | $2.63 \times 10^{-3}$ | 45 | $1.82 \times 10^{-1}$ |
| $5.84 \times 10^{-5}$ | $3.29 \times 10^{-3}$ | 56 | $2.34 \times 10^{-1}$ |

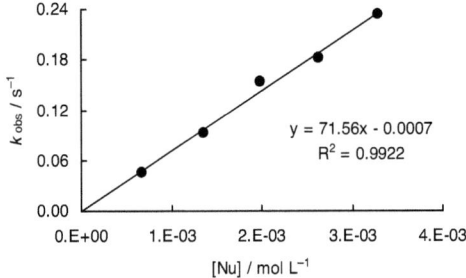

$k = 7.16 \times 10^1$ L mol$^{-1}$ s$^{-1}$

Reaction of Tr$^+$BF$_4^-$ with HSiEt$_3$ (stopped-flow UV-vis spectrometry, 430 nm, CH$_2$Cl$_2$, 20 °C).

| [E] / mol L$^{-1}$ | [Nu] / mol L$^{-1}$ | [Nu]/[E] | $k_{obs}$ / s$^{-1}$ |
|---|---|---|---|
| 9.36 × 10$^{-5}$ | 6.32 × 10$^{-3}$ | 68 | 1.04 |
| 9.36 × 10$^{-5}$ | 1.26 × 10$^{-2}$ | 135 | 1.85 |
| 9.36 × 10$^{-5}$ | 1.90 × 10$^{-2}$ | 203 | 2.82 |
| 9.36 × 10$^{-5}$ | 2.53 × 10$^{-2}$ | 270 | 3.69 |
| 9.36 × 10$^{-5}$ | 3.16 × 10$^{-2}$ | 338 | 4.43 |

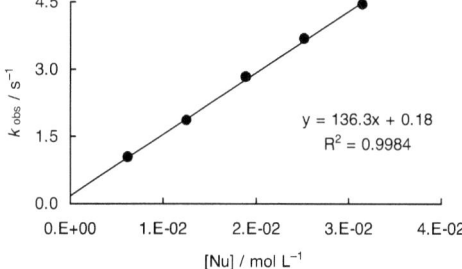

$k = 1.36 \times 10^2$ L mol$^{-1}$ s$^{-1}$

Reaction of Tr$^+$ with HSiEt$_3$ (conventional UV-vis spectrometry, 432 nm, CH$_2$Cl$_2$, 20 °C). The carbocation was generated in solution by mixing TrCl with variable amounts of GaCl$_3$.

| [E] / mol L$^{-1}$ | [GaCl$_3$]/[E] | [Nu] / mol L$^{-1}$ | [Nu]/[E] | $k_{obs}$ / s$^{-1}$ |
|---|---|---|---|---|
| 5.90 × 10$^{-5}$ | 4 | 9.86 × 10$^{-4}$ | 17 | 1.25 × 10$^{-1}$ |
| 5.82 × 10$^{-5}$ | 6 | 9.73 × 10$^{-4}$ | 17 | 1.27 × 10$^{-1}$ |
| 6.21 × 10$^{-5}$ | 8 | 1.04 × 10$^{-3}$ | 17 | 1.30 × 10$^{-1}$ |
| 5.56 × 10$^{-5}$ | 10 | 9.30 × 10$^{-4}$ | 17 | 1.28 × 10$^{-1}$ |
| 5.45 × 10$^{-5}$ | 13 | 9.12 × 10$^{-4}$ | 17 | 1.29 × 10$^{-1}$ |

Reaction of Tr$^+$ with HSiEt$_3$ (stopped-flow UV-vis spectrometry, 432 nm, CH$_2$Cl$_2$, 20 °C). The carbocation was generated in solution by mixing TrCl with trimethylsilyl triflate (TMSOTf).

| [E] / mol L$^{-1}$ | [TMSOTf]/[E] | [Nu] / mol L$^{-1}$ | [Nu]/[E] | $k_{obs}$ / s$^{-1}$ |
|---|---|---|---|---|
| 7.17 × 10$^{-5}$ | 5.77 | 1.86 × 10$^{-3}$ | 26 | 2.12 × 10$^{-1}$ |
| 7.17 × 10$^{-5}$ | 5.77 | 3.72 × 10$^{-3}$ | 52 | 4.78 × 10$^{-1}$ |
| 7.17 × 10$^{-5}$ | 5.77 | 5.58 × 10$^{-3}$ | 78 | 7.21 × 10$^{-1}$ |
| 7.17 × 10$^{-5}$ | 5.77 | 7.44 × 10$^{-3}$ | 104 | 9.31 × 10$^{-1}$ |
| 7.17 × 10$^{-5}$ | 5.77 | 9.31 × 10$^{-3}$ | 130 | 1.19 |

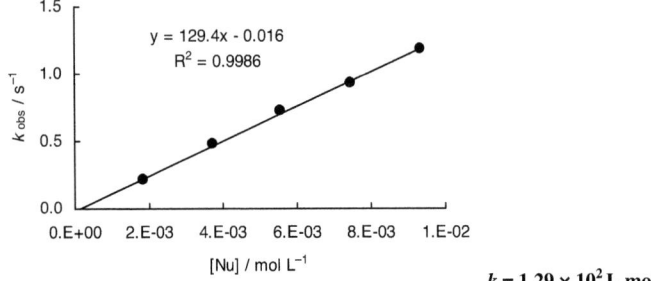

$k = 1.29 \times 10^2 \text{ L mol}^{-1} \text{ s}^{-1}$

Reaction of Tr$^+$BF$_4^-$ with HSiBu$_3$ (stopped-flow UV-vis spectrometry, 430 nm, CH$_2$Cl$_2$, 20 °C).

| [E] / mol L$^{-1}$ | [Nu] / mol L$^{-1}$ | [Nu]/[E] | $k_{obs}$ / s$^{-1}$ |
|---|---|---|---|
| 9.36 × 10$^{-5}$ | 5.23 × 10$^{-3}$ | 56 | 1.64 |
| 9.36 × 10$^{-5}$ | 1.05 × 10$^{-2}$ | 112 | 3.19 |
| 9.36 × 10$^{-5}$ | 1.57 × 10$^{-2}$ | 168 | 4.50 |
| 9.36 × 10$^{-5}$ | 2.09 × 10$^{-2}$ | 223 | 5.80 |
| 9.36 × 10$^{-5}$ | 2.61 × 10$^{-2}$ | 279 | 7.50 |

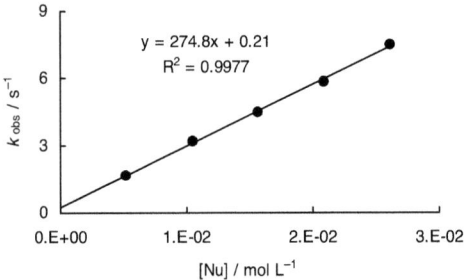

$k = 2.75 \times 10^2 \text{ L mol}^{-1} \text{ s}^{-1}$

Reaction of Tr$^+$BF$_4^-$ with Bu$_4$Sn (conventional UV-vis spectrometry, 412 nm, CH$_2$Cl$_2$, 20 °C, kinetics of low quality).

| [E] / mol L$^{-1}$ | [Nu] / mol L$^{-1}$ | [Nu]/[E] | $k_{obs}$ / s$^{-1}$ |
|---|---|---|---|
| 5.36 × 10$^{-5}$ | 2.95 × 10$^{-3}$ | 55 | 3.87 × 10$^{-4}$ |
| 5.46 × 10$^{-5}$ | 6.00 × 10$^{-3}$ | 110 | 6.04 × 10$^{-4}$ |
| 5.33 × 10$^{-5}$ | 8.78 × 10$^{-3}$ | 165 | 7.68 × 10$^{-4}$ |
| 5.49 × 10$^{-5}$ | 1.51 × 10$^{-2}$ | 275 | 1.18 × 10$^{-3}$ |
| 5.37 × 10$^{-5}$ | 2.11 × 10$^{-2}$ | 393 | 1.61 × 10$^{-3}$ |

Experimental Part 239

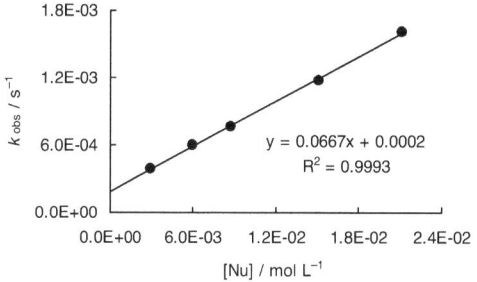

$k = 6.67 \times 10^{-2}$ L mol$^{-1}$ s$^{-1}$

Reaction of Tr$^+$BF$_4^-$ with HSiMe$_2$Ph (stopped-flow UV-vis spectrometry, 430 nm, CH$_2$Cl$_2$, 20 °C).

| [E] / mol L$^{-1}$ | [Nu] / mol L$^{-1}$ | [Nu]/[E] | $k_{obs}$ / s$^{-1}$ |
|---|---|---|---|
| 1.20 × 10$^{-4}$ | 1.37 × 10$^{-3}$ | 11 | 4.98 × 10$^{-1}$ |
| 1.20 × 10$^{-4}$ | 5.47 × 10$^{-3}$ | 46 | 1.23 |
| 1.20 × 10$^{-4}$ | 8.21 × 10$^{-3}$ | 68 | 1.83 |
| 1.20 × 10$^{-4}$ | 1.09 × 10$^{-2}$ | 91 | 2.27 |
| 1.20 × 10$^{-4}$ | 1.37 × 10$^{-2}$ | 114 | 2.71 |

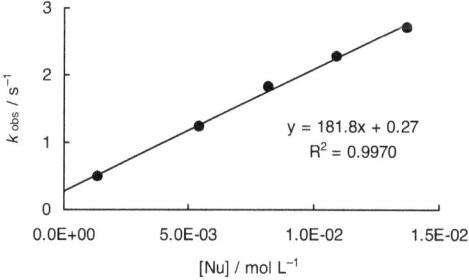

$k = 1.82 \times 10^2$ L mol$^{-1}$ s$^{-1}$

Reaction of Tr$^+$BF$_4^-$ with HSi(SiMe$_3$)$_3$ (stopped-flow UV-vis spectrometry, 430 nm, CH$_2$Cl$_2$, 20 °C).

| [E] / mol L$^{-1}$ | [Nu] / mol L$^{-1}$ | [Nu]/[E] | $k_{obs}$ / s$^{-1}$ |
|---|---|---|---|
| 1.19 × 10$^{-4}$ | 2.73 × 10$^{-3}$ | 23 | 3.17 |
| 1.19 × 10$^{-4}$ | 5.45 × 10$^{-3}$ | 46 | 6.47 |
| 1.19 × 10$^{-4}$ | 8.18 × 10$^{-3}$ | 69 | 9.00 |
| 1.19 × 10$^{-4}$ | 1.09 × 10$^{-2}$ | 92 | 1.14 × 10$^1$ |
| 1.19 × 10$^{-4}$ | 1.36 × 10$^{-2}$ | 114 | 1.40 × 10$^1$ |

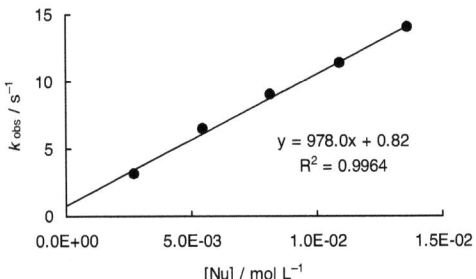

$k = 9.78 \times 10^2 \, \text{L mol}^{-1} \, \text{s}^{-1}$

Reaction of Tr$^+$BF$_4^-$ with HSiEt$_3$ (conventional UV-vis spectrometry, 432 nm, CH$_3$CN, 20 °C).

| [E] / mol L$^{-1}$ | [Nu] / mol L$^{-1}$ | [Nu]/[E] | $k_{obs}$ / s$^{-1}$ |
|---|---|---|---|
| 4.28 × 10$^{-5}$ | 4.97 × 10$^{-4}$ | 12 | 1.15 × 10$^{-2}$ |
| 3.82 × 10$^{-5}$ | 8.86 × 10$^{-4}$ | 23 | 1.81 × 10$^{-2}$ |
| 3.95 × 10$^{-5}$ | 1.37 × 10$^{-3}$ | 35 | 2.80 × 10$^{-2}$ |
| 4.22 × 10$^{-5}$ | 1.96 × 10$^{-3}$ | 46 | 4.46 × 10$^{-2}$ |
| 4.06 × 10$^{-5}$ | 2.36 × 10$^{-3}$ | 58 | 4.86 × 10$^{-2}$ |

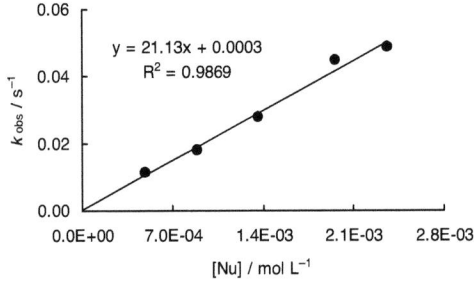

$k = 2.11 \times 10^1 \, \text{L mol}^{-1} \, \text{s}^{-1}$

Reaction of MeTr$^+$BF$_4^-$ with HSiPh$_3$ (conventional UV-vis spectrometry, 455 nm, CH$_2$Cl$_2$, 20 °C).

| [E] / mol L$^{-1}$ | [Nu] / mol L$^{-1}$ | [Nu]/[E] | $k_{obs}$ / s$^{-1}$ |
|---|---|---|---|
| 5.60 × 10$^{-5}$ | 1.23 × 10$^{-3}$ | 22 | 1.85 × 10$^{-2}$ |
| 5.69 × 10$^{-5}$ | 2.50 × 10$^{-3}$ | 44 | 3.88 × 10$^{-2}$ |
| 5.78 × 10$^{-5}$ | 3.81 × 10$^{-3}$ | 66 | 5.86 × 10$^{-2}$ |
| 5.44 × 10$^{-5}$ | 4.79 × 10$^{-3}$ | 88 | 7.40 × 10$^{-2}$ |
| 5.44 × 10$^{-5}$ | 5.99 × 10$^{-3}$ | 110 | 9.29 × 10$^{-2}$ |

Experimental Part 241

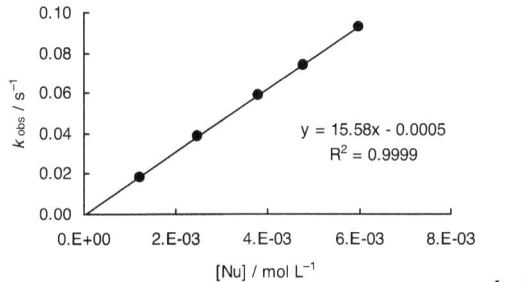

$k = 1.56 \times 10^1 \, \text{L mol}^{-1} \, \text{s}^{-1}$

Reaction of MeTr$^+$BF$_4^-$ with HSiEt$_3$ (conventional UV-vis spectrometry, 448 nm, CH$_2$Cl$_2$, 20 °C).

| [E] / mol L$^{-1}$ | [Nu] / mol L$^{-1}$ | [Nu]/[E] | $k_{obs}$ / s$^{-1}$ |
|---|---|---|---|
| 5.35 × 10$^{-5}$ | 9.44 × 10$^{-4}$ | 18 | 3.51 × 10$^{-2}$ |
| 5.18 × 10$^{-5}$ | 1.83 × 10$^{-3}$ | 35 | 6.73 × 10$^{-2}$ |
| 5.21 × 10$^{-5}$ | 2.76 × 10$^{-3}$ | 53 | 1.03 × 10$^{-1}$ |
| 5.22 × 10$^{-5}$ | 3.69 × 10$^{-3}$ | 71 | 1.39 × 10$^{-1}$ |
| 5.11 × 10$^{-5}$ | 4.51 × 10$^{-3}$ | 88 | 1.65 × 10$^{-1}$ |

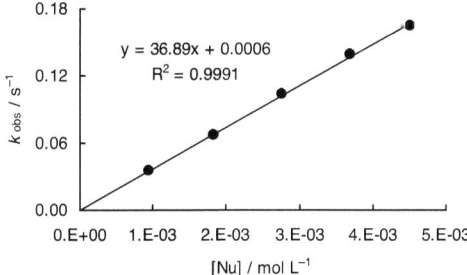

$k = 3.69 \times 10^1 \, \text{L mol}^{-1} \, \text{s}^{-1}$

Reaction of MeTr$^+$BF$_4^-$ with HSiBu$_3$ (stopped-flow UV-vis spectrometry, 448 nm, CH$_2$Cl$_2$, 20 °C).

| [E] / mol L$^{-1}$ | [Nu] / mol L$^{-1}$ | [Nu]/[E] | $k_{obs}$ / s$^{-1}$ |
|---|---|---|---|
| 1.23 × 10$^{-4}$ | 5.13 × 10$^{-3}$ | 42 | 4.87 × 10$^{-1}$ |
| 1.23 × 10$^{-4}$ | 1.03 × 10$^{-2}$ | 84 | 8.70 × 10$^{-1}$ |
| 1.23 × 10$^{-4}$ | 1.54 × 10$^{-2}$ | 125 | 1.36 |
| 1.23 × 10$^{-4}$ | 2.15 × 10$^{-2}$ | 175 | 1.75 |
| 1.23 × 10$^{-4}$ | 3.00 × 10$^{-2}$ | 244 | 2.26 |

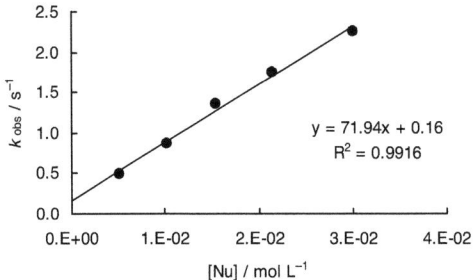

$k = 7.19 \times 10^1 \text{ L mol}^{-1} \text{ s}^{-1}$

Reaction of MeTr$^+$BF$_4^-$ with HSiMe$_2$Ph (stopped-flow UV-vis spectrometry, 448 nm, CH$_2$Cl$_2$, 20 °C).

| [E] / mol L$^{-1}$ | [Nu] / mol L$^{-1}$ | [Nu]/[E] | $k_{obs}$ / s$^{-1}$ |
|---|---|---|---|
| 1.28 × 10$^{-4}$ | 3.35 × 10$^{-3}$ | 26 | 1.94 × 10$^{-1}$ |
| 1.28 × 10$^{-4}$ | 6.69 × 10$^{-3}$ | 52 | 3.66 × 10$^{-1}$ |
| 1.28 × 10$^{-4}$ | 1.00 × 10$^{-2}$ | 78 | 5.17 × 10$^{-1}$ |
| 1.28 × 10$^{-4}$ | 1.34 × 10$^{-2}$ | 105 | 7.47 × 10$^{-1}$ |
| 1.28 × 10$^{-4}$ | 1.67 × 10$^{-2}$ | 131 | 8.98 × 10$^{-1}$ |

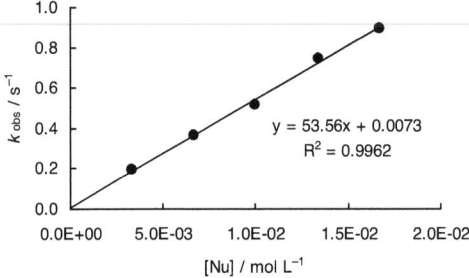

$k = 5.36 \times 10^1 \text{ L mol}^{-1} \text{ s}^{-1}$

Reaction of MeTr$^+$BF$_4^-$ with HSi(SiMe$_3$)$_3$ (stopped-flow UV-vis spectrometry, 455 nm, CH$_2$Cl$_2$, 20 °C).

| [E] / mol L$^{-1}$ | [Nu] / mol L$^{-1}$ | [Nu]/[E] | $k_{obs}$ / s$^{-1}$ |
|---|---|---|---|
| 6.74 × 10$^{-5}$ | 5.91 × 10$^{-3}$ | 88 | 1.96 |
| 6.74 × 10$^{-5}$ | 9.11 × 10$^{-3}$ | 135 | 2.99 |
| 6.74 × 10$^{-5}$ | 1.40 × 10$^{-2}$ | 208 | 4.52 |
| 6.74 × 10$^{-5}$ | 1.65 × 10$^{-2}$ | 245 | 5.28 |
| 6.74 × 10$^{-5}$ | 2.03 × 10$^{-2}$ | 301 | 6.59 |

# Experimental Part 243

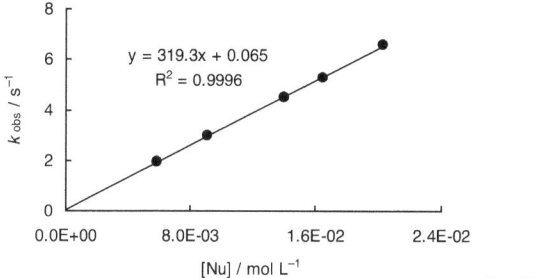

$k = 3.19 \times 10^2 \text{ L mol}^{-1} \text{ s}^{-1}$

Reaction of $Me_2Tr^+BF_4^-$ with $HSiPh_3$ (conventional UV-vis spectrometry, 460 nm, $CH_2Cl_2$, 20 °C).

| [E] / mol L$^{-1}$ | [Nu] / mol L$^{-1}$ | [Nu]/[E] | $k_{obs}$ / s$^{-1}$ |
|---|---|---|---|
| $7.24 \times 10^{-5}$ | $2.96 \times 10^{-3}$ | 41 | $1.17 \times 10^{-2}$ |
| $7.16 \times 10^{-5}$ | $5.86 \times 10^{-3}$ | 82 | $2.28 \times 10^{-2}$ |
| $7.04 \times 10^{-5}$ | $8.63 \times 10^{-3}$ | 123 | $3.46 \times 10^{-2}$ |
| $7.02 \times 10^{-5}$ | $1.15 \times 10^{-2}$ | 164 | $4.52 \times 10^{-2}$ |
| $1.67 \times 10^{-4}$ | $1.37 \times 10^{-2}$ | 82 | $5.51 \times 10^{-2}$ |

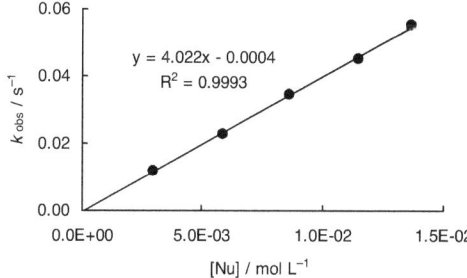

$k = 4.02 \text{ L mol}^{-1} \text{ s}^{-1}$

Reaction of $Me_2Tr^+BF_4^-$ with $HSiEt_3$ (conventional UV-vis spectrometry, 460 nm, $CH_2Cl_2$, 20 °C).

| [E] / mol L$^{-1}$ | [Nu] / mol L$^{-1}$ | [Nu]/[E] | $k_{obs}$ / s$^{-1}$ |
|---|---|---|---|
| $5.23 \times 10^{-5}$ | $1.13 \times 10^{-3}$ | 22 | $1.24 \times 10^{-2}$ |
| $4.84 \times 10^{-5}$ | $2.10 \times 10^{-3}$ | 43 | $2.27 \times 10^{-2}$ |
| $4.97 \times 10^{-5}$ | $3.23 \times 10^{-3}$ | 65 | $3.44 \times 10^{-2}$ |
| $4.79 \times 10^{-5}$ | $4.15 \times 10^{-3}$ | 87 | $4.51 \times 10^{-2}$ |
| $4.85 \times 10^{-5}$ | $5.25 \times 10^{-3}$ | 108 | $5.70 \times 10^{-2}$ |

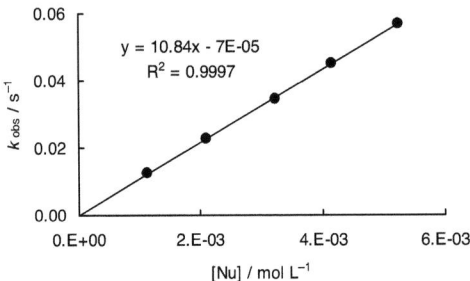

$k = 1.08 \times 10^1 \text{ L mol}^{-1} \text{ s}^{-1}$

Reaction of $Me_2Tr^+BF_4^-$ with $HSiBu_3$ (conventional UV-vis spectrometry, 460 nm, $CH_2Cl_2$, 20 °C).

| [E] / mol L$^{-1}$ | [Nu] / mol L$^{-1}$ | [Nu]/[E] | $k_{obs}$ / s$^{-1}$ |
|---|---|---|---|
| $5.79 \times 10^{-5}$ | $7.58 \times 10^{-4}$ | 13 | $1.78 \times 10^{-2}$ |
| $5.56 \times 10^{-5}$ | $1.46 \times 10^{-3}$ | 26 | $3.33 \times 10^{-2}$ |
| $5.30 \times 10^{-5}$ | $2.08 \times 10^{-3}$ | 39 | $4.89 \times 10^{-2}$ |
| $5.42 \times 10^{-5}$ | $2.84 \times 10^{-3}$ | 52 | $6.56 \times 10^{-2}$ |
| $5.20 \times 10^{-5}$ | $3.40 \times 10^{-3}$ | 66 | $7.89 \times 10^{-2}$ |

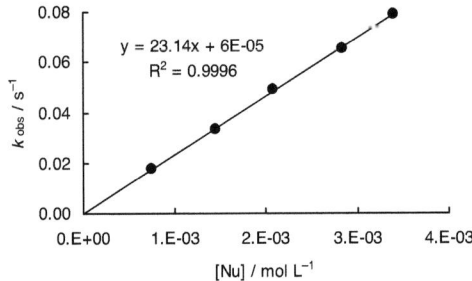

$k = 2.31 \times 10^1 \text{ L mol}^{-1} \text{ s}^{-1}$

Reaction of $Me_2Tr^+BF_4^-$ with $HSiMe_2Ph$ (conventional UV-vis spectrometry, 460 nm, 20 °C).

| [E] / mol L$^{-1}$ | [Nu] / mol L$^{-1}$ | [Nu]/[E] | $k_{obs}$ / s$^{-1}$ |
|---|---|---|---|
| $6.73 \times 10^{-5}$ | $1.49 \times 10^{-3}$ | 22 | $2.12 \times 10^{-2}$ |
| $5.12 \times 10^{-5}$ | $2.84 \times 10^{-3}$ | 56 | $4.13 \times 10^{-2}$ |
| $5.20 \times 10^{-5}$ | $4.33 \times 10^{-3}$ | 83 | $6.40 \times 10^{-2}$ |
| $5.30 \times 10^{-5}$ | $5.87 \times 10^{-3}$ | 111 | $8.59 \times 10^{-2}$ |
| $5.30 \times 10^{-5}$ | $7.34 \times 10^{-3}$ | 139 | $1.11 \times 10^{-1}$ |

# Experimental Part

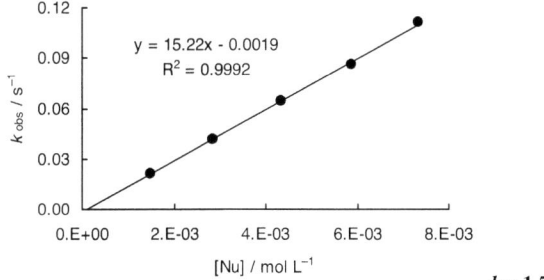

$k = 1.52 \times 10^1$ L mol$^{-1}$ s$^{-1}$

Reaction of Me$_2$Tr$^+$BF$_4^-$ with HSi(SiMe$_3$)$_3$ (stopped-flow UV-vis spectrometry, 460 nm, CH$_2$Cl$_2$, 20 °C, kinetics of low quality).

| [E] / mol L$^{-1}$ | [Nu] / mol L$^{-1}$ | [Nu]/[E] | $k_{obs}$ / s$^{-1}$ |
|---|---|---|---|
| 5.58 × 10$^{-5}$ | 5.91 × 10$^{-3}$ | 106 | 4.10 × 10$^{-1}$ |
| 5.58 × 10$^{-5}$ | 1.65 × 10$^{-2}$ | 296 | 5.97 × 10$^{-1}$ |
| 5.58 × 10$^{-5}$ | 2.03 × 10$^{-2}$ | 364 | 9.32 × 10$^{-1}$ |

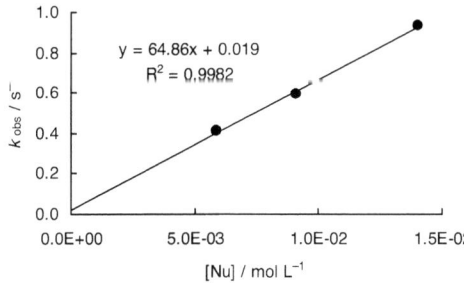

$k = 6.49 \times 10^1$ L mol$^{-1}$ s$^{-1}$

Reaction of Me$_2$Tr$^+$BF$_4^-$ with HSnBu$_3$ (stopped-flow UV-vis spectrometry, 460 nm, CH$_2$Cl$_2$, 20 °C).

| [E] / mol L$^{-1}$ | [Nu] / mol L$^{-1}$ | [Nu]/[E] | $k_{obs}$ / s$^{-1}$ |
|---|---|---|---|
| 4.75 × 10$^{-5}$ | 1.24 × 10$^{-3}$ | 26 | 7.46 × 10$^1$ |
| 4.75 × 10$^{-5}$ | 2.49 × 10$^{-3}$ | 52 | 1.52 × 10$^2$ |

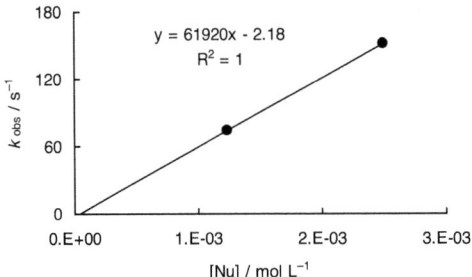

$k = 6.19 \times 10^4$ L mol$^{-1}$ s$^{-1}$

Reaction of Me$_2$Tr$^+$BF$_4^-$ with DSiBu$_3$ (CH$_2$Cl$_2$, conventional UV-vis spectrometry, 460 nm).

| [E] / mol L$^{-1}$ | [Nu] / mol L$^{-1}$ | [Nu]/[E] | $k_{obs}$ / s$^{-1}$ |
|---|---|---|---|
| 5.55 × 10$^{-5}$ | 6.05 × 10$^{-4}$ | 11 | 7.82 × 10$^{-3}$ |
| 5.56 × 10$^{-5}$ | 1.21 × 10$^{-3}$ | 22 | 1.62 × 10$^{-2}$ |
| 5.42 × 10$^{-5}$ | 1.77 × 10$^{-3}$ | 33 | 2.37 × 10$^{-2}$ |
| 5.46 × 10$^{-5}$ | 2.38 × 10$^{-3}$ | 44 | 3.27 × 10$^{-2}$ |
| 5.43 × 10$^{-5}$ | 2.96 × 10$^{-3}$ | 55 | 4.11 × 10$^{-2}$ |

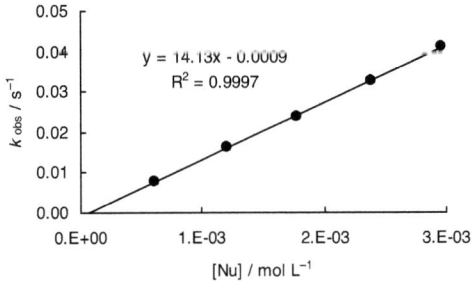

$k = 1.41 \times 10^1$ L mol$^{-1}$ s$^{-1}$

Reaction of Me$_3$Tr$^+$BF$_4^-$ with HSiPh$_3$ (conventional UV-vis spectrometry, 455 nm, CH$_2$Cl$_2$, 20 °C).

| [E] / mol L$^{-1}$ | [Nu] / mol L$^{-1}$ | [Nu]/[E] | $k_{obs}$ / s$^{-1}$ |
|---|---|---|---|
| 3.62 × 10$^{-5}$ | 6.24 × 10$^{-4}$ | 17 | 7.76 × 10$^{-4}$ |
| 3.81 × 10$^{-5}$ | 1.31 × 10$^{-3}$ | 35 | 1.60 × 10$^{-3}$ |
| 3.58 × 10$^{-5}$ | 1.85 × 10$^{-3}$ | 52 | 2.26 × 10$^{-3}$ |
| 3.57 × 10$^{-5}$ | 2.46 × 10$^{-3}$ | 69 | 3.01 × 10$^{-3}$ |
| 3.52 × 10$^{-5}$ | 3.03 × 10$^{-3}$ | 86 | 3.69 × 10$^{-3}$ |

# Experimental Part

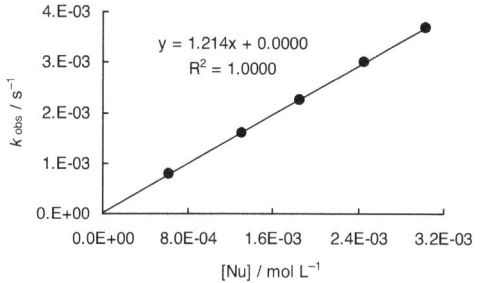

$k = 1.21$ L mol$^{-1}$ s$^{-1}$

Reaction of Me$_3$Tr$^+$BF$_4^-$ with HSiEt$_3$ (conventional UV-vis spectrometry, 455 nm, CH$_2$Cl$_2$, 20 °C).

| [E] / mol L$^{-1}$ | [Nu] / mol L$^{-1}$ | [Nu]/[E] | $k_{obs}$ / s$^{-1}$ |
|---|---|---|---|
| 3.62 × 10$^{-5}$ | 6.73 × 10$^{-4}$ | 19 | 2.34 × 10$^{-3}$ |
| 3.61 × 10$^{-5}$ | 1.34 × 10$^{-3}$ | 37 | 4.56 × 10$^{-3}$ |
| 3.68 × 10$^{-5}$ | 2.06 × 10$^{-3}$ | 56 | 7.08 × 10$^{-3}$ |
| 3.52 × 10$^{-5}$ | 2.62 × 10$^{-3}$ | 75 | 9.00 × 10$^{-3}$ |
| 3.53 × 10$^{-5}$ | 3.29 × 10$^{-3}$ | 93 | 1.13 × 10$^{-2}$ |

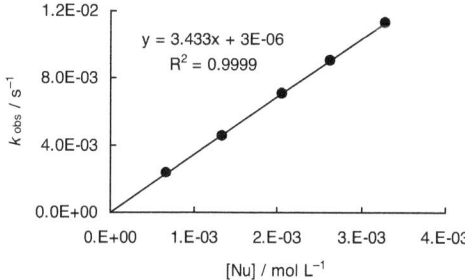

$k = 3.43$ L mol$^{-1}$ s$^{-1}$

Reaction of Me$_3$Tr$^+$BF$_4^-$ with HSiBu$_3$ (conventional UV-vis spectrometry, 455 nm, CH$_2$Cl$_2$, 20 °C).

| [E] / mol L$^{-1}$ | [Nu] / mol L$^{-1}$ | [Nu]/[E] | $k_{obs}$ / s$^{-1}$ |
|---|---|---|---|
| 3.74 × 10$^{-5}$ | 8.39 × 10$^{-4}$ | 22 | 5.08 × 10$^{-3}$ |
| 3.63 × 10$^{-5}$ | 1.62 × 10$^{-3}$ | 45 | 1.00 × 10$^{-2}$ |
| 3.67 × 10$^{-5}$ | 2.46 × 10$^{-3}$ | 67 | 1.48 × 10$^{-2}$ |
| 3.59 × 10$^{-5}$ | 3.21 × 10$^{-3}$ | 90 | 1.95 × 10$^{-2}$ |
| 3.63 × 10$^{-5}$ | 4.07 × 10$^{-3}$ | 112 | 2.48 × 10$^{-2}$ |

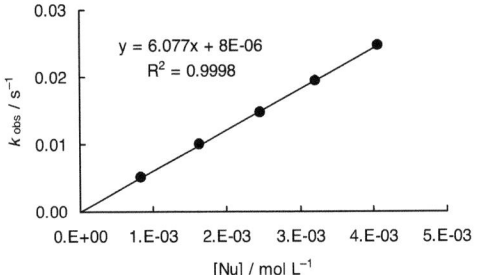

$k = 6.08 \text{ L mol}^{-1} \text{ s}^{-1}$

Reaction of Me$_3$Tr$^+$BF$_4^-$ with HSiMe$_2$Ph (conventional UV-vis spectrometry, 455 nm, CH$_2$Cl$_2$, 20 °C).

| [E] / mol L$^{-1}$ | [Nu] / mol L$^{-1}$ | [Nu]/[E] | $k_{obs}$ / s$^{-1}$ |
|---|---|---|---|
| 3.74 × 10$^{-5}$ | 7.25 × 10$^{-4}$ | 19 | 3.02 × 10$^{-3}$ |
| 3.61 × 10$^{-5}$ | 1.40 × 10$^{-3}$ | 39 | 6.49 × 10$^{-3}$ |
| 3.72 × 10$^{-5}$ | 2.16 × 10$^{-3}$ | 58 | 9.89 × 10$^{-3}$ |
| 3.50 × 10$^{-5}$ | 2.71 × 10$^{-3}$ | 78 | 1.25 × 10$^{-2}$ |
| 3.57 × 10$^{-5}$ | 3.46 × 10$^{-3}$ | 97 | 1.60 × 10$^{-2}$ |

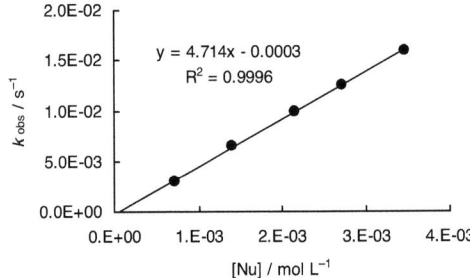

$k = 4.71 \text{ L mol}^{-1} \text{ s}^{-1}$

Reaction of Me$_3$Tr$^+$BF$_4^-$ with HSnBu$_3$ (stopped-flow UV-vis spectrometry, 455 nm, CH$_2$Cl$_2$, 20 °C).

| [E] / mol L$^{-1}$ | [Nu] / mol L$^{-1}$ | [Nu]/[E] | $k_{obs}$ / s$^{-1}$ |
|---|---|---|---|
| 5.37 × 10$^{-5}$ | 1.41 × 10$^{-3}$ | 26 | 3.80 × 10$^{1}$ |
| 5.37 × 10$^{-5}$ | 2.82 × 10$^{-3}$ | 53 | 7.49 × 10$^{1}$ |
| 5.37 × 10$^{-5}$ | 4.23 × 10$^{-3}$ | 79 | 1.17 × 10$^{2}$ |
| 5.37 × 10$^{-5}$ | 5.63 × 10$^{-3}$ | 105 | 1.59 × 10$^{2}$ |
| 5.37 × 10$^{-5}$ | 7.04 × 10$^{-3}$ | 131 | 2.04 × 10$^{2}$ |

Experimental Part 249

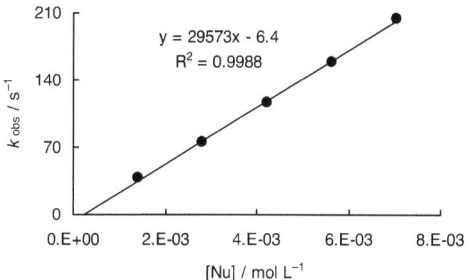

$k = 2.96 \times 10^4$ L mol$^{-1}$ s$^{-1}$

Reaction of (MeO)Tr$^+$BF$_4^-$ with HSiPh$_3$ (conventional UV-vis spectrometry, 478 nm, CH$_2$Cl$_2$, 20 °C).

| [E] / mol L$^{-1}$ | [Nu] / mol L$^{-1}$ | [Nu]/[E] | $k_{obs}$ / s$^{-1}$ |
|---|---|---|---|
| 3.89 × 10$^{-5}$ | 9.15 × 10$^{-4}$ | 24 | 5.87 × 10$^{-4}$ |
| 3.71 × 10$^{-5}$ | 1.75 × 10$^{-3}$ | 47 | 1.16 × 10$^{-3}$ |
| 3.67 × 10$^{-5}$ | 2.59 × 10$^{-3}$ | 71 | 1.71 × 10$^{-3}$ |
| 3.71 × 10$^{-5}$ | 3.61 × 10$^{-3}$ | 97 | 2.36 × 10$^{-3}$ |
| 3.93 × 10$^{-5}$ | 4.63 × 10$^{-3}$ | 118 | 2.98 × 10$^{-3}$ |

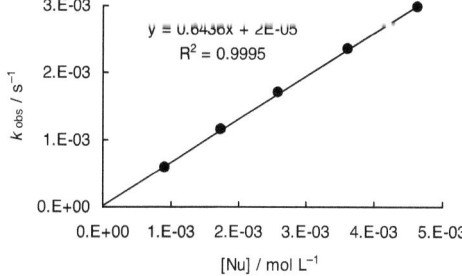

$k = 6.44 \times 10^{-1}$ L mol$^{-1}$ s$^{-1}$

Reaction of (MeO)Tr$^+$BF$_4^-$ with HSiEt$_3$ (conventional UV-vis spectrometry, 478 nm, CH$_2$Cl$_2$, 20 °C).

| [E] / mol L$^{-1}$ | [Nu] / mol L$^{-1}$ | [Nu]/[E] | $k_{obs}$ / s$^{-1}$ |
|---|---|---|---|
| 3.84 × 10$^{-5}$ | 1.36 × 10$^{-3}$ | 35 | 2.21 × 10$^{-3}$ |
| 3.87 × 10$^{-5}$ | 2.74 × 10$^{-3}$ | 71 | 4.41 × 10$^{-3}$ |
| 3.83 × 10$^{-5}$ | 4.07 × 10$^{-3}$ | 106 | 6.47 × 10$^{-3}$ |
| 3.90 × 10$^{-5}$ | 5.38 × 10$^{-3}$ | 138 | 8.53 × 10$^{-3}$ |
| 3.75 × 10$^{-5}$ | 6.65 × 10$^{-3}$ | 177 | 1.05 × 10$^{-2}$ |

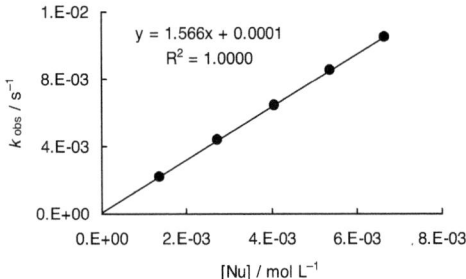

$k = 1.57 \text{ L mol}^{-1} \text{ s}^{-1}$

Reaction of (MeO)Tr$^+$BF$_4^-$ with HSiBu$_3$ (conventional UV-vis spectrometry, 478 nm, CH$_2$Cl$_2$, 20 °C).

| [E] / mol L$^{-1}$ | [Nu] / mol L$^{-1}$ | [Nu]/[E] | $k_{obs}$ / s$^{-1}$ |
|---|---|---|---|
| 3.19 × 10$^{-5}$ | 8.88 × 10$^{-4}$ | 28 | 3.05 × 10$^{-3}$ |
| 3.24 × 10$^{-5}$ | 1.80 × 10$^{-3}$ | 56 | 6.20 × 10$^{-3}$ |
| 3.23 × 10$^{-5}$ | 2.72 × 10$^{-3}$ | 84 | 9.24 × 10$^{-3}$ |
| 3.27 × 10$^{-5}$ | 3.63 × 10$^{-3}$ | 111 | 1.26 × 10$^{-2}$ |
| 3.19 × 10$^{-5}$ | 4.43 × 10$^{-3}$ | 139 | 1.52 × 10$^{-2}$ |

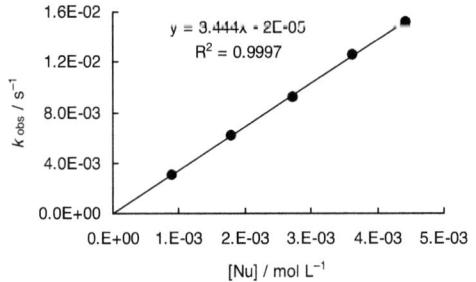

$k = 3.44 \text{ L mol}^{-1} \text{ s}^{-1}$

Reaction of (MeO)Tr$^+$BF$_4^-$ with HSiMe$_2$Ph (conventional UV-vis spectrometry, 478 nm, CH$_2$Cl$_2$, 20 °C).

| [E] / mol L$^{-1}$ | [Nu] / mol L$^{-1}$ | [Nu]/[E] | $k_{obs}$ / s$^{-1}$ |
|---|---|---|---|
| 3.33 × 10$^{-5}$ | 9.68 × 10$^{-4}$ | 29 | 2.33 × 10$^{-3}$ |
| 3.33 × 10$^{-5}$ | 1.94 × 10$^{-3}$ | 58 | 4.66 × 10$^{-3}$ |
| 3.15 × 10$^{-5}$ | 2.87 × 10$^{-3}$ | 91 | 6.63 × 10$^{-3}$ |
| 3.29 × 10$^{-5}$ | 3.83 × 10$^{-3}$ | 116 | 9.07 × 10$^{-3}$ |
| 3.30 × 10$^{-5}$ | 4.81 × 10$^{-3}$ | 145 | 1.18 × 10$^{-2}$ |

# Experimental Part 251

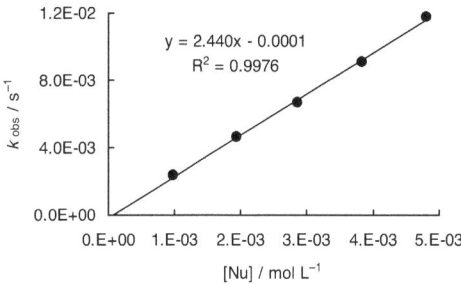

$k = 2.44$ L mol$^{-1}$ s$^{-1}$

Reaction of (MeO)Tr$^+$BF$_4^-$ with HSnBu$_3$ (stopped-flow UV-vis spectrometry, 478 nm, CH$_2$Cl$_2$, 20 °C).

| [E] / mol L$^{-1}$ | [Nu] / mol L$^{-1}$ | [Nu]/[E] | $k_{obs}$ / s$^{-1}$ |
|---|---|---|---|
| 4.89 × 10$^{-5}$ | 1.41 × 10$^{-3}$ | 29 | 1.85 × 10$^1$ |
| 4.89 × 10$^{-5}$ | 2.82 × 10$^{-3}$ | 58 | 3.77 × 10$^1$ |
| 4.89 × 10$^{-5}$ | 4.23 × 10$^{-3}$ | 87 | 5.44 × 10$^1$ |
| 4.89 × 10$^{-5}$ | 5.63 × 10$^{-3}$ | 115 | 7.33 × 10$^1$ |
| 4.89 × 10$^{-5}$ | 7.04 × 10$^{-3}$ | 144 | 9.37 × 10$^1$ |

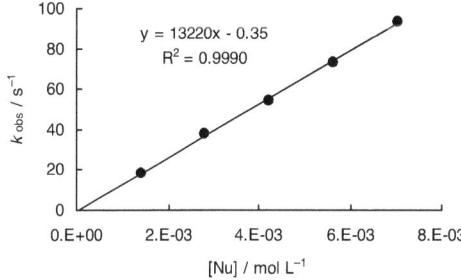

$k = 1.32 \times 10^4$ L mol$^{-1}$ s$^{-1}$

Reaction of (MeO)$_2$Tr$^+$BF$_4^-$ with HSiPh$_3$ (conventional UV-vis spectrometry, 504 nm, CH$_2$Cl$_2$, 20 °C).

| [E] / mol L$^{-1}$ | [Nu] / mol L$^{-1}$ | [Nu]/[E] | $k_{obs}$ / s$^{-1}$ |
|---|---|---|---|
| 3.01 × 10$^{-5}$ | 9.59 × 10$^{-4}$ | 32 | 2.47 × 10$^{-5}$ |
| 2.92 × 10$^{-5}$ | 1.70 × 10$^{-3}$ | 58 | 4.34 × 10$^{-5}$ |
| 2.95 × 10$^{-5}$ | 2.61 × 10$^{-3}$ | 88 | 6.81 × 10$^{-5}$ |
| 2.95 × 10$^{-5}$ | 3.69 × 10$^{-3}$ | 125 | 9.47 × 10$^{-5}$ |
| 3.06 × 10$^{-5}$ | 5.49 × 10$^{-3}$ | 180 | 1.43 × 10$^{-4}$ |

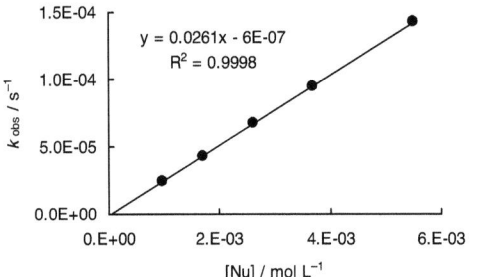

$k = 2.61 \times 10^{-2}$ L mol$^{-1}$ s$^{-1}$

Reaction of (MeO)$_2$Tr$^+$BF$_4^-$ with HSiEt$_3$ (conventional UV-vis spectrometry, 504 nm, CH$_2$Cl$_2$, 20 °C).

| [E] / mol L$^{-1}$ | [Nu] / mol L$^{-1}$ | [Nu]/[E] | $k_{obs}$ / s$^{-1}$ |
|---|---|---|---|
| 2.59 × 10$^{-5}$ | 1.76 × 10$^{-3}$ | 68 | 1.04 × 10$^{-4}$ |
| 2.67 × 10$^{-5}$ | 2.73 × 10$^{-3}$ | 102 | 1.60 × 10$^{-4}$ |
| 2.55 × 10$^{-5}$ | 3.48 × 10$^{-3}$ | 136 | 2.04 × 10$^{-4}$ |
| 2.45 × 10$^{-5}$ | 4.18 × 10$^{-3}$ | 171 | 1.04 × 10$^{-4}$ |

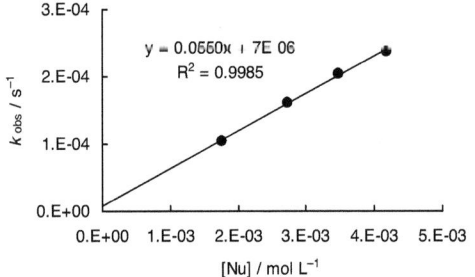

$k = 5.59 \times 10^{-2}$ L mol$^{-1}$ s$^{-1}$

Reaction of (MeO)$_2$Tr$^+$BF$_4^-$ with HSiBu$_3$ (conventional UV-vis spectrometry, 504 nm, CH$_2$Cl$_2$, 20 °C).

| [E] / mol L$^{-1}$ | [Nu] / mol L$^{-1}$ | [Nu]/[E] | $k_{obs}$ / s$^{-1}$ |
|---|---|---|---|
| 2.89 × 10$^{-5}$ | 1.84 × 10$^{-3}$ | 64 | 2.47 × 10$^{-4}$ |
| 2.89 × 10$^{-5}$ | 2.75 × 10$^{-3}$ | 95 | 3.68 × 10$^{-4}$ |
| 2.88 × 10$^{-5}$ | 3.67 × 10$^{-3}$ | 127 | 4.97 × 10$^{-4}$ |
| 2.92 × 10$^{-5}$ | 4.64 × 10$^{-3}$ | 159 | 6.05 × 10$^{-4}$ |

# Experimental Part

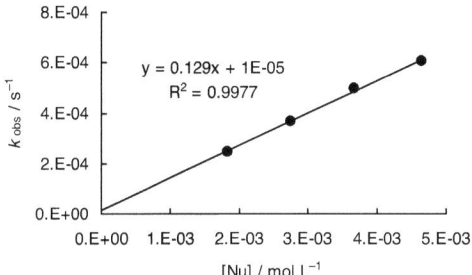

$k = 1.29 \times 10^{-1}$ L mol$^{-1}$ s$^{-1}$

Reaction of (MeO)$_2$Tr$^+$BF$_4^-$ with HSiMe$_2$Ph (conventional UV-vis spectrometry, 504 nm, CH$_2$Cl$_2$, 20 °C).

| [E] / mol L$^{-1}$ | [Nu] / mol L$^{-1}$ | [Nu]/[E] | $k_{obs}$ / s$^{-1}$ |
|---|---|---|---|
| 8.21 × 10$^{-5}$ | 9.66 × 10$^{-4}$ | 12 | 9.26 × 10$^{-5}$ |
| 7.75 × 10$^{-5}$ | 1.82 × 10$^{-3}$ | 23 | 1.78 × 10$^{-4}$ |
| 7.97 × 10$^{-5}$ | 2.94 × 10$^{-3}$ | 37 | 2.93 × 10$^{-4}$ |
| 7.62 × 10$^{-5}$ | 3.75 × 10$^{-3}$ | 49 | 3.93 × 10$^{-4}$ |
| 7.56 × 10$^{-5}$ | 4.65 × 10$^{-3}$ | 62 | 4.84 × 10$^{-4}$ |

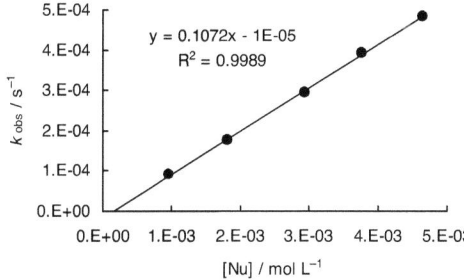

$k = 1.07 \times 10^{-1}$ L mol$^{-1}$ s$^{-1}$

Reaction of (MeO)$_2$Tr$^+$BF$_4^-$ with HSnBu$_3$ (stopped-flow UV-vis spectrometry, 504 nm, CH$_2$Cl$_2$, 20 °C).

| [E] / mol L$^{-1}$ | [Nu] / mol L$^{-1}$ | [Nu]/[E] | $k_{obs}$ / s$^{-1}$ |
|---|---|---|---|
| 3.59 × 10$^{-5}$ | 1.54 × 10$^{-3}$ | 43 | 2.10 |
| 3.59 × 10$^{-5}$ | 3.08 × 10$^{-3}$ | 86 | 4.33 |
| 3.59 × 10$^{-5}$ | 4.62 × 10$^{-3}$ | 129 | 6.44 |
| 3.59 × 10$^{-5}$ | 6.15 × 10$^{-3}$ | 171 | 8.64 |
| 3.59 × 10$^{-5}$ | 7.70 × 10$^{-3}$ | 214 | 1.09 × 10$^1$ |

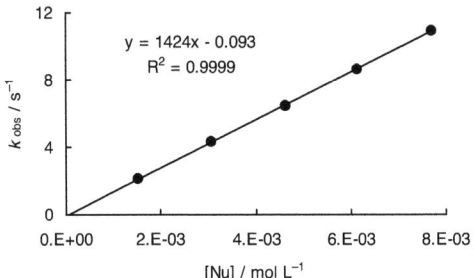

$k = 1.42 \times 10^3$ L mol$^{-1}$ s$^{-1}$

Reaction of (MeO)$_3$Tr$^+$BF$_4^-$ with HSnBu$_3$ (stopped-flow UV-vis spectrometry, 487 nm, CH$_2$Cl$_2$, 20 °C).

| [E] / mol L$^{-1}$ | [Nu] / mol L$^{-1}$ | [Nu]/[E] | $k_{obs}$ / s$^{-1}$ |
|---|---|---|---|
| 2.86 × 10$^{-5}$ | 1.24 × 10$^{-3}$ | 43 | 2.18 × 10$^{-1}$ |
| 2.86 × 10$^{-5}$ | 2.49 × 10$^{-3}$ | 87 | 5.25 × 10$^{-1}$ |
| 2.86 × 10$^{-5}$ | 3.73 × 10$^{-3}$ | 130 | 7.64 × 10$^{-1}$ |
| 2.86 × 10$^{-5}$ | 4.98 × 10$^{-3}$ | 174 | 1.07 |
| 2.86 × 10$^{-5}$ | 6.22 × 10$^{-3}$ | 217 | 1.36 |

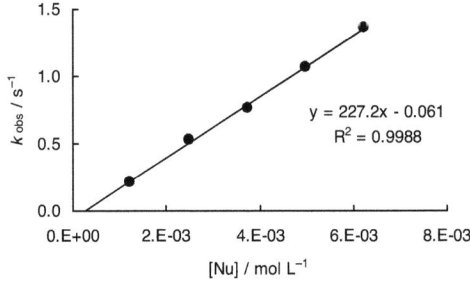

$k = 2.27 \times 10^2$ L mol$^{-1}$ s$^{-1}$

Reaction of (Me$_2$N)Tr$^+$BF$_4^-$ with HSnBu$_3$ (conventional UV-vis spectrometry, 482 nm, CH$_2$Cl$_2$, 20 °C).

| [E] / mol L$^{-1}$ | [Nu] / mol L$^{-1}$ | [Nu]/[E] | $k_{obs}$ / s$^{-1}$ |
|---|---|---|---|
| 4.96 × 10$^{-5}$ | 1.64 × 10$^{-3}$ | 33 | 1.86 × 10$^{-2}$ |
| 4.92 × 10$^{-5}$ | 3.26 × 10$^{-3}$ | 66 | 3.70 × 10$^{-2}$ |
| 4.94 × 10$^{-5}$ | 4.90 × 10$^{-3}$ | 99 | 5.84 × 10$^{-2}$ |
| 4.85 × 10$^{-5}$ | 6.42 × 10$^{-3}$ | 132 | 7.58 × 10$^{-2}$ |
| 4.79 × 10$^{-5}$ | 7.93 × 10$^{-3}$ | 165 | 9.16 × 10$^{-2}$ |

# Experimental Part 255

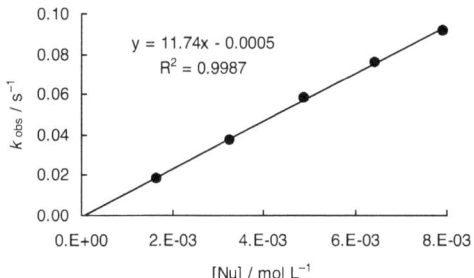

$k = 1.17 \times 10^1$ L mol$^{-1}$ s$^{-1}$

Reaction of (Me$_2$N)$_2$Tr$^+$BF$_4^-$ with HSnBu$_3$ (conventional UV-vis spectrometry, 623 nm, CH$_2$Cl$_2$, 20 °C).

| [E] / mol L$^{-1}$ | [Nu] / mol L$^{-1}$ | [Nu]/[E] | $k_{obs}$ / s$^{-1}$ |
|---|---|---|---|
| 1.66 × 10$^{-5}$ | 2.36 × 10$^{-3}$ | 142 | 7.03 × 10$^{-4}$ |
| 1.63 × 10$^{-5}$ | 2.78 × 10$^{-3}$ | 171 | 8.27 × 10$^{-4}$ |
| 1.60 × 10$^{-5}$ | 3.19 × 10$^{-3}$ | 199 | 9.44 × 10$^{-4}$ |
| 1.62 × 10$^{-5}$ | 7.81 × 10$^{-3}$ | 483 | 2.35 × 10$^{-3}$ |
| 1.72 × 10$^{-5}$ | 6.65 × 10$^{-3}$ | 386 | 2.03 × 10$^{-3}$ |

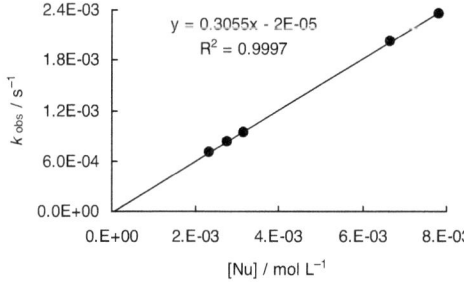

$k = 3.06 \times 10^{-1}$ L mol$^{-1}$ s$^{-1}$

Reaction of (Me$_2$N)$_2$Tr$^+$BF$_4^-$ with Bu$_4$N$^+$BH$_4^-$ (stopped-flow UV-vis spectrometry, 623 nm, CH$_2$Cl$_2$, 20 °C).

| [E] / mol L$^{-1}$ | [Nu] / mol L$^{-1}$ | [Nu]/[E] | $k_{obs}$ / s$^{-1}$ |
|---|---|---|---|
| 1.64 × 10$^{-5}$ | 9.02 × 10$^{-5}$ | 6 | 2.07 |
| 1.64 × 10$^{-5}$ | 1.80 × 10$^{-4}$ | 11 | 5.23 |
| 1.64 × 10$^{-5}$ | 2.71 × 10$^{-4}$ | 17 | 8.89 |
| 1.64 × 10$^{-5}$ | 3.61 × 10$^{-4}$ | 22 | 1.21 × 10$^1$ |

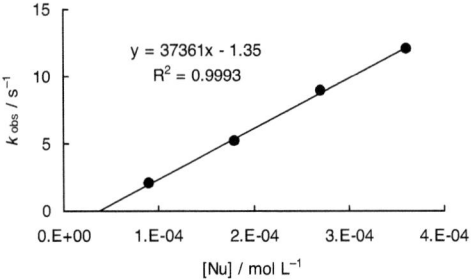

$k = 3.74 \times 10^4 \text{ L mol}^{-1} \text{ s}^{-1}$

Reaction of $(Me_2N)_3Tr^+Cl^-$ with $Bu_4N^+BH_4^-$ (stopped-flow UV-vis spectrometry, 620 nm, $CH_2Cl_2$, 20 °C).

| [E] / mol L$^{-1}$ | [Nu] / mol L$^{-1}$ | [Nu]/[E] | $k_{obs}$ / s$^{-1}$ |
|---|---|---|---|
| $2.25 \times 10^{-5}$ | $8.74 \times 10^{-4}$ | 39 | 1.63 |
| $2.25 \times 10^{-5}$ | $1.88 \times 10^{-3}$ | 84 | 3.31 |
| $2.25 \times 10^{-5}$ | $3.48 \times 10^{-3}$ | 155 | 5.90 |
| $2.25 \times 10^{-5}$ | $5.56 \times 10^{-3}$ | 247 | 9.16 |

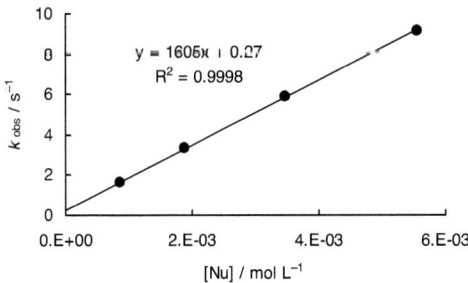

$k = 1.61 \times 10^3 \text{ L mol}^{-1} \text{ s}^{-1}$

Experimental Part 257

## 4.7. Reactions of benzhydrylium ions with hydride donors

Reaction of (ani)PhCH$^+$ with Bu$_4$Sn (conventional UV-vis spectrometry, 469 nm, CH$_2$Cl$_2$, 20 °C). The carbocation was generated in solution by mixing (ani)PhCHCl with GaCl$_3$.

| [E] / mol L$^{-1}$ | [GaCl$_3$]/[E] | [Nu] / mol L$^{-1}$ | [Nu]/[E] | $k_{obs}$ / s$^{-1}$ |
|---|---|---|---|---|
| 4.32 × 10$^{-5}$ | 17 | 1.40 × 10$^{-3}$ | 32 | 8.62 × 10$^{-2}$ |
| 4.29 × 10$^{-5}$ | 17 | 2.78 × 10$^{-3}$ | 65 | 2.22 × 10$^{-1}$ |
| 4.06 × 10$^{-5}$ | 17 | 3.95 × 10$^{-3}$ | 97 | 3.22 × 10$^{-1}$ |
| 4.11 × 10$^{-5}$ | 17 | 5.32 × 10$^{-3}$ | 130 | 4.40 × 10$^{-1}$ |

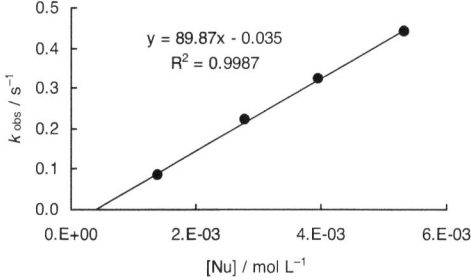

$k = 8.99 \times 10^1$ L mol$^{-1}$ s$^{-1}$

Reaction of (ani)(tol)CH$^+$ with Bu$_4$Sn (conventional UV-vis spectrometry, 488 nm, CH$_2$Cl$_2$, 20 °C). The carbocation was generated in solution by mixing (ani)(tol)CHCl with GaCl$_3$.

| [E] / mol L$^{-1}$ | [GaCl$_3$]/[E] | [Nu] / mol L$^{-1}$ | [Nu]/[E] | $k_{obs}$ / s$^{-1}$ |
|---|---|---|---|---|
| 5.31 × 10$^{-5}$ | 20 | 1.31 × 10$^{-3}$ | 25 | 1.77 × 10$^{-2}$ |
| 3.71 × 10$^{-5}$ | 20 | 2.75 × 10$^{-3}$ | 74 | 4.32 × 10$^{-2}$ |
| 3.29 × 10$^{-5}$ | 20 | 4.89 × 10$^{-3}$ | 149 | 8.03 × 10$^{-2}$ |
| 3.43 × 10$^{-5}$ | 20 | 6.38 × 10$^{-3}$ | 186 | 1.06 × 10$^{-1}$ |

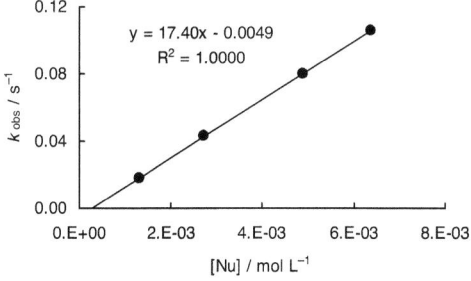

$k = 1.74 \times 10^1$ L mol$^{-1}$ s$^{-1}$

Reaction of (ani)$_2$CH$^+$ with Bu$_4$Sn (conventional UV-vis spectrometry, 512 nm, CH$_2$Cl$_2$, 20 °C). The carbocation was generated in solution by mixing (ani)$_2$CHCl with GaCl$_3$.

| [E] / mol L$^{-1}$ | [GaCl$_3$]/[E] | [Nu] / mol L$^{-1}$ | [Nu]/[E] | $k_{obs}$ / s$^{-1}$ |
|---|---|---|---|---|
| 2.10 × 10$^{-5}$ | 32 | 2.48 × 10$^{-3}$ | 118 | 1.85 × 10$^{-3}$ |
| 2.10 × 10$^{-5}$ | 32 | 3.72 × 10$^{-3}$ | 177 | 2.41 × 10$^{-3}$ |
| 2.16 × 10$^{-5}$ | 32 | 5.09 × 10$^{-3}$ | 236 | 2.90 × 10$^{-3}$ |
| 2.05 × 10$^{-5}$ | 32 | 6.06 × 10$^{-3}$ | 295 | 3.67 × 10$^{-3}$ |

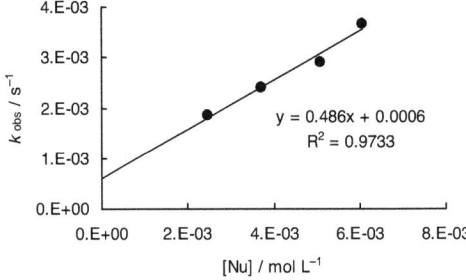

$k = 4.86 \times 10^{-1}$ L mol$^{-1}$ s$^{-1}$

Plot of log $k$ for the reactions of Bu$_4$Sn with substituted benzhydrylium ions versus the corresponding $E$-parameters (CH$_2$Cl$_2$, 20 °C).

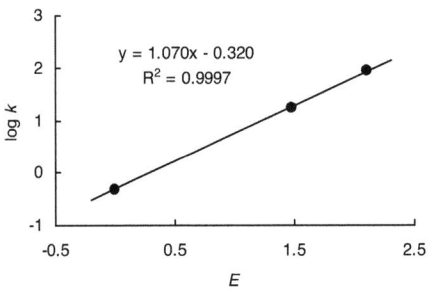

$N = -0.30, s_N = 1.07$

Reaction of (ani)PhCH$^+$ with 2-propyl-1,3-dioxolane (conventional UV-vis spectrometry, 469 nm, CH$_2$Cl$_2$, 20 °C). The carbocation was generated in solution by mixing (ani)PhCHCl with GaCl$_3$.

| [E] / mol L$^{-1}$ | [GaCl$_3$]/[E] | [Nu] / mol L$^{-1}$ | [Nu]/[E] | $k_{obs}$ / s$^{-1}$ |
|---|---|---|---|---|
| 2.91 × 10$^{-5}$ | 14 | 7.55 × 10$^{-4}$ | 26 | 4.63 × 10$^{-3}$ |
| 2.89 × 10$^{-5}$ | 14 | 1.50 × 10$^{-3}$ | 52 | 9.75 × 10$^{-3}$ |
| 2.88 × 10$^{-5}$ | 14 | 2.23 × 10$^{-3}$ | 78 | 1.47 × 10$^{-2}$ |
| 2.86 × 10$^{-5}$ | 14 | 2.96 × 10$^{-3}$ | 104 | 2.10 × 10$^{-2}$ |

Experimental Part 259

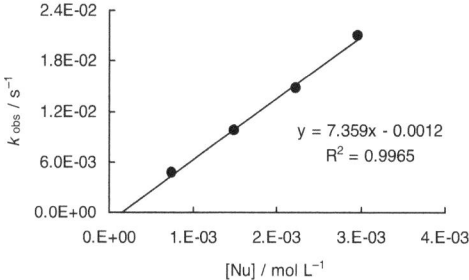

$k = 7.36 \text{ L mol}^{-1} \text{ s}^{-1}$

Reaction of (ani)(tol)CH$^+$ with 2-propyl-1,3-dioxolane (conventional UV-vis spectrometry, 488 nm, CH$_2$Cl$_2$, 20 °C). The carbocation was generated in solution by mixing (ani)(tol)CHCl with GaCl$_3$.

| [E] / mol L$^{-1}$ | [GaCl$_3$]/[E] | [Nu] / mol L$^{-1}$ | [Nu]/[E] | $k_{obs}$ / s$^{-1}$ |
|---|---|---|---|---|
| 3.60 × 10$^{-5}$ | 22 | 1.48 × 10$^{-3}$ | 41 | 2.61 × 10$^{-3}$ |
| 3.59 × 10$^{-5}$ | 22 | 2.21 × 10$^{-3}$ | 62 | 3.93 × 10$^{-3}$ |
| 3.59 × 10$^{-5}$ | 22 | 2.95 × 10$^{-3}$ | 82 | 6.00 × 10$^{-3}$ |
| 3.55 × 10$^{-5}$ | 22 | 3.64 × 10$^{-3}$ | 103 | 7.11 × 10$^{-3}$ |

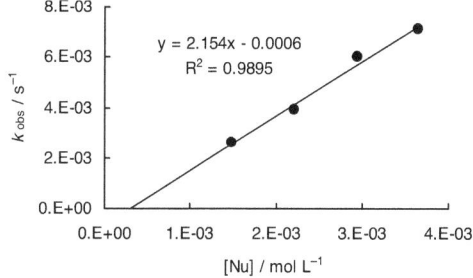

$k = 2.15 \text{ L mol}^{-1} \text{ s}^{-1}$

Reaction of (ani)$_2$CH$^+$ with 2-propyl-1,3-dioxolane (conventional UV-vis spectrometry, 512 nm, CH$_2$Cl$_2$, 20 °C). The carbocation was generated in solution by mixing (ani)$_2$CHCl with GaCl$_3$.

| [E] / mol L$^{-1}$ | [GaCl$_3$]/[E] | [Nu] / mol L$^{-1}$ | [Nu]/[E] | $k_{obs}$ / s$^{-1}$ |
|---|---|---|---|---|
| 2.96 × 10$^{-5}$ | 27 | 2.93 × 10$^{-3}$ | 99 | 4.13 × 10$^{-4}$ |
| 1.46 × 10$^{-5}$ | 27 | 4.35 × 10$^{-3}$ | 297 | 6.88 × 10$^{-4}$ |
| 1.46 × 10$^{-5}$ | 27 | 5.78 × 10$^{-3}$ | 396 | 8.83 × 10$^{-4}$ |
| 1.44 × 10$^{-5}$ | 27 | 7.15 × 10$^{-3}$ | 496 | 1.00 × 10$^{-4}$ |

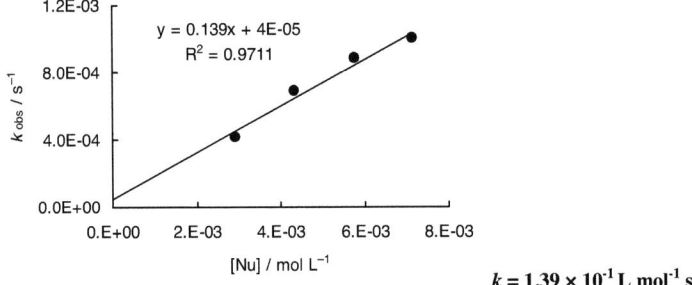

$k = 1.39 \times 10^{-1} \text{ L mol}^{-1} \text{ s}^{-1}$

Plot of log $k$ for the reactions of 2-propyl-1,3-dioxolane with substituted benzhy-drylium ions versus the corresponding $E$-parameters (CH$_2$Cl$_2$, 20 °C).

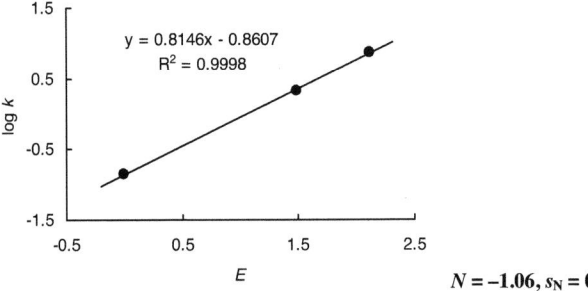

$N = -1.06, s_N = 0.81$

Reaction of (fur)$_2$CH$^+$ with HSiEt$_3$ (stopped-flow UV-vis spectrometry, 512 nm, CH$_2$Cl$_2$, 20 °C). The carbocation was generated in solution by mixing (fur)$_2$CHCl with GaCl$_3$.

| [E] / mol L$^{-1}$ | [GaCl$_3$]/[E] | [Nu] / mol L$^{-1}$ | [Nu]/[E] | $k_{obs}$ / s$^{-1}$ |
|---|---|---|---|---|
| $3.03 \times 10^{-5}$ | 21 | $7.10 \times 10^{-3}$ | 239 | $2.97 \times 10^{-1}$ |
| $3.03 \times 10^{-5}$ | 21 | $1.55 \times 10^{-2}$ | 521 | $6.47 \times 10^{-1}$ |
| $3.03 \times 10^{-5}$ | 21 | $2.53 \times 10^{-2}$ | 851 | 1.01 |
| $3.03 \times 10^{-5}$ | 21 | $3.61 \times 10^{-2}$ | $1.22 \times 10^{3}$ | 1.39 |
| $3.03 \times 10^{-5}$ | 21 | $4.69 \times 10^{-2}$ | $1.58 \times 10^{3}$ | 1.81 |

# Experimental Part

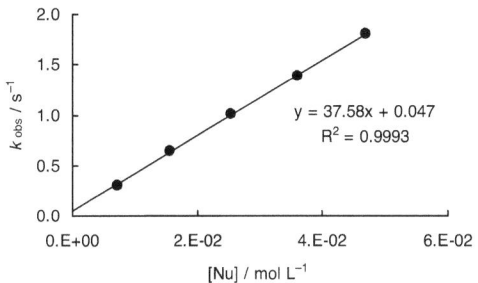

$k = 3.76 \times 10^1 \, \text{L mol}^{-1} \, \text{s}^{-1}$

Reaction of (ani)(tol)CH$^+$ with HSiEt$_3$ (stopped-flow UV-vis spectrometry, 488 nm, CH$_2$Cl$_2$, 20 °C). The carbocation was generated in solution by mixing (ani)(tol)CHCl with GaCl$_3$.

| [E] / mol L$^{-1}$ | [GaCl$_3$]/[E] | [Nu] / mol L$^{-1}$ | [Nu]/[E] | $k_{obs}$ / s$^{-1}$ |
|---|---|---|---|---|
| 3.65 × 10$^{-5}$ | 40 | 1.09 × 10$^{-2}$ | 299 | 5.11 × 10$^1$ |
| 3.65 × 10$^{-5}$ | 40 | 1.59 × 10$^{-2}$ | 435 | 7.72 × 10$^1$ |
| 3.65 × 10$^{-5}$ | 40 | 2.61 × 10$^{-2}$ | 715 | 1.27 × 10$^2$ |
| 3.65 × 10$^{-5}$ | 40 | 3.37 × 10$^{-2}$ | 923 | 1.61 × 10$^2$ |
| 3.65 × 10$^{-5}$ | 40 | 4.90 × 10$^{-2}$ | 1.34 × 10$^3$ | 2.38 × 10$^2$ |

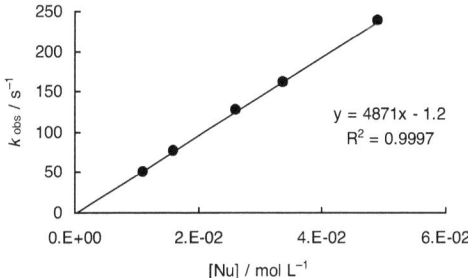

$k = 4.87 \times 10^3 \, \text{L mol}^{-1} \, \text{s}^{-1}$

Reaction of (ani)$_2$CH$^+$ with HSiEt$_3$ (stopped-flow UV-vis spectrometry, 512 nm, CH$_2$Cl$_2$, 20 °C). The carbocation was generated in solution by mixing (ani)$_2$CHCl with GaCl$_3$.

| [E] / mol L$^{-1}$ | [GaCl$_3$]/[E] | [Nu] / mol L$^{-1}$ | [Nu]/[E] | $k_{obs}$ / s$^{-1}$ |
|---|---|---|---|---|
| 3.54 × 10$^{-5}$ | 22 | 7.10 × 10$^{-3}$ | 200 | 2.59 |
| 3.54 × 10$^{-5}$ | 22 | 1.55 × 10$^{-2}$ | 437 | 5.70 |
| 3.54 × 10$^{-5}$ | 22 | 2.53 × 10$^{-2}$ | 714 | 9.37 |
| 3.54 × 10$^{-5}$ | 22 | 3.61 × 10$^{-2}$ | 1.02 × 10$^3$ | 1.41 × 10$^1$ |
| 3.54 × 10$^{-5}$ | 22 | 4.69 × 10$^{-2}$ | 1.32 × 10$^3$ | 1.83 × 10$^1$ |

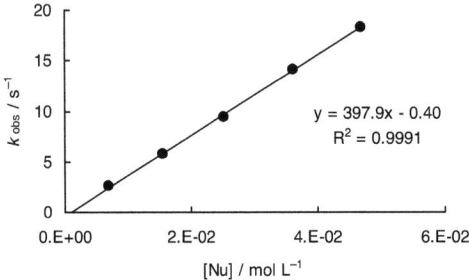

$k = 3.98 \times 10^2 \text{ L mol}^{-1} \text{ s}^{-1}$

Plot of log $k$ for the reactions of HSiEt$_3$ with substituted benzhydrylium ions versus the corresponding $E$-parameters (CH$_2$Cl$_2$, 20 °C).

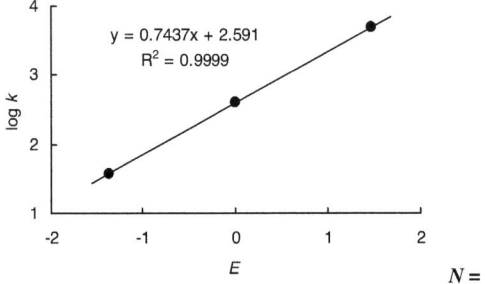

$N = 3.48, s_N = 0.74$

Reaction of (ani)(tol)CH$^+$ with HSiMe$_2$Ph (stopped-flow UV-vis spectrometry, 488 nm, CH$_2$Cl$_2$, 20 °C). The carbocation was generated in solution by mixing (ani)(tol)CHCl with GaCl$_3$.

| [E] / mol L$^{-1}$ | [GaCl$_3$]/[E] | [Nu] / mol L$^{-1}$ | [Nu]/[E] | $k_{obs}$ / s$^{-1}$ |
|---|---|---|---|---|
| 3.65 × 10$^{-5}$ | 40 | 3.93 × 10$^{-3}$ | 108 | 4.49 × 10$^1$ |
| 3.65 × 10$^{-5}$ | 40 | 7.87 × 10$^{-3}$ | 216 | 6.80 × 10$^1$ |
| 3.65 × 10$^{-5}$ | 40 | 1.18 × 10$^{-2}$ | 323 | 9.09 × 10$^1$ |
| 3.65 × 10$^{-5}$ | 40 | 1.57 × 10$^{-2}$ | 430 | 1.16 × 10$^2$ |

# Experimental Part 263

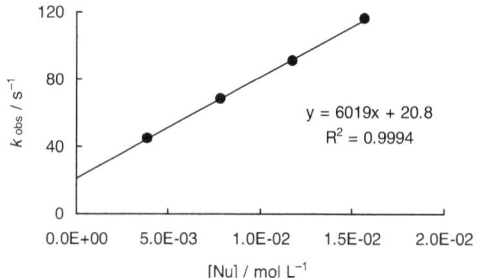

$k = 6.02 \times 10^3$ L mol$^{-1}$ s$^{-1}$

Reaction of (fur)$_2$CH$^+$ with HSiMe$_2$Ph (stopped-flow UV-vis spectrometry, 535 nm, CH$_2$Cl$_2$, 20 °C). The carbocation was generated in solution by mixing (fur)$_2$CHCl with GaCl$_3$.

| [E] / mol L$^{-1}$ | [GaCl$_3$]/[E] | [Nu] / mol L$^{-1}$ | [Nu]/[E] | $k_{obs}$ / s$^{-1}$ |
|---|---|---|---|---|
| 3.03 × 10$^{-5}$ | 21 | 3.93 × 10$^{-3}$ | 130 | 1.72 × 10$^{-1}$ |
| 3.03 × 10$^{-5}$ | 21 | 7.87 × 10$^{-3}$ | 260 | 3.47 × 10$^{-1}$ |
| 3.03 × 10$^{-5}$ | 21 | 1.18 × 10$^{-2}$ | 389 | 5.31 × 10$^{-1}$ |
| 3.03 × 10$^{-5}$ | 21 | 1.57 × 10$^{-2}$ | 518 | 6.99 × 10$^{-1}$ |
| 3.03 × 10$^{-5}$ | 21 | 1.97 × 10$^{-2}$ | 650 | 8.80 × 10$^{-1}$ |

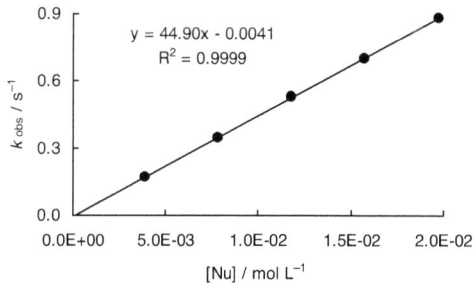

$k = 4.49 \times 10^1$ L mol$^{-1}$ s$^{-1}$

Reaction of (ani)$_2$CH$^+$ with HSiMe$_2$Ph (stopped-flow UV-vis spectrometry, 512 nm, CH$_2$Cl$_2$, 20 °C). The carbocation was generated in solution by mixing (ani)$_2$CHCl with GaCl$_3$.

| [E] / mol L$^{-1}$ | [GaCl$_3$]/[E] | [Nu] / mol L$^{-1}$ | [Nu]/[E] | $k_{obs}$ / s$^{-1}$ |
|---|---|---|---|---|
| 3.54 × 10$^{-5}$ | 22 | 3.93 × 10$^{-3}$ | 111 | 1.60 |
| 3.54 × 10$^{-5}$ | 22 | 7.87 × 10$^{-3}$ | 222 | 3.34 |
| 3.54 × 10$^{-5}$ | 22 | 1.18 × 10$^{-2}$ | 333 | 5.00 |
| 3.54 × 10$^{-5}$ | 22 | 1.57 × 10$^{-2}$ | 444 | 6.74 |
| 3.54 × 10$^{-5}$ | 22 | 1.97 × 10$^{-2}$ | 556 | 8.44 |

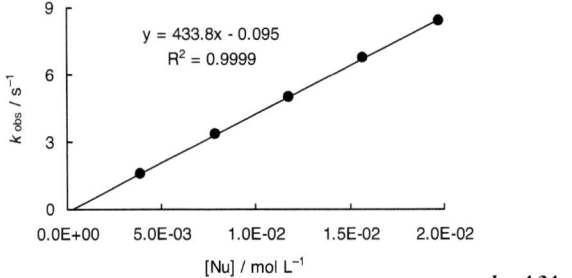

$$k = 4.34 \times 10^2 \text{ L mol}^{-1} \text{ s}^{-1}$$

Plot of log $k$ for the reactions of HSiMe$_2$Ph with substituted benzhydrylium ions versus the corresponding $E$-parameters (CH$_2$Cl$_2$, 20 °C).

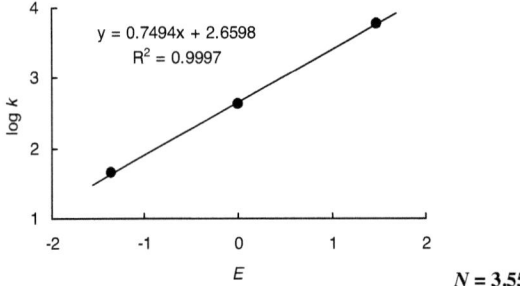

$$N = 3.55, s_N = 0.75$$

Reaction of (ani)(tol)CH$^+$ with HSiBu$_3$ (stopped-flow UV-vis spectrometry, 488 nm, CH$_2$Cl$_2$, 20 °C). The carbocation was generated in solution by mixing (ani)(tol)CHCl with GaCl$_3$.

| [E] / mol L$^{-1}$ | [GaCl$_3$]/[E] | [Nu] / mol L$^{-1}$ | [Nu]/[E] | $k_{obs}$ / s$^{-1}$ |
|---|---|---|---|---|
| $3.40 \times 10^{-5}$ | 53 | $3.47 \times 10^{-3}$ | 102 | $3.87 \times 10^1$ |
| $3.40 \times 10^{-5}$ | 53 | $4.94 \times 10^{-3}$ | 145 | $5.20 \times 10^1$ |
| $3.40 \times 10^{-5}$ | 53 | $8.28 \times 10^{-3}$ | 244 | $8.95 \times 10^1$ |
| $3.40 \times 10^{-5}$ | 53 | $9.35 \times 10^{-3}$ | 275 | $1.04 \times 10^2$ |
| $3.40 \times 10^{-5}$ | 53 | $1.11 \times 10^{-2}$ | 326 | $1.23 \times 10^2$ |

## Experimental Part

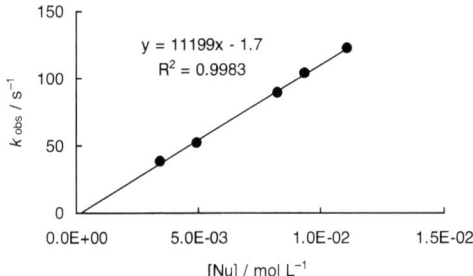

$k = 1.12 \times 10^4$ L mol$^{-1}$ s$^{-1}$

Reaction of (ani)$_2$CH$^+$ with HSiBu$_3$ (stopped-flow UV-vis spectrometry, 512 nm, CH$_2$Cl$_2$, 20 °C). The carbocation was generated in solution by mixing (ani)$_2$CHCl with GaCl$_3$.

| [E] / mol L$^{-1}$ | [GaCl$_3$]/[E] | [Nu] / mol L$^{-1}$ | [Nu]/[E] | $k_{obs}$ / s$^{-1}$ |
|---|---|---|---|---|
| 2.59 × 10$^{-5}$ | 57 | 3.47 × 10$^{-3}$ | 134 | 2.94 |
| 2.59 × 10$^{-5}$ | 57 | 8.28 × 10$^{-3}$ | 320 | 7.10 |
| 2.59 × 10$^{-5}$ | 57 | 9.35 × 10$^{-3}$ | 361 | 8.17 |
| 2.59 × 10$^{-5}$ | 57 | 1.11 × 10$^{-2}$ | 428 | 9.70 |

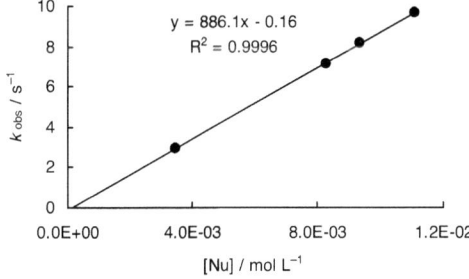

$k = 8.86 \times 10^2$ L mol$^{-1}$ s$^{-1}$

Reaction of (fur)$_2$CH$^+$ with HSiBu$_3$ (stopped-flow UV-vis spectrometry, 535 nm, CH$_2$Cl$_2$, 20 °C). The carbocation was generated in solution by mixing (fur)$_2$CHCl with GaCl$_3$.

| [E] / mol L$^{-1}$ | [GaCl$_3$]/[E] | [Nu] / mol L$^{-1}$ | [Nu]/[E] | $k_{obs}$ / s$^{-1}$ |
|---|---|---|---|---|
| 2.62 × 10$^{-5}$ | 52 | 3.47 × 10$^{-3}$ | 132 | 2.75 × 10$^{-1}$ |
| 2.62 × 10$^{-5}$ | 52 | 4.94 × 10$^{-3}$ | 189 | 4.11 × 10$^{-1}$ |
| 2.62 × 10$^{-5}$ | 52 | 8.28 × 10$^{-3}$ | 316 | 7.14 × 10$^{-1}$ |
| 2.62 × 10$^{-5}$ | 52 | 9.35 × 10$^{-3}$ | 357 | 8.16 × 10$^{-1}$ |
| 2.62 × 10$^{-5}$ | 52 | 1.11 × 10$^{-2}$ | 423 | 9.40 × 10$^{-1}$ |

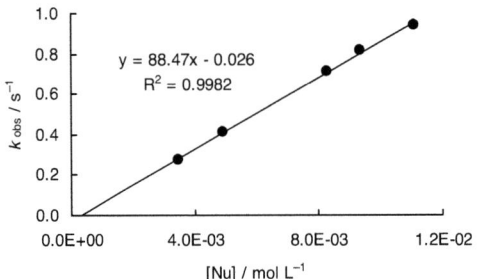

$k = 8.85 \times 10^1$ L mol$^{-1}$ s$^{-1}$

Plot of log $k$ for the reactions of HSiBu$_3$ with substituted benzhydrylium ions versus the corresponding $E$-parameters (CH$_2$Cl$_2$, 20 °C).

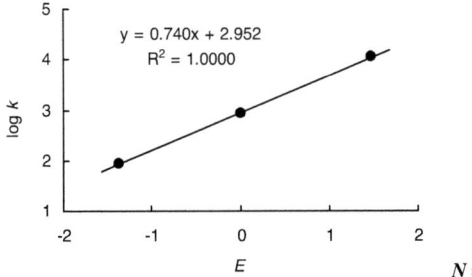

$N = 3.99, s_N = 0.74$

Reaction of (ani)(tol)CH$^+$ with HSiPh$_3$ (stopped-flow UV-vis spectrometry, 488 nm, CH$_2$Cl$_2$, 20 °C). The carbocation was generated in solution by mixing (ani)(tol)CHCl with GaCl$_3$.

| [E] / mol L$^{-1}$ | [GaCl$_3$]/[E] | [Nu] / mol L$^{-1}$ | [Nu]/[E] | $k_{obs}$ / s$^{-1}$ |
|---|---|---|---|---|
| 3.40 × 10$^{-5}$ | 53 | 4.24 × 10$^{-3}$ | 125 | 3.92 |
| 3.40 × 10$^{-5}$ | 53 | 9.02 × 10$^{-3}$ | 265 | 8.70 |
| 3.40 × 10$^{-5}$ | 53 | 1.42 × 10$^{-2}$ | 417 | 1.38 × 10$^1$ |
| 3.40 × 10$^{-5}$ | 53 | 1.65 × 10$^{-2}$ | 485 | 1.64 × 10$^1$ |
| 3.40 × 10$^{-5}$ | 53 | 1.89 × 10$^{-2}$ | 556 | 1.89 × 10$^1$ |

Experimental Part 267

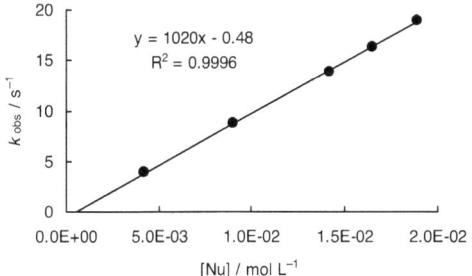

$k = 1.02 \times 10^3$ L mol$^{-1}$ s$^{-1}$

Reaction of (ani)$_2$CH$^+$ with HSiPh$_3$ (stopped-flow UV-vis spectrometry, 512 nm, CH$_2$Cl$_2$, 20 °C). The carbocation was generated in solution by mixing (ani)$_2$CHCl with GaCl$_3$.

| [E] / mol L$^{-1}$ | [GaCl$_3$]/[E] | [Nu] / mol L$^{-1}$ | [Nu]/[E] | $k_{obs}$ / s$^{-1}$ |
|---|---|---|---|---|
| 2.59 × 10$^{-5}$ | 57 | 4.24 × 10$^{-3}$ | 164 | 3.11 × 10$^{-1}$ |
| 2.59 × 10$^{-5}$ | 57 | 9.02 × 10$^{-3}$ | 348 | 6.72 × 10$^{-1}$ |
| 2.59 × 10$^{-5}$ | 57 | 1.42 × 10$^{-2}$ | 547 | 1.03 |
| 2.59 × 10$^{-5}$ | 57 | 1.65 × 10$^{-2}$ | 637 | 1.24 |
| 2.59 × 10$^{-5}$ | 57 | 1.89 × 10$^{-2}$ | 730 | 1.39 |

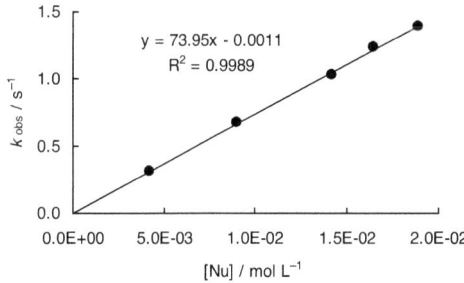

$k = 7.40 \times 10^1$ L mol$^{-1}$ s$^{-1}$

Reaction of (fur)$_2$CH$^+$ with HSiPh$_3$ (stopped-flow UV-vis spectrometry, 535 nm, CH$_2$Cl$_2$, 20 °C). The carbocation was generated in solution by mixing (fur)$_2$CHCl with GaCl$_3$.

| [E] / mol L$^{-1}$ | [GaCl$_3$]/[E] | [Nu] / mol L$^{-1}$ | [Nu]/[E] | $k_{obs}$ / s$^{-1}$ |
|---|---|---|---|---|
| 2.62 × 10$^{-5}$ | 52 | 4.24 × 10$^{-3}$ | 162 | 3.81 × 10$^{-2}$ |
| 2.62 × 10$^{-5}$ | 52 | 9.02 × 10$^{-3}$ | 344 | 8.19 × 10$^{-2}$ |
| 2.62 × 10$^{-5}$ | 52 | 1.42 × 10$^{-2}$ | 541 | 1.29 × 10$^{-1}$ |
| 2.62 × 10$^{-5}$ | 52 | 1.65 × 10$^{-2}$ | 630 | 1.49 × 10$^{-1}$ |
| 2.62 × 10$^{-5}$ | 52 | 1.89 × 10$^{-2}$ | 722 | 1.70 × 10$^{-1}$ |

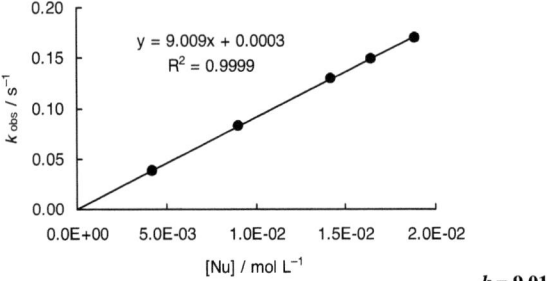

$k = 9.01$ L mol$^{-1}$ s$^{-1}$

Plot of log $k$ for the reactions of HSiPh$_3$ with substituted benzhydrylium ions versus the corresponding $E$-parameters (CH$_2$Cl$_2$, 20 °C).

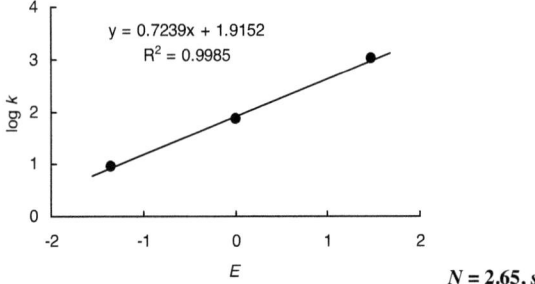

$N = 2.65$, $s_N = 0.72$

Reaction of (fur)$_2$CH$^+$ with HSi(SiMe$_3$)$_3$ (stopped-flow UV-vis spectrometry, 535 nm, CH$_2$Cl$_2$, 20 °C). The carbocation was generated in solution by mixing (fur)$_2$CHCl with trimethylsilyl triflate (TMSOTf).

| [E] / mol L$^{-1}$ | [TMSOTf]/[E] | [Nu] / mol L$^{-1}$ | [Nu]/[E] | $k_{obs}$ / s$^{-1}$ |
|---|---|---|---|---|
| 4.39 × 10$^{-5}$ | 14 | 1.99 × 10$^{-3}$ | 45 | 1.55 × 10$^{-1}$ |
| 4.39 × 10$^{-5}$ | 14 | 3.68 × 10$^{-3}$ | 84 | 3.02 × 10$^{-1}$ |
| 4.39 × 10$^{-5}$ | 14 | 7.32 × 10$^{-3}$ | 167 | 6.50 × 10$^{-1}$ |
| 4.39 × 10$^{-5}$ | 14 | 1.15 × 10$^{-2}$ | 262 | 9.94 × 10$^{-1}$ |

# Experimental Part 269

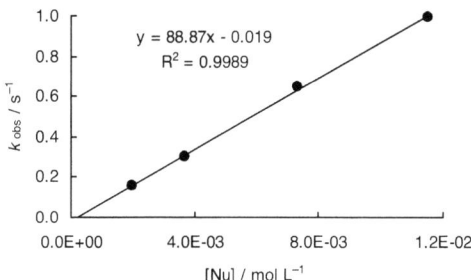

$k = 8.89 \times 10^1$ L mol$^{-1}$ s$^{-1}$

Reaction of (ani)$_2$CH$^+$ with HSi(SiMe$_3$)$_3$ (stopped-flow UV-vis spectrometry, 512 nm, CH$_2$Cl$_2$, 20 °C). The carbocation was generated in solution by mixing (ani)$_2$CHCl with trimethylsilyl triflate (TMSOTf).

| [E] / mol L$^{-1}$ | [TMSOTf]/[E] | [Nu] / mol L$^{-1}$ | [Nu]/[E] | $k_{obs}$ / s$^{-1}$ |
|---|---|---|---|---|
| 2.71 × 10$^{-5}$ | 6 | 1.89 × 10$^{-3}$ | 70 | 1.27 |
| 2.71 × 10$^{-5}$ | 6 | 4.02 × 10$^{-3}$ | 148 | 2.58 |
| 2.71 × 10$^{-5}$ | 6 | 7.04 × 10$^{-3}$ | 260 | 5.71 |
| 2.71 × 10$^{-5}$ | 6 | 1.07 × 10$^{-2}$ | 395 | 9.20 |
| 2.71 × 10$^{-5}$ | 6 | 1.27 × 10$^{-2}$ | 469 | 1.18 × 10$^1$ |

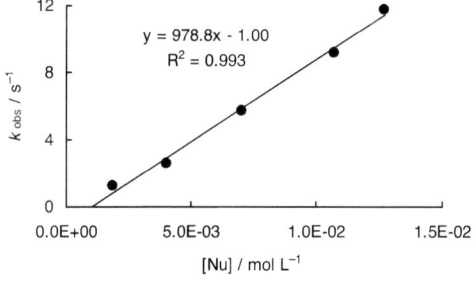

$k = 9.79 \times 10^2$ L mol$^{-1}$ s$^{-1}$

Reaction of (ani)(tol)CH$^+$ with HSi(SiMe$_3$)$_3$ (stopped-flow UV-vis spectrometry, 482 nm, CH$_2$Cl$_2$, 20 °C). The carbocation was generated in solution by mixing (ani)(tol)CHCl with trimethylsilyl triflate (TMSOTf).

| [E] / mol L$^{-1}$ | [TMSOTf]/[E] | [Nu] / mol L$^{-1}$ | [Nu]/[E] | $k_{obs}$ / s$^{-1}$ |
|---|---|---|---|---|
| 7.21 × 10$^{-5}$ | 9 | 1.99 × 10$^{-3}$ | 28 | 2.39 × 10$^1$ |
| 7.21 × 10$^{-5}$ | 9 | 3.68 × 10$^{-3}$ | 51 | 3.85 × 10$^1$ |
| 7.21 × 10$^{-5}$ | 9 | 7.32 × 10$^{-3}$ | 102 | 7.65 × 10$^1$ |
| 7.21 × 10$^{-5}$ | 9 | 8.97 × 10$^{-3}$ | 124 | 8.25 × 10$^1$ |
| 7.21 × 10$^{-5}$ | 9 | 1.15 × 10$^{-2}$ | 160 | 1.07 × 10$^2$ |

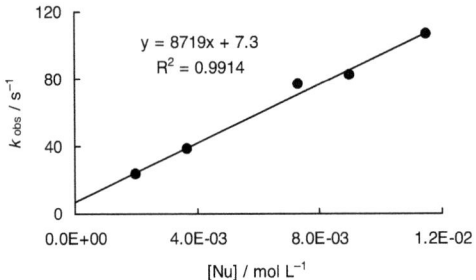

$k = 8.72 \times 10^3$ L mol$^{-1}$ s$^{-1}$

Reaction of (mfa)$_2$CH$^+$BF$_4^-$ with HSi(SiMe$_3$)$_3$ (conventional UV-vis spectrometry, 593 nm, CH$_2$Cl$_2$, 20 °C).

| [E] / mol L$^{-1}$ | [Nu] / mol L$^{-1}$ | [Nu]/[E] | $k_{obs}$ / s$^{-1}$ |
|---|---|---|---|
| 9.56 × 10$^{-6}$ | 2.75 × 10$^{-3}$ | 288 | 1.41 × 10$^{-3}$ |
| 1.01 × 10$^{-5}$ | 4.38 × 10$^{-3}$ | 432 | 2.18 × 10$^{-3}$ |
| 9.91 × 10$^{-6}$ | 5.71 × 10$^{-3}$ | 576 | 2.79 × 10$^{-3}$ |
| 9.68 × 10$^{-6}$ | 8.37 × 10$^{-3}$ | 864 | 4.40 × 10$^{-3}$ |

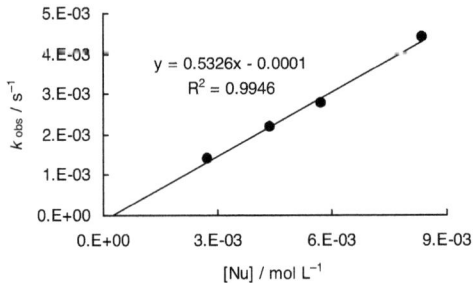

$k = 5.33 \times 10^{-1}$ L mol$^{-1}$ s$^{-1}$

Reaction of (pfa)$_2$CH$^+$BF$_4^-$ with HSi(SiMe$_3$)$_3$ (conventional UV-vis spectrometry, 601 nm, CH$_2$Cl$_2$, 20 °C).

| [E] / mol L$^{-1}$ | [Nu] / mol L$^{-1}$ | [Nu]/[E] | $k_{obs}$ / s$^{-1}$ |
|---|---|---|---|
| 2.33 × 10$^{-5}$ | 2.83 × 10$^{-3}$ | 121 | 6.08 × 10$^{-3}$ |
| 2.31 × 10$^{-5}$ | 4.21 × 10$^{-3}$ | 182 | 7.82 × 10$^{-3}$ |
| 2.32 × 10$^{-5}$ | 5.65 × 10$^{-3}$ | 243 | 1.13 × 10$^{-2}$ |
| 2.39 × 10$^{-5}$ | 8.71 × 10$^{-3}$ | 365 | 1.75 × 10$^{-2}$ |

# Experimental Part 271

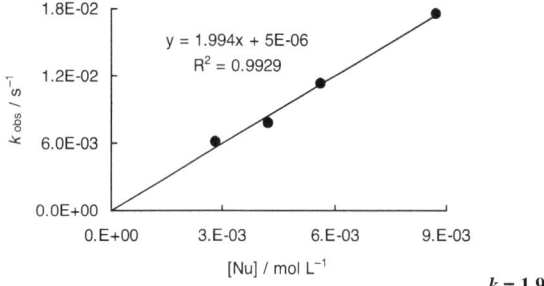

$k = 1.99 \text{ L mol}^{-1} \text{ s}^{-1}$

Plot of log $k$ for the reactions of HSi(SiMe$_3$)$_3$ with substituted benzhydrylium ions versus the corresponding $E$-parameters (CH$_2$Cl$_2$, 20 °C).

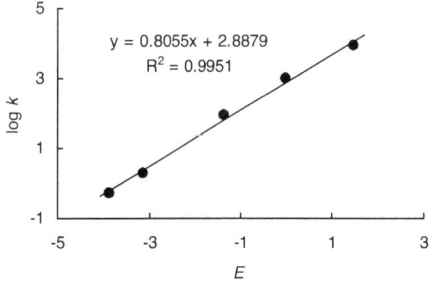

$N = 3.59, s_N = 0.81$

Reaction of (dma)$_2$CH$^+$BF$_4^-$ with carbene borane **1** (conventional UV-vis spectrometry, 612 nm, CH$_2$Cl$_2$, 20 °C).

| [E] / mol L$^{-1}$ | [Nu] / mol L$^{-1}$ | [Nu]/[E] | $k_{\text{obs}}$ / s$^{-1}$ |
|---|---|---|---|
| $1.02 \times 10^{-5}$ | $1.06 \times 10^{-4}$ | 10 | $1.35 \times 10^{-2}$ |
| $1.05 \times 10^{-5}$ | $2.18 \times 10^{-4}$ | 21 | $2.75 \times 10^{-2}$ |
| $1.07 \times 10^{-5}$ | $3.33 \times 10^{-4}$ | 31 | $3.98 \times 10^{-2}$ |

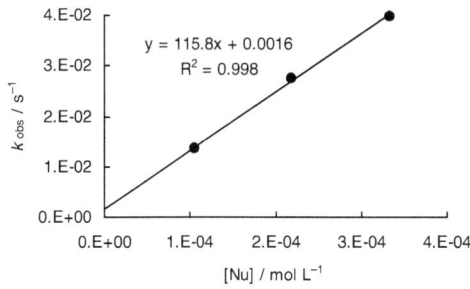

$k = 1.16 \times 10^2 \text{ L mol}^{-1} \text{ s}^{-1}$

Reaction of (thq)$_2$CH$^+$BF$_4^-$ with carbene borane **1** (conventional UV-vis spectrometry, 627 nm, CH$_2$Cl$_2$, 20 °C).

| [E] / mol L$^{-1}$ | [Nu] / mol L$^{-1}$ | [Nu]/[E] | $k_{obs}$ / s$^{-1}$ |
|---|---|---|---|
| 1.09 × 10$^{-5}$ | 1.07 × 10$^{-4}$ | 10 | 1.53 × 10$^{-3}$ |
| 1.09 × 10$^{-5}$ | 2.14 × 10$^{-4}$ | 20 | 2.80 × 10$^{-3}$ |
| 2.18 × 10$^{-5}$ | 3.23 × 10$^{-4}$ | 15 | 3.78 × 10$^{-3}$ |

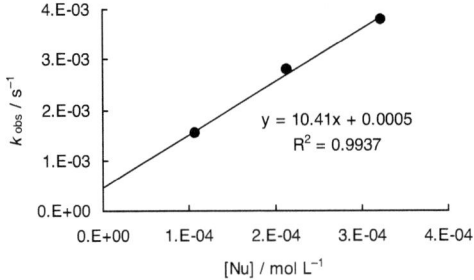

$k = 1.04 \times 10^1$ L mol$^{-1}$ s$^{-1}$

Reaction of (jul)$_2$CH$^+$BF$_4^-$ with carbene borane **1** (conventional UV-vis spectrometry, 643 nm, CH$_2$Cl$_2$, 20 °C).

| [E] / mol L$^{-1}$ | [Nu] / mol L$^{-1}$ | [Nu]/[E] | $k_{obs}$ / s$^{-1}$ |
|---|---|---|---|
| 1.38 × 10$^{-5}$ | 2.17 × 10$^{-4}$ | 16 | 2.73 × 10$^{-4}$ |
| 1.40 × 10$^{-5}$ | 3.29 × 10$^{-4}$ | 24 | 4.15 × 10$^{-4}$ |

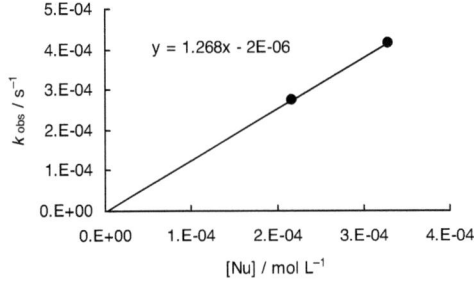

$k = 1.27$ L mol$^{-1}$ s$^{-1}$

Reaction of (thq)$_2$CH$^+$BF$_4^-$ with carbene borane **2** (conventional UV-vis spectrometry, 627 nm, CH$_2$Cl$_2$, 20 °C).

| [E] / mol L$^{-1}$ | [Nu] / mol L$^{-1}$ | [Nu]/[E] | $k_{obs}$ / s$^{-1}$ |
|---|---|---|---|
| 1.10 × 10$^{-5}$ | 1.61 × 10$^{-4}$ | 15 | 5.73 × 10$^{-2}$ |
| 1.12 × 10$^{-5}$ | 3.29 × 10$^{-4}$ | 29 | 1.28 × 10$^{-1}$ |
| 1.62 × 10$^{-5}$ | 4.74 × 10$^{-4}$ | 29 | 1.84 × 10$^{-1}$ |

Experimental Part 273

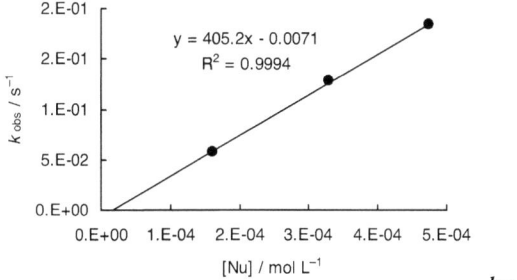

$k = 4.05 \times 10^2$ L mol$^{-1}$ s$^{-1}$

Reaction of (jul)$_2$CH$^+$BF$_4^-$ with carbene borane **2** (conventional UV-vis spectrometry, 643 nm, CH$_2$Cl$_2$, 20 °C).

| [E] / mol L$^{-1}$ | [Nu] / mol L$^{-1}$ | [Nu]/[E] | $k_{obs}$ / s$^{-1}$ |
|---|---|---|---|
| 1.44 × 10$^{-5}$ | 1.68 × 10$^{-4}$ | 12 | 6.89 × 10$^{-3}$ |
| 1.32 × 10$^{-5}$ | 3.08 × 10$^{-4}$ | 23 | 1.50 × 10$^{-2}$ |
| 1.38 × 10$^{-5}$ | 4.84 × 10$^{-4}$ | 35 | 2.45 × 10$^{-2}$ |
| 1.38 × 10$^{-5}$ | 6.44 × 10$^{-4}$ | 47 | 3.23 × 10$^{-2}$ |
| 1.34 × 10$^{-5}$ | 7.78 × 10$^{-4}$ | 58 | 3.75 × 10$^{-2}$ |

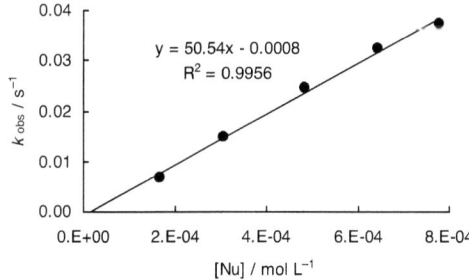

$k = 5.05 \times 10^1$ L mol$^{-1}$ s$^{-1}$

### 4.8. Reactions of tritylium ions with imidazole

Reaction of (MeO)$_2$Tr$^+$BF$_4^-$ with imidazole (stopped-flow UV-vis spectrometry, 497 nm, CH$_3$CN, 20 °C).

| [E] / mol L$^{-1}$ | [Nu] / mol L$^{-1}$ | [Nu]/[E] | $k_{obs}$ / s$^{-1}$ |
|---|---|---|---|
| 4.36 × 10$^{-5}$ | 6.56 × 10$^{-4}$ | 15 | 6.56 × 10$^{-4}$ |
| 4.36 × 10$^{-5}$ | 1.31 × 10$^{-3}$ | 30 | 1.31 × 10$^{-3}$ |
| 4.36 × 10$^{-5}$ | 1.97 × 10$^{-3}$ | 45 | 1.97 × 10$^{-3}$ |

# Experimental Part

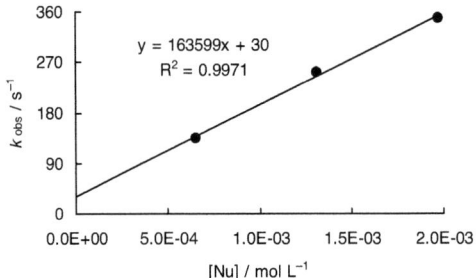

$k = 1.64 \times 10^5$ L mol$^{-1}$ s$^{-1}$

Reaction of (MeO)$_3$Tr$^+$BF$_4^-$ with imidazole (stopped-flow UV-vis spectrometry, 484 nm, CH$_3$CN, 20 °C).

| [E] / mol L$^{-1}$ | [Nu] / mol L$^{-1}$ | [Nu]/[E] | $k_{obs}$ / s$^{-1}$ |
|---|---|---|---|
| 3.24 × 10$^{-5}$ | 6.56 × 10$^{-4}$ | 20 | 1.89 × 10$^1$ |
| 3.24 × 10$^{-5}$ | 1.31 × 10$^{-3}$ | 40 | 3.69 × 10$^1$ |
| 3.24 × 10$^{-5}$ | 1.97 × 10$^{-3}$ | 61 | 5.36 × 10$^1$ |
| 3.24 × 10$^{-5}$ | 2.62 × 10$^{-3}$ | 81 | 6.98 × 10$^1$ |
| 3.24 × 10$^{-5}$ | 3.28 × 10$^{-3}$ | 101 | 8.70 × 10$^1$ |

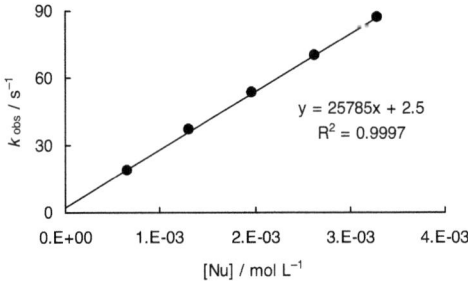

$k = 2.58 \times 10^4$ L mol$^{-1}$ s$^{-1}$

Reaction of (Me$_2$N)Tr$^+$BF$_4^-$ with imidazole (stopped-flow UV-vis spectrometry, 461 nm, CH$_3$CN, 20 °C).

| [E] / mol L$^{-1}$ | [Nu] / mol L$^{-1}$ | [Nu]/[E] | $k_{obs}$ / s$^{-1}$ |
|---|---|---|---|
| 6.30 × 10$^{-5}$ | 1.20 × 10$^{-3}$ | 19 | 6.20 × 10$^{-2}$ |
| 6.30 × 10$^{-5}$ | 2.57 × 10$^{-3}$ | 41 | 1.31 × 10$^{-1}$ |
| 6.30 × 10$^{-5}$ | 4.99 × 10$^{-3}$ | 79 | 2.49 × 10$^{-1}$ |
| 6.30 × 10$^{-5}$ | 6.67 × 10$^{-3}$ | 106 | 3.49 × 10$^{-1}$ |

Experimental Part

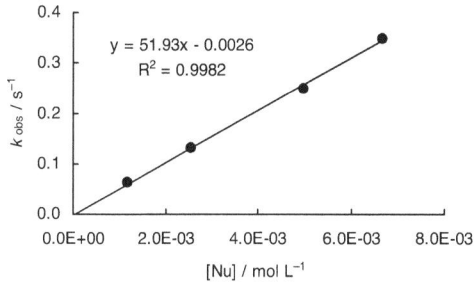

$k = 5.19 \times 10^1$ L mol$^{-1}$ s$^{-1}$

### 4.9. Ionizations of trityl halides and esters

Ionization rates of Me$_3$TrCl ($c_0$ = 1.12 × 10$^{-3}$ mol L$^{-1}$) in the presence of various concentrations of piperidine; 100AN, 25 C°, stopped-flow conductimetry.

| [pip] / mol L$^{-1}$ | [pip] / [substrate] | $k_{obs}$ / s$^{-1}$ |
|---|---|---|
| 2.22 × 10$^{-2}$ | 20 | 2.12 × 10$^2$ |
| 3.43 × 10$^{-2}$ | 31 | 2.27 × 10$^2$ |
| 6.66 × 10$^{-2}$ | 59 | 2.50 × 10$^2$ |
| 8.17 × 10$^{-2}$ | 73 | 2.20 × 10$^2$ |
| 1.05 × 10$^{-1}$ | 94 | 2.38 × 10$^2$ |
| 1.21 × 10$^{-1}$ | 108 | 2.50 × 10$^2$ |
| $k_{ion}$ = 2.50 × 10$^2$ s$^{-1}$ | | |

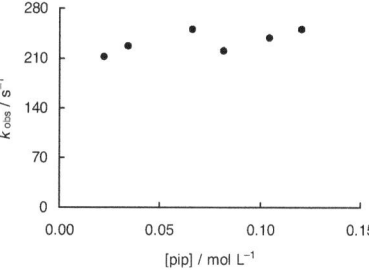

Ionization rates of Me$_3$TrCl ($c_0$ = 8.73 × 10$^{-4}$ mol L$^{-1}$) in the presence of various concentrations of piperidine; 100A, 25 C°, stopped-flow conductimetry.

| [pip] / mol L$^{-1}$ | [pip] / [substrate] | $k_{obs}$ / s$^{-1}$ |
|---|---|---|
| 1.37 × 10$^{-2}$ | 16 | 6.97 × 10$^{-1}$ |
| 2.11 × 10$^{-2}$ | 24 | 6.70 × 10$^{-1}$ |
| 5.07 × 10$^{-2}$ | 58 | 7.68 × 10$^{-1}$ |
| 6.43 × 10$^{-2}$ | 74 | 7.60 × 10$^{-1}$ |
| 8.77 × 10$^{-2}$ | 100 | 7.15 × 10$^{-1}$ |
| 1.12 × 10$^{-1}$ | 128 | 6.89 × 10$^{-1}$ |
| $k_{ion}$ = 7.68 × 10$^{-1}$ s$^{-1}$ | | |

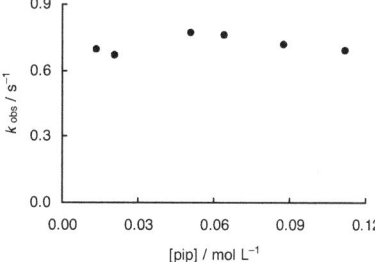

Ionization rates of Me$_2$TrCl ($c_0$ = 1.05 × 10$^{-3}$ mol L$^{-1}$) in the presence of various concentrations of piperidine; 100AN, 25 C°, stopped-flow conductimetry.

| [pip] / mol L$^{-1}$ | [pip] / [substrate] | $k_{obs}$ / s$^{-1}$ |
|---|---|---|
| 1.97 × 10$^{-2}$ | 19 | 4.40 × 10$^1$ |
| 6.63 × 10$^{-2}$ | 63 | 4.95 × 10$^1$ |
| 9.46 × 10$^{-2}$ | 90 | 5.19 × 10$^1$ |
| 1.20 × 10$^{-1}$ | 114 | 4.80 × 10$^1$ |
| 2.25 × 10$^{-1}$ | 214 | 4.80 × 10$^1$ |
| $k_{ion}$ = 5.19 × 10$^1$ s$^{-1}$ | | |

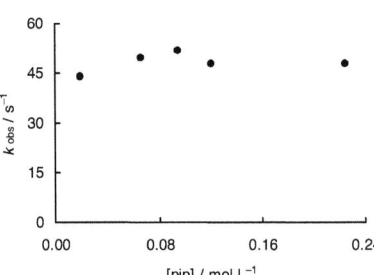

Ionization rates of Me$_2$TrCl ($c_0$ = 8.47 × 10$^{-4}$ mol L$^{-1}$) in the presence of various concentrations of piperidine; 100A, 25 C°, stopped-flow conductimetry.

| [pip] / mol L$^{-1}$ | [pip] / [substrate] | $k_{obs}$ / s$^{-1}$ |
|---|---|---|
| 1.48 × 10$^{-2}$ | 17 | 7.11 × 10$^{-2}$ |
| 2.64 × 10$^{-2}$ | 31 | 7.86 × 10$^{-2}$ |
| 4.76 × 10$^{-2}$ | 56 | 8.11 × 10$^{-2}$ |
| 8.67 × 10$^{-2}$ | 102 | 8.04 × 10$^{-2}$ |
| 1.15 × 10$^{-1}$ | 136 | 7.99 × 10$^{-2}$ |
| 1.28 × 10$^{-1}$ | 151 | 7.99 × 10$^{-2}$ |
| $k_{ion}$ = 8.11 × 10$^{-2}$ s$^{-1}$ | | |

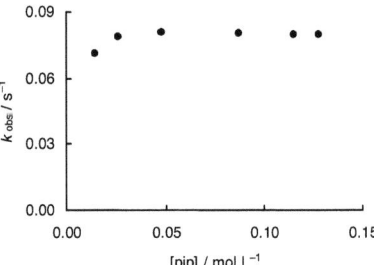

Ionization rates of MeTrCl ($c_0$ = 1.07 × 10$^{-3}$ mol L$^{-1}$) in the presence of various concentrations of piperidine; 100AN, 25 C°, stopped-flow conductimetry.

| [pip] / mol L$^{-1}$ | [pip] / [substrate] | $k_{obs}$ / s$^{-1}$ |
|---|---|---|
| 6.82 × 10$^{-3}$ | 6 | 4.07 |
| 1.36 × 10$^{-2}$ | 13 | 5.02 |
| 6.82 × 10$^{-2}$ | 64 | 5.91 |
| 1.36 × 10$^{-1}$ | 127 | 5.80 |
| $k_{ion}$ = 5.91 s$^{-1}$ | | |

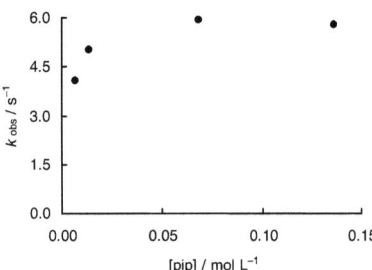

Experimental Part 277

Ionization rates of MeTrCl ($c_0 = 1.00 \times 10^{-3}$ mol L$^{-1}$) in the presence of various concentrations of piperidine; 90AN10W, 25 C°, stopped-flow conductimetry.

| [pip] / mol L$^{-1}$ | [pip] / [substrate] | $k_{obs}$ / s$^{-1}$ |
|---|---|---|
| $4.50 \times 10^{-3}$ | 5 | $3.99 \times 10^2$ |
| $9.01 \times 10^{-3}$ | 9 | $4.35 \times 10^2$ |
| $3.64 \times 10^{-2}$ | 36 | $4.77 \times 10^2$ |
| $7.21 \times 10^{-2}$ | 72 | $4.66 \times 10^2$ |
| $1.11 \times 10^{-1}$ | 111 | $4.50 \times 10^2$ |
| $1.44 \times 10^{-1}$ | 144 | $4.48 \times 10^2$ |
| $k_{ion} = 4.77 \times 10^2$ s$^{-1}$ | | |

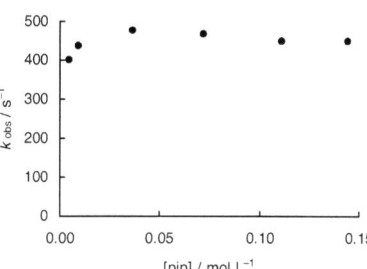

Ionization rates of MeTrBr ($c_0 = 9.78 \times 10^{-4}$ mol L$^{-1}$) in the presence of various concentrations of piperidine; 100A, 25 C°, stopped-flow conductimetry.

| [pip] / mol L$^{-1}$ | [pip] / [substrate] | $k_{obs}$ / s$^{-1}$ |
|---|---|---|
| $1.42 \times 10^{-2}$ | 15 | $2.70 \times 10^1$ |
| $2.84 \times 10^{-2}$ | 29 | $3.15 \times 10^1$ |
| $5.69 \times 10^{-2}$ | 58 | $3.49 \times 10^1$ |
| $8.53 \times 10^{-2}$ | 87 | $3.91 \times 10^1$ |
| $1.41 \times 10^{-1}$ | 144 | $3.91 \times 10^1$ |
| $2.84 \times 10^{-1}$ | 290 | $3.72 \times 10^1$ |
| $k_{ion} = 3.91 \times 10^1$ s$^{-1}$ | | |

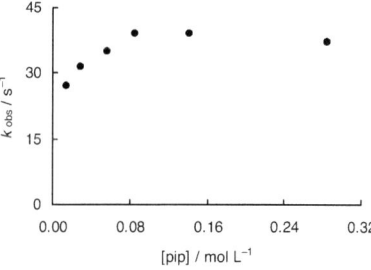

Ionization rates of TrCl ($c_0 = 1.65 \times 10^{-3}$ mol L$^{-1}$) in the presence of various concentrations of piperidine; 100AN, 25 C°, stopped-flow conductimetry.

| [pip] / mol L$^{-1}$ | [pip] / [substrate] | $k_{obs}$ / s$^{-1}$ |
|---|---|---|
| $1.29 \times 10^{-2}$ | 8 | $4.34 \times 10^{-1}$ |
| $2.59 \times 10^{-2}$ | 16 | $4.72 \times 10^{-1}$ |
| $5.18 \times 10^{-2}$ | 31 | $4.91 \times 10^{-1}$ |
| $1.04 \times 10^{-1}$ | 63 | $4.85 \times 10^{-1}$ |
| $k_{ion} = 4.91 \times 10^{-1}$ s$^{-1}$ | | |

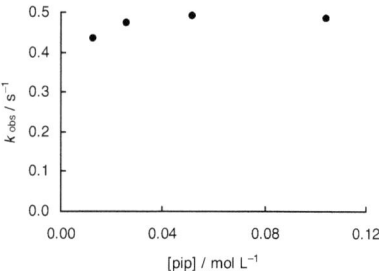

Ionization rates of TrCl ($c_0 = 1.00 \times 10^{-3}$ mol L$^{-1}$) in the presence of various concentrations of piperidine; 90AN10W, 25 C°, stopped-flow conductimetry.

| [pip] / mol L$^{-1}$ | [pip] / [substrate] | $k_{obs}$ / s$^{-1}$ |
|---|---|---|
| 4.54 × 10$^{-3}$ | 5 | 6.65 × 10$^{1}$ |
| 9.08 × 10$^{-3}$ | 9 | 7.46 × 10$^{1}$ |
| 3.63 × 10$^{-2}$ | 36 | 7.92 × 10$^{1}$ |
| 7.27 × 10$^{-2}$ | 73 | 8.09 × 10$^{1}$ |
| 1.09 × 10$^{-1}$ | 109 | 7.87 × 10$^{1}$ |
| 1.46 × 10$^{-1}$ | 146 | 7.66 × 10$^{1}$ |
| $k_{ion} = 8.09 \times 10^{1}$ s$^{-1}$ | | |

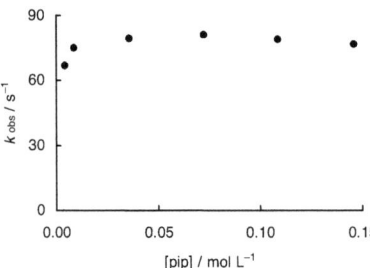

Ionization rates of TrCl ($c_0 = 9.97 \times 10^{-4}$ mol L$^{-1}$) in the presence of various concentrations of piperidine; 80AN20W, 25 C°, stopped-flow conductimetry.

| [pip] / mol L$^{-1}$ | [pip] / [substrate] | $k_{obs}$ / s$^{-1}$ |
|---|---|---|
| 5.00 × 10$^{-3}$ | 5 | 2.39 × 10$^{2}$ |
| 1.00 × 10$^{-2}$ | 10 | 2.46 × 10$^{2}$ |
| 4.00 × 10$^{-2}$ | 40 | 2.52 × 10$^{2}$ |
| 8.00 × 10$^{-2}$ | 80 | 2.45 × 10$^{2}$ |
| 1.18 × 10$^{-1}$ | 119 | 2.47 × 10$^{2}$ |
| 1.65 × 10$^{-1}$ | 165 | 2.35 × 10$^{2}$ |
| $k_{ion} = 2.52 \times 10^{2}$ s$^{-1}$ | | |

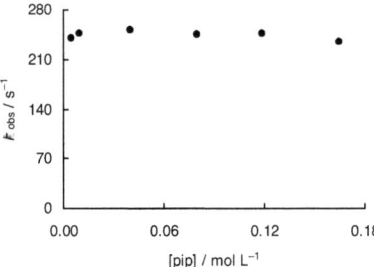

Ionization rates of TrBr ($c_0 = 9.28 \times 10^{-4}$ mol L$^{-1}$) in the presence of various concentrations of piperidine; 100AN, 25 C°, stopped-flow conductimetry.

| [pip] / mol L$^{-1}$ | [pip] / [substrate] | $k_{obs}$ / s$^{-1}$ |
|---|---|---|
| 1.08 × 10$^{-3}$ | 1 | 5.56 × 10$^{1}$ |
| 5.39 × 10$^{-3}$ | 6 | 2.68 × 10$^{2}$ |
| 1.08 × 10$^{-2}$ | 12 | 3.88 × 10$^{2}$ |
| 2.16 × 10$^{-2}$ | 23 | 4.97 × 10$^{2}$ |
| 4.31 × 10$^{-2}$ | 46 | 5.60 × 10$^{2}$ |
| 1.63 × 10$^{-1}$ | 176 | 6.04 × 10$^{2}$ |
| $k_{ion} = 6.04 \times 10^{2}$ s$^{-1}$ | | |

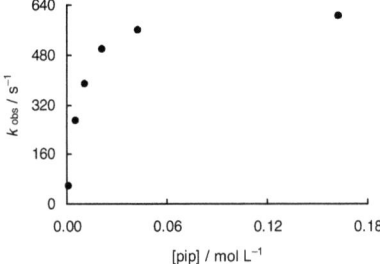

Experimental Part 279

Ionization rates of TrF ($c_0 = 6.71 \times 10^{-4}$ mol L$^{-1}$) in the presence of various concentrations of amines; aqueous acetonitrile, 25 C°, conventional conductimetry.

| solvent | amine | [amine] / mol L$^{-1}$ | $k_{ion}$ / s$^{-1}$ | average $k_{ion}$ / s$^{-1}$ |
|---|---|---|---|---|
| 80AN20W | piperidine | 6.61 × 10$^{-3}$ | 3.98 × 10$^{-5}$ | 4.02 × 10$^{-5}$ |
|  |  |  | 4.06 × 10$^{-5}$ |  |
| 60AN40W | piperidine | 6.61 × 10$^{-3}$ | 4.35 × 10$^{-4}$ | 4.37 × 10$^{-4}$ |
|  | triethylamine | 5.64 × 10$^{-3}$ | 4.38 × 10$^{-4}$ |  |
| 50AN50W | piperidine | 6.61 × 10$^{-3}$ | 1.30 × 10$^{-3}$ | 1.30 × 10$^{-3}$ |
|  | triethylamine | 5.64 × 10$^{-3}$ | 1.29 × 10$^{-3}$ |  |

Ionization rates of ($p$Cl)TrCl in the presence of various concentrations of piperidine; 80AN20W, 25 C°, stopped-flow conductimetry.

| [substrate] / mol L$^{-1}$ | [pip] / mol L$^{-1}$ | [pip] / [substrate] | $k_{obs}$ / s$^{-1}$ |
|---|---|---|---|
| 8.62 × 10$^{-4}$ | 0 | 0 | 1.00 × 10$^2$ |
| 8.97 × 10$^{-4}$ | 1.81 × 10$^{-2}$ | 20 | 1.25 × 10$^2$ |
| 8.97 × 10$^{-4}$ | 2.35 × 10$^{-2}$ | 26 | 1.15 × 10$^2$ |
| 8.97 × 10$^{-4}$ | 3.74 × 10$^{-2}$ | 42 | 1.30 × 10$^2$ |
| 8.97 × 10$^{-4}$ | 5.34 × 10$^{-2}$ | 60 | 1.21 × 10$^2$ |
| 8.97 × 10$^{-4}$ | 7.15 × 10$^{-2}$ | 80 | 1.19 × 10$^2$ |
| $k_{ion} = 1.30 \times 10^2$ s$^{-1}$ | | | |

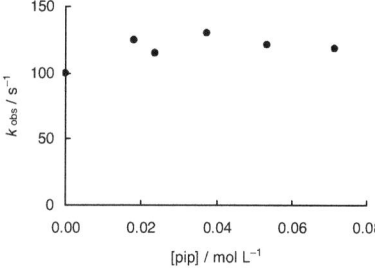

Ionization rates of ($p$Cl)TrCl ($c_0 = 8.62 \times 10^{-4}$ mol L$^{-1}$) in aqueous acetonitrile without amine, 25 C°, stopped-flow conductimetry.

| solvent | $k_{ion}$ / s$^{-1}$ |
|---|---|
| 60AN40W | 4.89 × 10$^2$ |
| 50AN50W | 9.52 × 10$^2$ |

Ionization rates of (pCl)TrCl in the presence of various concentrations of piperidine; 90A10W, 25 C°, stopped-flow conductimetry.

| [substrate] / mol L$^{-1}$ | [pip] / mol L$^{-1}$ | [pip] / [substrate] | $k_{obs}$ / s$^{-1}$ |
|---|---|---|---|
| 8.75 × 10$^{-4}$ | 0 | 0 | 8.32 × 10$^{-1}$ |
| 8.30 × 10$^{-4}$ | 4.17 × 10$^{-2}$ | 50 | 1.52 |
| 8.30 × 10$^{-4}$ | 9.26 × 10$^{-2}$ | 112 | 1.59 |
| 8.30 × 10$^{-4}$ | 1.40 × 10$^{-1}$ | 169 | 1.56 |
| 8.30 × 10$^{-4}$ | 1.82 × 10$^{-1}$ | 219 | 1.52 |
| 8.30 × 10$^{-4}$ | 2.74 × 10$^{-1}$ | 330 | 1.45 |
| 8.30 × 10$^{-4}$ | 3.85 × 10$^{-1}$ | 464 | 1.34 |
| 8.30 × 10$^{-4}$ | 4.87 × 10$^{-1}$ | 587 | 1.21 |
| 8.30 × 10$^{-4}$ | 5.62 × 10$^{-1}$ | 677 | 1.16 |
| | $k_{ion} = 1.59$ s$^{-1}$ | | |

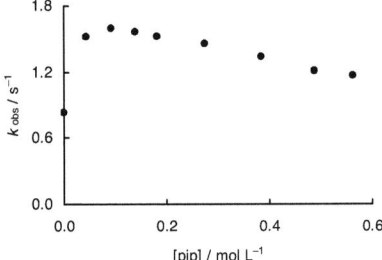

Ionization rates of (pCl)TrCl ($c_0$ = 8.75 × 10$^{-4}$ mol L$^{-1}$) in the presence of various concentrations of piperidine; 80A20W, 25 C°, stopped-flow conductimetry.

| [pip] / mol L$^{-1}$ | [pip] / [substrate] | $k_{obs}$ / s$^{-1}$ |
|---|---|---|
| 0 | 0 | 1.12 × 10$^{1}$ |
| 5.34 × 10$^{-3}$ | 6 | 1.22 × 10$^{1}$ |
| 1.39 × 10$^{-2}$ | 16 | 1.22 × 10$^{1}$ |
| 2.14 × 10$^{-2}$ | 24 | 1.24 × 10$^{1}$ |
| 3.10 × 10$^{-2}$ | 35 | 1.28 × 10$^{1}$ |
| 4.38 × 10$^{-2}$ | 50 | 1.27 × 10$^{1}$ |
| $k_{ion} = 1.28 \times 10^{1}$ s$^{-1}$ | | |

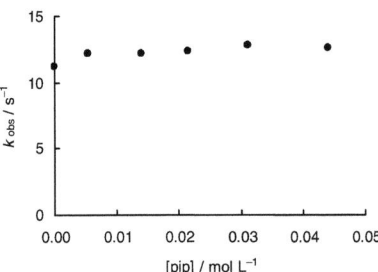

# Experimental Part

Ionization rates of $(p\text{Cl})\text{TrCl}$ ($c_0 = 8.75 \times 10^{-4}$ mol L$^{-1}$) in aqueous acetone without amine, 25 C°, stopped-flow conductimetry.

| solvent | $k_{ion}$ / s$^{-1}$ |
|---|---|
| 60A40W | $1.66 \times 10^2$ |
| 50A50W | $5.90 \times 10^2$ |

Ionization rates of $(p\text{Cl})\text{TrPNB}$ in aqueous acetonitrile with triethylamine as additive, 25 C°, conventional conductimetry.

| solvent | [substrate] / mol L$^{-1}$ | [TEA] / mol L$^{-1}$ | $k_{ion}$ / s$^{-1}$ | average $k_{ion}$ / s$^{-1}$ |
|---|---|---|---|---|
| 90AN10W | $7.51 \times 10^{-4}$ | $3.62 \times 10^{-3}$ | $7.41 \times 10^{-4}$ | $7.24 \times 10^{-4}$ |
|  | $5.26 \times 10^{-4}$ | $4.42 \times 10^{-3}$ | $7.07 \times 10^{-4}$ |  |
| 80AN20W | $6.16 \times 10^{-4}$ | $3.70 \times 10^{-3}$ | $2.22 \times 10^{-3}$ | $2.22 \times 10^{-3}$ |
|  | $6.16 \times 10^{-4}$ | $3.51 \times 10^{-3}$ | $2.22 \times 10^{-3}$ |  |
| 60AN40W | $6.16 \times 10^{-4}$ | $3.33 \times 10^{-3}$ | $6.18 \times 10^{-3}$ | $6.10 \times 10^{-3}$ |
|  | $6.16 \times 10^{-4}$ | $3.14 \times 10^{-3}$ | $6.02 \times 10^{-4}$ |  |
| 50AN50W | $6.16 \times 10^{-4}$ | $4.37 \times 10^{-3}$ | $1.05 \times 10^{-2}$ | $1.04 \times 10^{-2}$ |
|  | $6.16 \times 10^{-4}$ | $3.39 \times 10^{-3}$ | $1.02 \times 10^{-2}$ |  |

Ionization rates of $(p\text{Cl})\text{TrOBz}$ ($c_0 = 9.86 \times 10^{-4}$ mol L$^{-1}$) in the presence of piperidine; aqueous acetone, 25 C°, conventional conductimetry.

| solvent | [pip] / mol L$^{-1}$ | [pip] / [substrate] | $k_{ion}$ / s$^{-1}$ |
|---|---|---|---|
| 60A40W | $1.17 \times 10^{-2}$ | 12 | $1.27 \times 10^{-4}$ |
| 50A50W | $1.17 \times 10^{-2}$ | 12 | $3.58 \times 10^{-4}$ |

Ionization rates of $(p\text{F})\text{TrCl}$ ($c_0 = 8.42 \times 10^{-4}$ mol L$^{-1}$) in the presence of various concentrations of piperidine; 100AN, 25 C°, stopped-flow conductimetry.

| [pip] / mol L$^{-1}$ | [pip] / [substrate] | $k_{obs}$ / s$^{-1}$ |
|---|---|---|
| $5.81 \times 10^{-3}$ | 7 | $5.14 \times 10^{-1}$ |
| $1.16 \times 10^{-2}$ | 14 | $5.85 \times 10^{-1}$ |
| $5.81 \times 10^{-2}$ | 69 | $6.35 \times 10^{-1}$ |
| $1.16 \times 10^{-1}$ | 138 | $6.47 \times 10^{-1}$ |
| $k_{ion} = 6.47 \times 10^{-1}$ s$^{-1}$ |  |  |

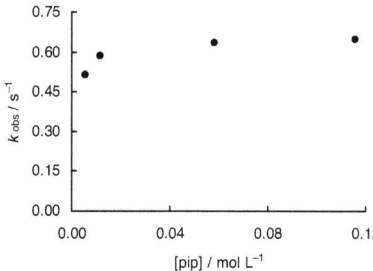

Ionization rates of (pF)TrCl ($c_0 = 1.00 \times 10^{-3}$ mol L$^{-1}$) in the presence of various concentrations of piperidine; 90AN10W, 25 C°, stopped-flow conductimetry.

| [pip] / mol L$^{-1}$ | [pip] / [substrate] | $k_{obs}$ / s$^{-1}$ |
|---|---|---|
| 0 | 0 | $4.29 \times 10^1$ |
| $4.56 \times 10^{-3}$ | 5 | $9.46 \times 10^1$ |
| $9.13 \times 10^{-3}$ | 9 | $1.03 \times 10^2$ |
| $3.65 \times 10^{-2}$ | 37 | $1.09 \times 10^2$ |
| $7.30 \times 10^{-2}$ | 73 | $1.12 \times 10^2$ |
| $1.08 \times 10^{-1}$ | 108 | $1.09 \times 10^2$ |
| $1.48 \times 10^{-1}$ | 148 | $1.04 \times 10^2$ |
| $k_{ion} = 1.12 \times 10^2$ s$^{-1}$ | | |

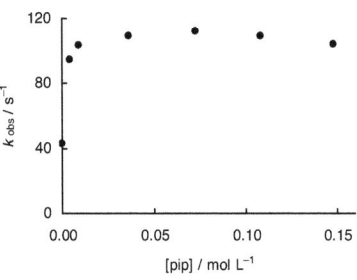

Ionization rates of (pF)TrCl ($c_0 = 1.02 \times 10^{-3}$ mol L$^{-1}$) in the presence of various concentrations of piperidine; 80AN20W, 25 C°, stopped-flow conductimetry.

| [pip] / mol L$^{-1}$ | [pip] / [substrate] | $k_{obs}$ / s$^{-1}$ |
|---|---|---|
| 0 | 0 | $2.58 \times 10^2$ |
| $1.45 \times 10^{-2}$ | 14 | $2.78 \times 10^2$ |
| $2.67 \times 10^{-2}$ | 26 | $2.87 \times 10^2$ |
| $3.74 \times 10^{-2}$ | 37 | $2.91 \times 10^2$ |
| $5.12 \times 10^{-2}$ | 51 | $3.11 \times 10^2$ |
| $7.26 \times 10^{-2}$ | 72 | $2.91 \times 10^2$ |
| $k_{ion} = 3.11 \times 10^2$ s$^{-1}$ | | |

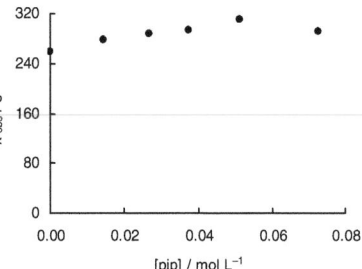

Ionization rates of (pF)TrCl in aqueous solvents without amine, 25 C°, stopped-flow conductimetry.

| solvent | [substrate] / mol L$^{-1}$ | $k_{obs}$ / s$^{-1}$ |
|---|---|---|
| 60AN40W | $1.02 \times 10^{-3}$ | $1.11 \times 10^3$ |
| 60A40W | $8.75 \times 10^{-4}$ | $4.46 \times 10^2$ |

# Experimental Part 283

Ionization rates of $(p\text{F})\text{TrCl}$ in the presence of various concentrations of piperidine; 90A10W, 25 C°, stopped-flow conductimetry.

| [substrate] / mol $L^{-1}$ | [pip] / mol $L^{-1}$ | [pip] / [substrate] | $k_{obs}$ / $s^{-1}$ |
|---|---|---|---|
| $8.16 \times 10^{-4}$ | 0 | 0 | 2.27 |
| $1.05 \times 10^{-3}$ | $4.17 \times 10^{-2}$ | 40 | 5.26 |
| $1.05 \times 10^{-3}$ | $9.26 \times 10^{-2}$ | 88 | 5.35 |
| $1.05 \times 10^{-3}$ | $1.40 \times 10^{-1}$ | 133 | 5.29 |
| $1.05 \times 10^{-3}$ | $1.82 \times 10^{-1}$ | 173 | 5.17 |
| $1.05 \times 10^{-3}$ | $2.74 \times 10^{-1}$ | 261 | 4.94 |
| $1.05 \times 10^{-3}$ | $3.85 \times 10^{-1}$ | 367 | 4.53 |
| $1.05 \times 10^{-3}$ | $4.87 \times 10^{-1}$ | 464 | 4.23 |
| $1.05 \times 10^{-3}$ | $5.62 \times 10^{-1}$ | 535 | 3.94 |
| $k_{ion} = 5.35$ $s^{-1}$ | | | |

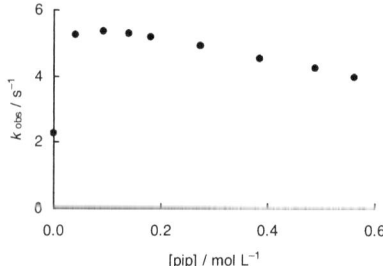

Ionization rates of $(p\text{F})\text{TrCl}$ in the presence of various concentrations of piperidine; 80A20W, 25 C°, stopped-flow conductimetry.

| [substrate] / mol $L^{-1}$ | [pip] / mol $L^{-1}$ | [pip] / [substrate] | $k_{obs}$ / $s^{-1}$ |
|---|---|---|---|
| $8.75 \times 10^{-4}$ | 0 | 0 | $3.06 \times 10^{1}$ |
| $1.01 \times 10^{-3}$ | $4.65 \times 10^{-3}$ | 5 | $2.77 \times 10^{1}$ |
| $1.01 \times 10^{-3}$ | $6.98 \times 10^{-3}$ | 7 | $2.83 \times 10^{1}$ |
| $1.01 \times 10^{-3}$ | $9.30 \times 10^{-3}$ | 9 | $2.92 \times 10^{1}$ |
| $k_{ion} = 3.06 \times 10^{1}$ $s^{-1}$ | | | |

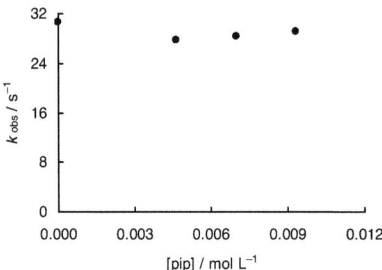

Ionization rates of $(pF)_2TrCl$ ($c_0 = 1.02 \times 10^{-3}$ mol L$^{-1}$) in the presence of various concentrations of piperidine; 90AN10W, 25 C°, stopped-flow conductimetry.

| [pip] / mol L$^{-1}$ | [pip] / [substrate] | $k_{obs}$ / s$^{-1}$ |
|---|---|---|
| $5.03 \times 10^{-3}$ | 5 | $1.30 \times 10^2$ |
| $1.01 \times 10^{-2}$ | 10 | $1.47 \times 10^2$ |
| $4.03 \times 10^{-2}$ | 40 | $1.55 \times 10^2$ |
| $8.05 \times 10^{-2}$ | 79 | $1.51 \times 10^2$ |
| $1.20 \times 10^{-1}$ | 118 | $1.50 \times 10^2$ |
| $1.62 \times 10^{-1}$ | 159 | $1.45 \times 10^2$ |
| $k_{ion} = 1.55 \times 10^2$ s$^{-1}$ | | |

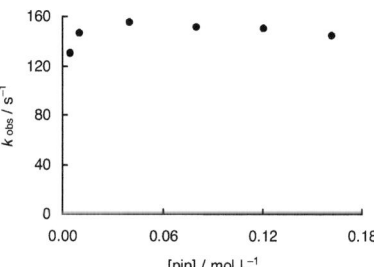

Ionization rates of $(pF)_3TrCl$ ($c_0 = 4.81 \times 10^{-4}$ mol L$^{-1}$) in the presence of various concentrations of piperidine; 100AN, 25 C°, stopped-flow conductimetry.

| [pip] / mol L$^{-1}$ | [pip] / [substrate] | $k_{obs}$ / s$^{-1}$ |
|---|---|---|
| $1.06 \times 10^{-3}$ | 2 | $3.30 \times 10^{-1}$ |
| $5.07 \times 10^{-3}$ | 11 | $7.08 \times 10^{-1}$ |
| $1.06 \times 10^{-2}$ | 22 | $8.33 \times 10^{-1}$ |
| $5.07 \times 10^{-2}$ | 105 | $9.97 \times 10^{-1}$ |
| $1.06 \times 10^{-1}$ | 220 | $1.02$ |
| $k_{ion} = 1.02$ s$^{-1}$ | | |

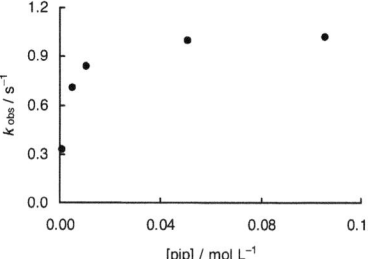

# Experimental Part

Ionization rates of $(p\text{F})_3\text{TrCl}$ ($c_0 = 9.62 \times 10^{-4}$ mol $\text{L}^{-1}$) in the presence of various concentrations of piperidine; 90AN10W, 25 C°, stopped-flow conductimetry.

| [pip] / mol $\text{L}^{-1}$ | [pip] / [substrate] | $k_{\text{obs}}$ / $\text{s}^{-1}$ |
|---|---|---|
| $2.33 \times 10^{-3}$ | 2 | $1.27 \times 10^2$ |
| $4.65 \times 10^{-3}$ | 5 | $1.40 \times 10^2$ |
| $6.98 \times 10^{-3}$ | 7 | $1.44 \times 10^2$ |
| $9.31 \times 10^{-3}$ | 10 | $1.65 \times 10^2$ |
| $1.16 \times 10^{-2}$ | 12 | $1.70 \times 10^2$ |
| $k_{\text{ion}} = 1.70 \times 10^2 \text{ s}^{-1}$ | | |

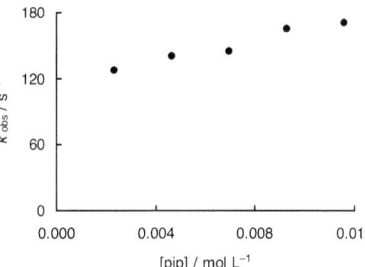

Ionization rates of $(m\text{F})\text{TrCl}$ in the presence of various concentrations of piperidine; 100AN, 25 C°, conventional conductimetry.

| [substrate] / mol $\text{L}^{-1}$ | [pip] / mol $\text{L}^{-1}$ | [pip] / [substrate] | $k_{\text{obs}}$ / $\text{s}^{-1}$ |
|---|---|---|---|
| $1.74 \times 10^{-3}$ | $1.86 \times 10^{-2}$ | 11 | $2.88 \times 10^{-2}$ |
| $1.53 \times 10^{-3}$ | $3.27 \times 10^{-2}$ | 21 | $2.87 \times 10^{-2}$ |
| $1.57 \times 10^{-3}$ | $6.73 \times 10^{-2}$ | 43 | $2.95 \times 10^{-2}$ |
| $k_{\text{ion}} = 2.95 \times 10^{-2} \text{ s}^{-1}$ | | | |

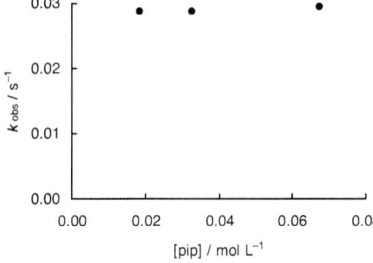

Ionization rates of ($m$F)TrCl ($c_0$ = 1.00 × $10^{-3}$ mol $L^{-1}$) in the presence of various concentrations of piperidine; 90AN10W, 25 C°, stopped-flow conductimetry.

| [pip] / mol $L^{-1}$ | [pip] / [substrate] | $k_{obs}$ / $s^{-1}$ |
|---|---|---|
| 0 | 0 | 5.41 |
| 4.50 × $10^{-3}$ | 5 | 7.61 |
| 9.01 × $10^{-3}$ | 9 | 8.42 |
| 3.64 × $10^{-2}$ | 36 | 9.49 |
| 7.21 × $10^{-2}$ | 72 | 9.51 |
| 1.11 × $10^{-1}$ | 111 | 9.28 |
| 1.44 × $10^{-1}$ | 144 | 9.05 |
| $k_{ion}$ = 9.51 $s^{-1}$ | | |

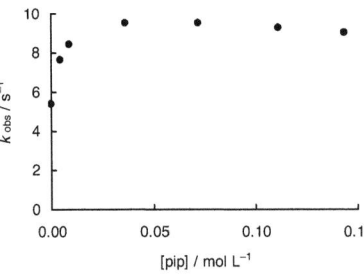

Ionization rates of ($m$F)TrCl ($c_0$ = 9.77 × $10^{-4}$ mol $L^{-1}$) in the presence of various concentrations of piperidine; 80AN20W, 25 C°, stopped-flow conductimetry.

| [pip] / mol $L^{-1}$ | [pip] / [substrate] | $k_{obs}$ / $s^{-1}$ |
|---|---|---|
| 0 | 0 | 2.82 × $10^{1}$ |
| 4.98 × $10^{-3}$ | 5 | 3.09 × $10^{1}$ |
| 9.97 × $10^{-3}$ | 10 | 3.27 × $10^{1}$ |
| 3.99 × $10^{-2}$ | 41 | 3.26 × $10^{1}$ |
| 7.97 × $10^{-2}$ | 82 | 3.23 × $10^{1}$ |
| 1.24 × $10^{-1}$ | 127 | 3.11 × $10^{1}$ |
| 1.62 × $10^{-1}$ | 166 | 3.00 × $10^{1}$ |
| $k_{ion}$ = 3.27 × $10^{1}$ $s^{-1}$ | | |

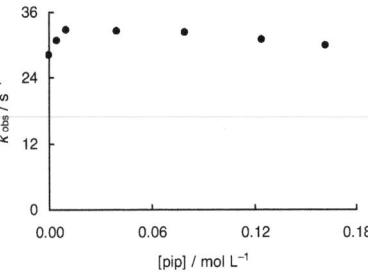

Ionization rates of ($m$F)TrCl in aqueous acetonitrile without amine; 25 C°, stopped-flow conductimetry.

| solvent | [substrate] / mol $L^{-1}$ | $k_{obs}$ / $s^{-1}$ |
|---|---|---|
| 90AN10W | 1.06 × $10^{-3}$ | 5.14 |
| | 1.06 × $10^{-4}$ | 9.22 |
| 80AN20W | 1.06 × $10^{-3}$ | 2.86 × $10^{1}$ |
| | 1.06 × $10^{-4}$ | 3.18 × $10^{1}$ |
| 60AN40W | 1.06 × $10^{-3}$ | 1.32 × $10^{2}$ |
| 50AN50W | 1.06 × $10^{-3}$ | 2.54 × $10^{2}$ |

Experimental Part 287

Ionization rates of $(m\text{F})\text{TrBr}$ ($c_0 = 9.96 \times 10^{-4}$ mol L$^{-1}$) in the presence of various concentrations of piperidine; 100AN, 25 C°, stopped-flow conductimetry.

| [pip] / mol L$^{-1}$ | [pip] / [substrate] | $k_{obs}$ / s$^{-1}$ |
|---|---|---|
| $1.08 \times 10^{-3}$ | 1 | 6.28 |
| $5.39 \times 10^{-3}$ | 5 | $1.81 \times 10^1$ |
| $1.08 \times 10^{-2}$ | 11 | $2.53 \times 10^1$ |
| $2.16 \times 10^{-2}$ | 22 | $2.89 \times 10^1$ |
| $4.31 \times 10^{-2}$ | 43 | $3.15 \times 10^1$ |
| $1.63 \times 10^{-1}$ | 164 | $3.45 \times 10^1$ |
| $k_{ion} = 3.45 \times 10^1$ s$^{-1}$ | | |

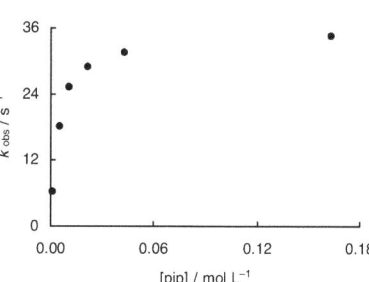

Ionization rates of $(m\text{F})\text{TrOAc}$ ($c_0 = 9.16 \times 10^{-4}$ mol L$^{-1}$) in the presence of triethylamine (c = $6.00 \times 10^{-3}$ mol L$^{-1}$); 60AN40W, 25 C°, conventional conductimetry.

| $k_{ion}$ / s$^{-1}$ | average $k_{ion}$ / s$^{-1}$ |
|---|---|
| $3.47 \times 10^{-5}$ | $3.45 \times 10^{-5}$ |
| $3.43 \times 10^{-5}$ | |

Ionization rates of $(m\text{F})(m\text{F})'\text{TrCl}$ in the presence of various concentrations of piperidine; 100AN, 25 C°, conventional conductimetry.

| [substrate] / mol L$^{-1}$ | [pip] / mol L$^{-1}$ | [pip] / [substrate] | $k_{obs}$ / s$^{-1}$ |
|---|---|---|---|
| $1.30 \times 10^{-3}$ | $8.33 \times 10^{-3}$ | 6 | $1.24 \times 10^{-3}$ |
| $1.39 \times 10^{-3}$ | $4.43 \times 10^{-2}$ | 32 | $1.46 \times 10^{-3}$ |
| $1.43 \times 10^{-3}$ | $1.17 \times 10^{-1}$ | 82 | $1.46 \times 10^{-3}$ |
| $k_{ion} = 1.46 \times 10^{-3}$ s$^{-1}$ | | | |

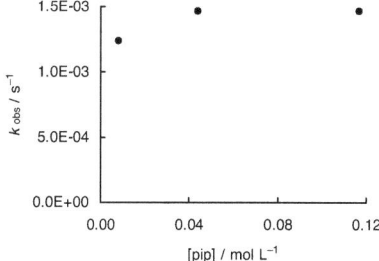

288                                                                                              Experimental Part

Ionization rates of $(m\text{F})(m\text{F})'\text{TrCl}$ ($c_0$ = 1.00 × 10$^{-3}$ mol L$^{-1}$) in the presence of various concentrations of piperidine; 90AN10W, 25 C°, stopped-flow conductimetry.

| [pip] / mol L$^{-1}$ | [pip] / [substrate] | $k_{obs}$ / s$^{-1}$ |
|---|---|---|
| 4.54 × 10$^{-3}$ | 5 | 6.52 × 10$^{-1}$ |
| 9.08 × 10$^{-3}$ | 9 | 7.78 × 10$^{-1}$ |
| 3.63 × 10$^{-2}$ | 36 | 8.26 × 10$^{-1}$ |
| 7.27 × 10$^{-2}$ | 73 | 8.15 × 10$^{-1}$ |
| 1.09 × 10$^{-1}$ | 109 | 8.32 × 10$^{-1}$ |
| 1.46 × 10$^{-1}$ | 146 | 8.02 × 10$^{-1}$ |
| $k_{ion}$ = 8.26 × 10$^{-1}$ s$^{-1}$ | | |

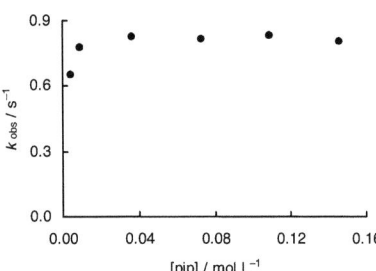

Ionization rates of $(m\text{F})(m\text{F})'\text{TrCl}$ ($c_0$ = 1.02 × 10$^{-3}$ mol L$^{-1}$) in the presence of various concentrations of piperidine; 80AN20W, 25 C°, stopped-flow conductimetry.

| [pip] / mol L$^{-1}$ | [pip] / [substrate] | $k_{obs}$ / s$^{-1}$ |
|---|---|---|
| 4.99 × 10$^{-3}$ | 5 | 3.54 |
| 9.97 × 10$^{-3}$ | 10 | 3.60 |
| 3.99 × 10$^{-2}$ | 39 | 3.50 |
| 7.98 × 10$^{-2}$ | 78 | 3.44 |
| 1.26 × 10$^{-1}$ | 123 | 3.32 |
| 1.60 × 10$^{-1}$ | 157 | 3.28 |
| $k_{ion}$ = 3.60 s$^{-1}$ | | |

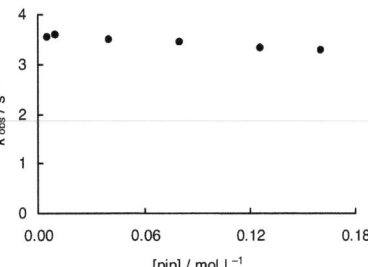

Ionization rates of $(m\text{F})(m\text{F})'\text{TrCl}$ in aqueous acetonitrile without amine; 25 C°, stopped-flow conductimetry.

| solvent | [substrate] / mol L$^{-1}$ | $k_{obs}$ / s$^{-1}$ |
|---|---|---|
| 90AN10W | 1.10 × 10$^{-3}$ | 7.16 × 10$^{-1}$ |
|  | 1.10 × 10$^{-4}$ | 8.74 × 10$^{-1}$ |
| 80AN20W | 1.10 × 10$^{-3}$ | 3.41 |
| 60AN40W | 1.10 × 10$^{-3}$ | 1.60 × 10$^{1}$ |
| 50AN50W | 1.10 × 10$^{-3}$ | 3.22 × 10$^{1}$ |

Ionization rates of $(m\text{F})(m\text{F})'\text{TrBr}$ ($c_0 = 1.01 \times 10^{-3}$ mol L$^{-1}$) in the presence of various concentrations of piperidine; 100AN, 25 C°, stopped-flow conductimetry.

| [pip] / mol L$^{-1}$ | [pip] / [substrate] | $k_{obs}$ / s$^{-1}$ |
|---|---|---|
| 4.78 × 10$^{-3}$ | 5 | 8.63 × 10$^{-1}$ |
| 9.55 × 10$^{-3}$ | 9 | 1.21 |
| 1.43 × 10$^{-2}$ | 14 | 1.42 |
| 1.91 × 10$^{-2}$ | 19 | 1.49 |
| 3.82 × 10$^{-2}$ | 38 | 1.67 |
| 7.64 × 10$^{-2}$ | 76 | 1.80 |
| 1.10 × 10$^{-1}$ | 109 | 1.81 |
| 1.66 × 10$^{-1}$ | 164 | 1.81 |
| $k_{ion}$ = **1.81 s$^{-1}$** | | |

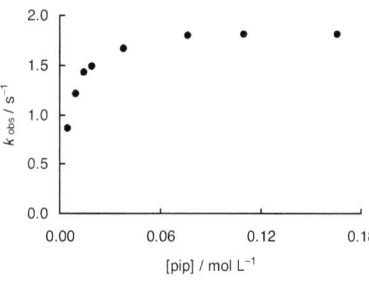

Ionization rates of $(m\text{F})(m\text{F})'\text{TrBr}$ ($c_0 = 1.01 \times 10^{-3}$ mol L$^{-1}$) in the presence of various concentrations of diethylamine; 100AN, 25 C°, stopped-flow conductimetry.

| [Et$_2$NH] / mol L$^{-1}$ | [Et$_2$NH] / [substrate] | $k_{obs}$ / s$^{-1}$ |
|---|---|---|
| 4.54 × 10$^{-3}$ | 5 | 1.60 × 10$^{-1}$ |
| 9.09 × 10$^{-3}$ | 9 | 3.10 × 10$^{-1}$ |
| 1.36 × 10$^{-2}$ | 14 | 4.37 × 10$^{-1}$ |
| 1.82 × 10$^{-2}$ | 18 | 5.90 × 10$^{-1}$ |
| 3.63 × 10$^{-2}$ | 36 | 9.12 × 10$^{-1}$ |
| 7.25 × 10$^{-2}$ | 72 | 1.17 |
| 1.10 × 10$^{-1}$ | 109 | 1.29 |
| 1.52 × 10$^{-1}$ | 151 | 1.36 |
| $k_{ion}$ = **1.36 s$^{-1}$** | | |

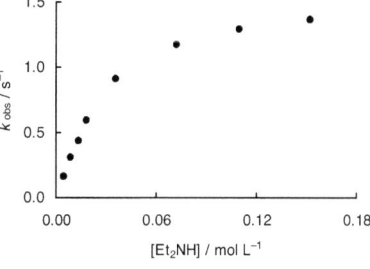

Ionization rates of $(mF)(mF)'TrBr$ ($c_0 = 1.02 \times 10^{-3}$ mol L$^{-1}$) in the presence of various concentrations of butylamine; 100AN, 25 C°, stopped-flow conductimetry.

| [BuNH$_2$] / mol L$^{-1}$ | [BuNH$_2$] / [substrate] | $k_{obs}$ / s$^{-1}$ |
|---|---|---|
| 4.55 × 10$^{-3}$ | 4 | 9.49 × 10$^{-1}$ |
| 9.09 × 10$^{-3}$ | 9 | 1.29 |
| 1.36 × 10$^{-2}$ | 13 | 1.46 |
| 1.82 × 10$^{-2}$ | 18 | 1.58 |
| 3.64 × 10$^{-2}$ | 36 | 1.73 |
| 7.27 × 10$^{-2}$ | 71 | 1.77 |
| 1.06 × 10$^{-1}$ | 104 | 1.78 |
| 1.57 × 10$^{-1}$ | 154 | 1.78 |
| $k_{ion} = 1.78$ s$^{-1}$ | | |

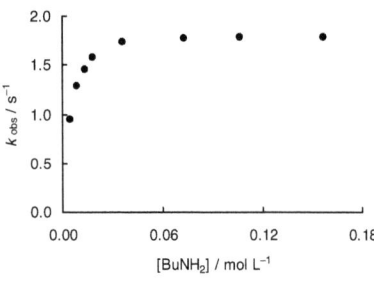

Ionization rates of $(mF)(mF)'TrBr$ ($c_0 = 1.01 \times 10^{-3}$ mol L$^{-1}$) in the presence of various concentrations of piperidine; 90AN10W, 25 C°, stopped-flow conductimetry.

| [pip] / mol L$^{-1}$ | [pip] / [substrate] | $k_{obs}$ / s$^{-1}$ |
|---|---|---|
| 4.49 × 10$^{-3}$ | 4 | 8.04 × 10$^{1}$ |
| 8.98 × 10$^{-3}$ | 9 | 9.63 × 10$^{1}$ |
| 3.59 × 10$^{-2}$ | 36 | 1.32 × 10$^{2}$ |
| 7.19 × 10$^{-2}$ | 71 | 1.40 × 10$^{2}$ |
| 1.11 × 10$^{-1}$ | 109 | 1.44 × 10$^{2}$ |
| 1.42 × 10$^{-1}$ | 140 | 1.31 × 10$^{2}$ |
| $k_{ion} = 1.44 \times 10^{2}$ s$^{-1}$ | | |

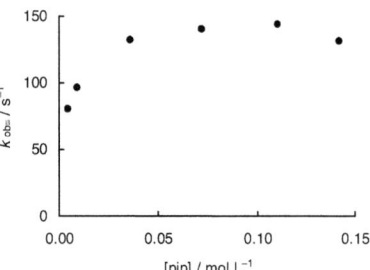

Ionization rates of $(mF)(mF)'TrBr$ ($c_0 = 7.85 \times 10^{-4}$ mol L$^{-1}$) in the presence of various concentrations of piperidine; 90A10W, 25 C°, stopped-flow conductimetry.

| [pip] / mol L$^{-1}$ | [pip] / [substrate] | $k_{obs}$ / s$^{-1}$ |
|---|---|---|
| 1.68 × 10$^{-2}$ | 21 | 5.41 |
| 9.29 × 10$^{-2}$ | 118 | 5.10 |
| 1.44 × 10$^{-1}$ | 183 | 5.26 |
| 2.05 × 10$^{-1}$ | 261 | 5.32 |
| $k_{ion} = 5.41$ s$^{-1}$ | | |

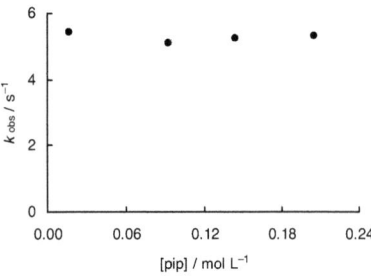

# Experimental Part

Ionization rates of $(m\text{F})_2\text{TrCl}$ in the presence of various concentrations of piperidine; 100AN, 25 C°, conventional conductimetry.

| [substrate] / mol L$^{-1}$ | [pip] / mol L$^{-1}$ | [pip] / [substrate] | $k_{obs}$ / s$^{-1}$ |
|---|---|---|---|
| 7.85 × 10$^{-4}$ | 9.00 × 10$^{-3}$ | 12 | 1.92 × 10$^{-3}$ |
| 7.39 × 10$^{-4}$ | 1.69 × 10$^{-2}$ | 23 | 2.02 × 10$^{-3}$ |
| 7.84 × 10$^{-4}$ | 3.59 × 10$^{-2}$ | 46 | 2.12 × 10$^{-3}$ |
| 6.66 × 10$^{-4}$ | 6.89 × 10$^{-2}$ | 104 | 2.06 × 10$^{-3}$ |
| $k_{ion} = 2.12 \times 10^{-3}$ s$^{-1}$ | | | |

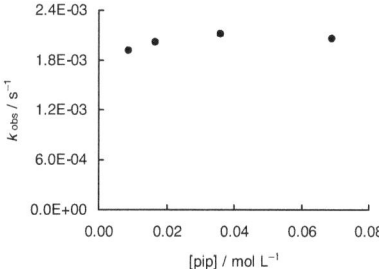

Ionization rates of $(m\text{F})_2\text{TrCl}$ ($c_0 = 9.91 \times 10^{-4}$ mol L$^{-1}$) in the presence of various concentrations of piperidine; 90AN10W, 25 C°, stopped-flow conductimetry.

| [pip] / mol L$^{-1}$ | [pip] / [substrate] | $k_{obs}$ / s$^{-1}$ |
|---|---|---|
| 5.03 × 10$^{-3}$ | 5 | 1.10 |
| 1.01 × 10$^{-2}$ | 10 | 1.15 |
| 4.03 × 10$^{-2}$ | 41 | 1.25 |
| 8.05 × 10$^{-2}$ | 81 | 1.22 |
| 1.20 × 10$^{-1}$ | 121 | 1.19 |
| 1.62 × 10$^{-1}$ | 163 | 1.14 |
| $k_{ion} = 1.25$ s$^{-1}$ | | |

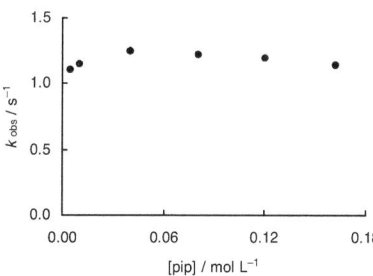

Ionization rates of $(mF)_2TrCl$ ($c_0 = 1.01 \times 10^{-3}$ mol $L^{-1}$) in the presence of various concentrations of piperidine; 80AN20W, 25 C°, stopped-flow conductimetry.

| [pip] / mol $L^{-1}$ | [pip] / [substrate] | $k_{obs}$ / $s^{-1}$ |
|---|---|---|
| $4.98 \times 10^{-3}$ | 5 | 4.80 |
| $9.97 \times 10^{-3}$ | 10 | 4.78 |
| $3.99 \times 10^{-2}$ | 40 | 4.70 |
| $7.97 \times 10^{-2}$ | 79 | 4.57 |
| $1.24 \times 10^{-1}$ | 123 | 4.43 |
| $1.62 \times 10^{-1}$ | 160 | 4.31 |
| $k_{ion} = 4.80$ $s^{-1}$ | | |

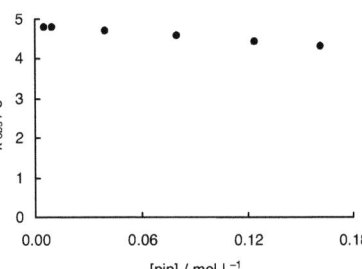

Ionization rates of $(mF)(mF)'(mF)''TrCl$ in the presence of various concentrations of piperidine; 100AN, 25 C°, conventional conductimetry.

| [substrate] / mol $L^{-1}$ | [pip] / mol $L^{-1}$ | [pip] / [substrate] | $k_{obs}$ / $s^{-1}$ |
|---|---|---|---|
| $1.22 \times 10^{-3}$ | $6.63 \times 10^{-3}$ | 5 | $7.10 \times 10^{-5}$ |
| $9.43 \times 10^{-4}$ | $1.84 \times 10^{-2}$ | 20 | $6.98 \times 10^{-5}$ |
| $9.32 \times 10^{-4}$ | $2.42 \times 10^{-2}$ | 26 | $7.10 \times 10^{-5}$ |
| $9.36 \times 10^{-4}$ | $1.39 \times 10^{-1}$ | 149 | $6.61 \times 10^{-5}$ |
| $9.36 \times 10^{-4}$ | $2.78 \times 10^{-1}$ | 297 | $5.89 \times 10^{-5}$ |
| $k_{ion} = 7.10 \times 10^{-5}$ $s^{-1}$ | | | |

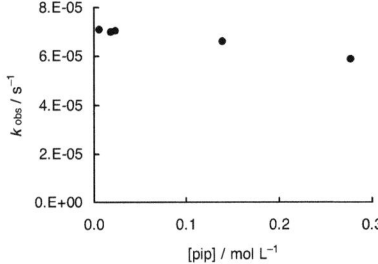

# Experimental Part 293

Ionization rates of $(mF)(mF)'(mF)''TrCl$ ($c_0 = 8.41 \times 10^{-4}$ mol L$^{-1}$) in the presence of various concentrations of piperidine; 90AN10W, 25 C°, conventional conductimetry.

| [pip] / mol L$^{-1}$ | [pip] / [substrate] | $k_{obs}$ / s$^{-1}$ |
|---|---|---|
| 0 | 0 | $5.99 \times 10^{-2}$ |
| $1.48 \times 10^{-2}$ | 18 | $6.34 \times 10^{-2}$ |
| $2.96 \times 10^{-2}$ | 35 | $6.38 \times 10^{-2}$ |
| $k_{ion} = 6.38 \times 10^{-2}$ s$^{-1}$ | | |

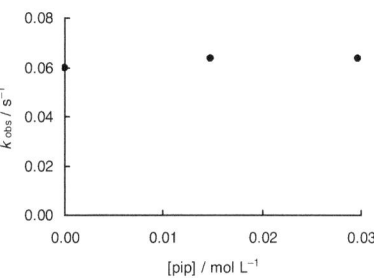

Ionization rates of $(mF)(mF)'(mF)''TrCl$ ($c_0 = 9.68 \times 10^{-4}$ mol L$^{-1}$) in the presence of various concentrations of piperidine; 80AN20W, 25 C°, stopped-flow conductimetry.

| [pip] / mol L$^{-1}$ | [pip] / [substrate] | $k_{obs}$ / s$^{-1}$ |
|---|---|---|
| $4.99 \times 10^{-3}$ | 5 | $2.74 \times 10^{-1}$ |
| $9.97 \times 10^{-3}$ | 10 | $2.71 \times 10^{-1}$ |
| $3.99 \times 10^{-2}$ | 41 | $2.64 \times 10^{-1}$ |
| $7.98 \times 10^{-2}$ | 81 | $2.64 \times 10^{-1}$ |
| $1.26 \times 10^{-1}$ | 127 | $2.42 \times 10^{-1}$ |
| $1.60 \times 10^{-1}$ | 163 | $2.41 \times 10^{-1}$ |
| $k_{ion} = 2.74 \times 10^{-1}$ s$^{-1}$ | | |

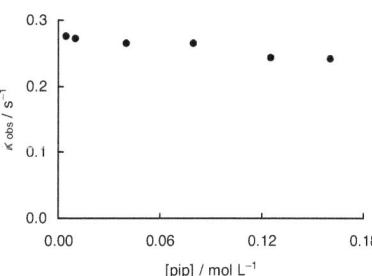

Ionization rates of $(mF)(mF)'(mF)''TrCl$ ($c_0 = 8.62 \times 10^{-4}$ mol L$^{-1}$) in the presence of various concentrations of piperidine; 90A10W, 25 C°, conventional conductimetry.

| [pip] / mol L$^{-1}$ | [pip] / [substrate] | $k_{obs}$ / s$^{-1}$ |
|---|---|---|
| 0 | 0 | $1.33 \times 10^{-3}$ |
| $1.43 \times 10^{-2}$ | 17 | $1.53 \times 10^{-3}$ |
| $2.87 \times 10^{-2}$ | 33 | $1.48 \times 10^{-3}$ |
| $k_{ion} = 1.53 \times 10^{-3}$ s$^{-1}$ | | |

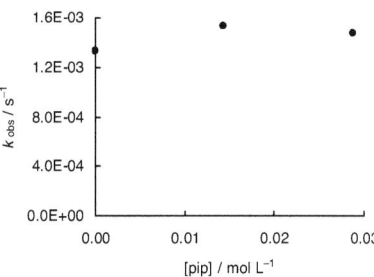

Ionization rates of $(mF)(mF)'(mF)''TrCl$ ($c_0 = 8.62 \times 10^{-4}$ mol L$^{-1}$) in the presence of various concentrations of piperidine; 80A20W, 25 °C, conventional conductimetry.

| [substrate] / mol L$^{-1}$ | [pip] / mol L$^{-1}$ | [pip] / [substrate] | $k_{obs}$ / s$^{-1}$ |
|---|---|---|---|
| $8.62 \times 10^{-4}$ | 0 | 0 | $1.52 \times 10^{-2}$ |
| $8.21 \times 10^{-4}$ | $1.43 \times 10^{-2}$ | 17 | $1.61 \times 10^{-2}$ |
| $8.21 \times 10^{-4}$ | $2.87 \times 10^{-2}$ | 35 | $1.65 \times 10^{-2}$ |
| $k_{ion} = 1.65 \times 10^{-2}$ s$^{-1}$ | | | |

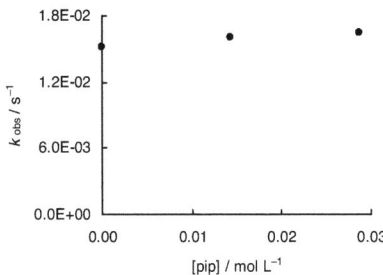

Ionization rates of $(mF)(mF)'(mF)''TrBr$ ($c_0 = 1.00 \times 10^{-3}$ mol L$^{-1}$) in the presence of various concentrations of piperidine; 100AN, 25 C°, stopped-flow conductimetry.

| [pip] / mol L$^{-1}$ | [pip] / [substrate] | $k_{obs}$ / s$^{-1}$ |
|---|---|---|
| $4.78 \times 10^{-3}$ | 5 | $4.69 \times 10^{-2}$ |
| $9.55 \times 10^{-3}$ | 10 | $5.76 \times 10^{-2}$ |
| $1.43 \times 10^{-2}$ | 14 | $6.56 \times 10^{-2}$ |
| $1.91 \times 10^{-2}$ | 19 | $7.11 \times 10^{-2}$ |
| $3.82 \times 10^{-2}$ | 38 | $7.66 \times 10^{-2}$ |
| $7.64 \times 10^{-2}$ | 76 | $8.04 \times 10^{-2}$ |
| $1.10 \times 10^{-1}$ | 110 | $8.18 \times 10^{-2}$ |
| $1.66 \times 10^{-1}$ | 166 | $8.27 \times 10^{-2}$ |
| $k_{ion} = 8.27 \times 10^{-2}$ s$^{-1}$ | | |

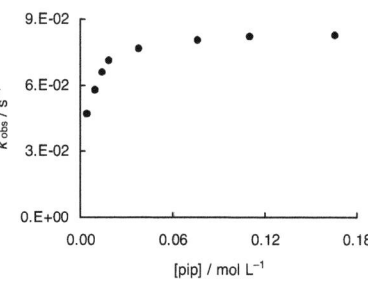

Experimental Part 295

Ionization rates of $(m\text{F})(m\text{F})'(m\text{F})''\text{TrBr}$ ($c_0 = 9.97 \times 10^{-4}$ mol L$^{-1}$) in the presence of various concentrations of piperidine; 90AN10W, 25 C°, stopped-flow conductimetry.

| [pip] / mol L$^{-1}$ | [pip] / [substrate] | $k_{obs}$ / s$^{-1}$ |
|---|---|---|
| $4.53 \times 10^{-3}$ | 5 | 7.40 |
| $9.05 \times 10^{-3}$ | 9 | 7.57 |
| $3.62 \times 10^{-2}$ | 36 | 9.23 |
| $7.24 \times 10^{-2}$ | 73 | 9.79 |
| $1.07 \times 10^{-1}$ | 108 | 9.91 |
| $1.46 \times 10^{-1}$ | 147 | 9.58 |
| $k_{ion} = 9.91$ s$^{-1}$ | | |

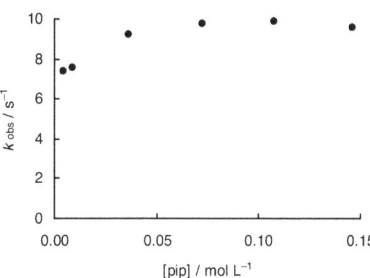

Ionization rates of $(m\text{F})(m\text{F})'(m\text{F})''\text{TrBr}$ ($c_0 = 9.97 \times 10^{-4}$ mol L$^{-1}$) in the presence of various concentrations of piperidine; 80AN20W, 25 C°, stopped-flow conductimetry.

| [pip] / mol L$^{-1}$ | [pip] / [substrate] | $k_{obs}$ / s$^{-1}$ |
|---|---|---|
| $1.00 \times 10^{-2}$ | 10 | $3.70 \times 10^{1}$ |
| $2.00 \times 10^{-2}$ | 20 | $3.90 \times 10^{1}$ |
| $4.00 \times 10^{-2}$ | 40 | $3.89 \times 10^{1}$ |
| $6.00 \times 10^{-2}$ | 60 | $3.90 \times 10^{1}$ |
| $k_{ion} = 3.90 \times 10^{1}$ s$^{-1}$ | | |

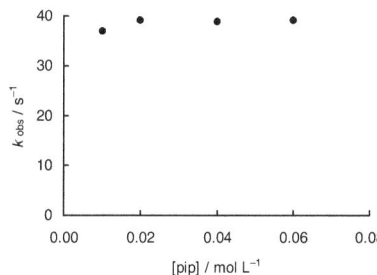

Ionization rates of $(m\text{F})(m\text{F})'(m\text{F})''\text{TrBr}$ ($c_0 = 9.60 \times 10^{-4}$ mol L$^{-1}$) in the presence of various concentrations of piperidine; 60AN40W, 25 C°, stopped-flow conductimetry.

| [pip] / mol L$^{-1}$ | [pip] / [substrate] | $k_{obs}$ / s$^{-1}$ |
|---|---|---|
| $2.73 \times 10^{-2}$ | 28 | $1.50 \times 10^{2}$ |
| $5.58 \times 10^{-2}$ | 58 | $1.42 \times 10^{2}$ |
| $1.95 \times 10^{-1}$ | 203 | $1.50 \times 10^{2}$ |
| $k_{ion} = 1.50 \times 10^{2}$ s$^{-1}$ | | |

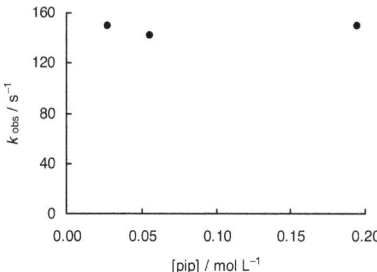

Ionization rate of $(mF)_2(mF)'_2TrCl$ ($c_0 = 9.18 \times 10^{-4}$ mol L$^{-1}$) in the presence of piperidine; 100AN, 25 C°, conventional conductimetry.

| [pip] / mol L$^{-1}$ | [pip] / [substrate] | $k_{ion}$ / s$^{-1}$ |
|---|---|---|
| 7.61 × 10$^{-2}$ | 83 | 6.3 × 10$^{-6}$ |

Ionization rates of $(mF)_2(mF)'_2TrCl$ ($c_0 = 7.94 \times 10^{-4}$ mol L$^{-1}$) in the presence of various concentrations of piperidine; 90AN10W, 25 C°, conventional conductimetry.

| [pip] / mol L$^{-1}$ | [pip] / [substrate] | $k_{obs}$ / s$^{-1}$ |
|---|---|---|
| 5.04 × 10$^{-3}$ | 6 | 9.49 × 10$^{-3}$ |
| 2.01 × 10$^{-2}$ | 25 | 9.62 × 10$^{-3}$ |
| 7.05 × 10$^{-2}$ | 89 | 1.01 × 10$^{-2}$ |
| 1.01 × 10$^{-1}$ | 126 | 9.89 × 10$^{-3}$ |
| 2.02 × 10$^{-1}$ | 253 | 8.79 × 10$^{-3}$ |
| $k_{ion} = 1.01 \times 10^{-2}$ s$^{-1}$ | | |

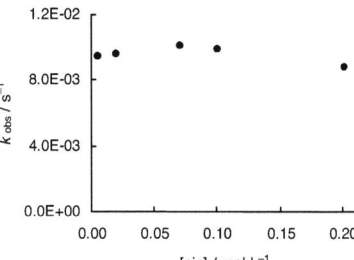

Ionization rates of $(mF)_2(mF)'_2TrCl$ ($c_0 = 8.55 \times 10^{-4}$ mol L$^{-1}$) in the presence of various concentrations of piperidine; 80AN20W, 25 C°, conventional conductimetry.

| [pip] / mol L$^{-1}$ | [pip] / [substrate] | $k_{obs}$ / s$^{-1}$ |
|---|---|---|
| 1.46 × 10$^{-2}$ | 17 | 4.76 × 10$^{-2}$ |
| 2.93 × 10$^{-2}$ | 34 | 4.54 × 10$^{-2}$ |
| $k_{ion} = 4.76 \times 10^{-2}$ s$^{-1}$ | | |

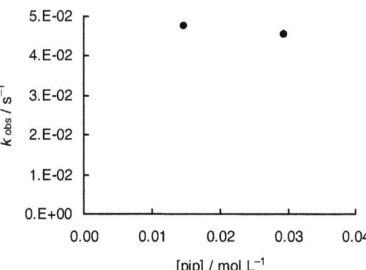

Experimental Part                                                                             297

Ionization rates of $(mF)_2(mF)'(mF)''TrBr$ in the presence of various concentrations of piperidine; 100AN, 25 C°, conventional conductimetry.

| [substrate] / mol L$^{-1}$ | [pip] / mol L$^{-1}$ | [pip] / [substrate] | $k_{obs}$ / s$^{-1}$ |
|---|---|---|---|
| $1.23 \times 10^{-3}$ | $6.21 \times 10^{-3}$ | 5 | $3.06 \times 10^{-3}$ |
| $1.13 \times 10^{-3}$ | $1.14 \times 10^{-2}$ | 10 | $3.73 \times 10^{-3}$ |
| $1.17 \times 10^{-3}$ | $2.96 \times 10^{-2}$ | 25 | $4.52 \times 10^{-3}$ |
| $1.10 \times 10^{-3}$ | $5.54 \times 10^{-2}$ | 51 | $4.98 \times 10^{-3}$ |
| $1.11 \times 10^{-3}$ | $1.12 \times 10^{-1}$ | 101 | $5.19 \times 10^{-3}$ |
| $1.07 \times 10^{-3}$ | $1.62 \times 10^{-1}$ | 152 | $5.20 \times 10^{-3}$ |
| $k_{ion} = 5.20 \times 10^{-3}$ s$^{-1}$ | | | |

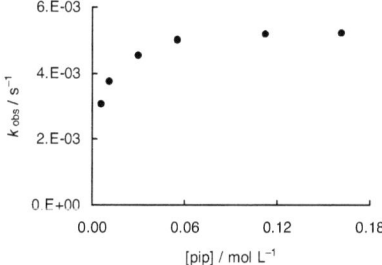

Ionization rates of $(mF)_2(mF)'(mF)''TrBr$ ($c_0 = 1.02 \times 10^{-3}$ mol L$^{-1}$) in the presence of various concentrations of piperidine; 90AN10W, 25 C°, stopped-flow conductimetry.

| [pip] / mol L$^{-1}$ | [pip] / [substrate] | $k_{obs}$ / s$^{-1}$ |
|---|---|---|
| $4.53 \times 10^{-3}$ | 4 | $6.79 \times 10^{-1}$ |
| $9.05 \times 10^{-3}$ | 9 | $7.15 \times 10^{-1}$ |
| $3.62 \times 10^{-2}$ | 35 | $7.51 \times 10^{-1}$ |
| $7.24 \times 10^{-2}$ | 71 | $7.85 \times 10^{-1}$ |
| $1.07 \times 10^{-1}$ | 105 | $7.73 \times 10^{-1}$ |
| $1.46 \times 10^{-1}$ | 143 | $7.30 \times 10^{-1}$ |
| $k_{ion} = 7.85 \times 10^{-1}$ s$^{-1}$ | | |

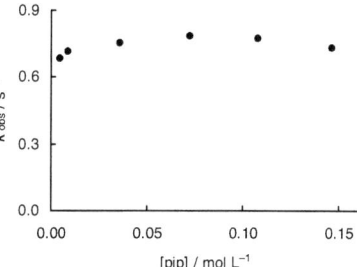

Ionization rates of $(mF)_2(mF)'(mF)''TrBr$ ($c_0 = 9.77 \times 10^{-4}$ mol L$^{-1}$) in the presence of various concentrations of piperidine; 80AN20W, 25 C°, stopped-flow conductimetry.

| [pip] / mol L$^{-1}$ | [pip] / [substrate] | $k_{obs}$ / s$^{-1}$ |
|---|---|---|
| $1.00 \times 10^{-2}$ | 10 | 3.48 |
| $2.00 \times 10^{-2}$ | 20 | 3.48 |
| $4.00 \times 10^{-2}$ | 41 | 3.48 |
| $6.00 \times 10^{-2}$ | 61 | 3.42 |
| $k_{ion} = 3.48$ s$^{-1}$ | | |

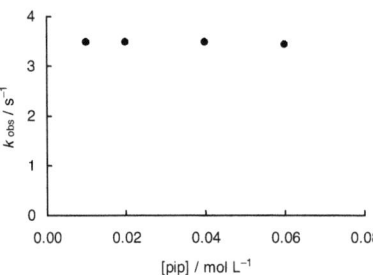

Ionization rates of $(mF)_2(mF)'(mF)''TrBr$ ($c_0 = 1.16 \times 10^{-3}$ mol L$^{-1}$) in the presence of various concentrations of piperidine; 60AN40W, 25 C°, stopped-flow conductimetry.

| [pip] / mol L$^{-1}$ | [pip] / [substrate] | $k_{obs}$ / s$^{-1}$ |
|---|---|---|
| $2.73 \times 10^{-2}$ | 24 | $1.62 \times 10^1$ |
| $5.58 \times 10^{-2}$ | 48 | $1.63 \times 10^1$ |
| $7.88 \times 10^{-2}$ | 68 | $1.57 \times 10^1$ |
| $1.95 \times 10^{-1}$ | 168 | $1.56 \times 10^1$ |
| $k_{ion} = 1.63 \times 10^1$ s$^{-1}$ | | |

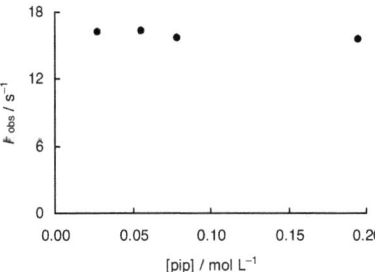

Ionization rates of $(mF)_2(mF)'(mF)''TrBr$ ($c_0 = 1.05 \times 10^{-3}$ mol L$^{-1}$) in the presence of various concentrations of piperidine; 90A10W, 25 C°, conventional conductimetry.

| [pip] / mol L$^{-1}$ | [pip] / [substrate] | $k_{obs}$ / s$^{-1}$ |
|---|---|---|
| $1.46 \times 10^{-2}$ | 14 | $1.51 \times 10^{-2}$ |
| $2.93 \times 10^{-2}$ | 28 | $1.49 \times 10^{-2}$ |
| $k_{ion} = 1.51 \times 10^{-2}$ s$^{-1}$ | | |

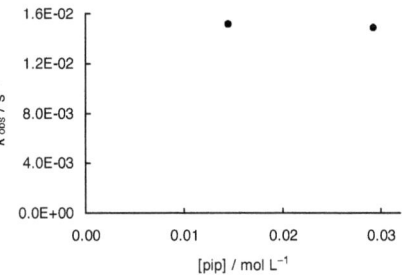

Experimental Part                                                                                           299

Ionization rate of $(mF)_6TrBr$ ($c_0 = 5.03 \times 10^{-4}$ mol L$^{-1}$) in the presence of piperidine; 100AN, 25 C°, conventional conductimetry.

| [pip] / mol L$^{-1}$ | [pip] / [substrate] | $k_{ion}$ / s$^{-1}$ |
|---|---|---|
| $1.06 \times 10^{-1}$ | 211 | $2.1 \times 10^{-5}$ |

Ionization rates of $(mF)_6TrBr$ in the presence of various concentrations of piperidine; 90AN10W, 25 C°, conventional conductimetry.

| [substrate] / mol L$^{-1}$ | [pip] / mol L$^{-1}$ | [pip] / [substrate] | $k_{obs}$ / s$^{-1}$ |
|---|---|---|---|
| $8.02 \times 10^{-4}$ | $5.00 \times 10^{-3}$ | 6 | $3.39 \times 10^{-3}$ |
| $7.99 \times 10^{-4}$ | $1.99 \times 10^{-2}$ | 25 | $3.35 \times 10^{-3}$ |
| $8.02 \times 10^{-4}$ | $7.00 \times 10^{-2}$ | 87 | $3.19 \times 10^{-3}$ |
| $8.02 \times 10^{-4}$ | $1.00 \times 10^{-1}$ | 125 | $3.23 \times 10^{-3}$ |
| $8.02 \times 10^{-4}$ | $2.00 \times 10^{-1}$ | 249 | $3.09 \times 10^{-3}$ |
| $k_{ion} = 3.39 \times 10^{-3}$ s$^{-1}$ | | | |

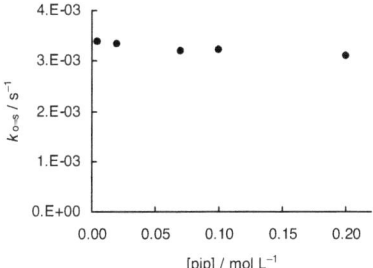

Ionization rates of $(mF)_6TrBr$ ($c_0 = 7.94 \times 10^{-4}$ mol L$^{-1}$) in the presence of various concentrations of piperidine; 80AN20W, 25 C°, conventional conductimetry.

| [pip] / mol L$^{-1}$ | [pip] / [substrate] | $k_{obs}$ / s$^{-1}$ |
|---|---|---|
| $8.03 \times 10^{-3}$ | 10 | $1.68 \times 10^{-2}$ |
| $1.61 \times 10^{-3}$ | 20 | $1.68 \times 10^{-2}$ |
| $3.21 \times 10^{-2}$ | 40 | $1.64 \times 10^{-2}$ |
| $4.82 \times 10^{-2}$ | 61 | $1.60 \times 10^{-2}$ |
| $k_{ion} = 1.68 \times 10^{-2}$ s$^{-1}$ | | |

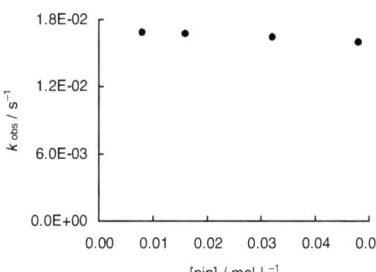

## 5. Computational Data

### 5.1. Methyl anion and hydroxide affinities of tritylium ions

Geometries of the tritylium ions, triarylmethanols and triarylethanes were optimized at the B3LYP/6-31G(d,p) level of theory. Thermochemical corrections (B3LYP/6-31G(d,p)) were combined with single point energies on the MP2(FC)/6-31+G(2d,p) level to give $H_{298}$ and $G_{298}$.

For some of the tritylium ions, triarylmethanols and triarylethanes different conformational isomers had to be considered.

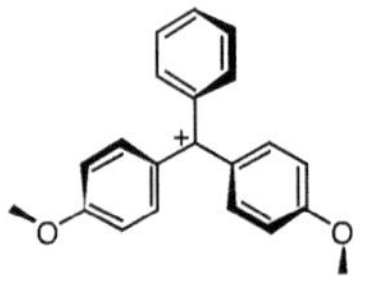

0.00 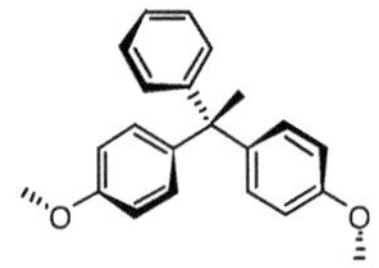        0.00

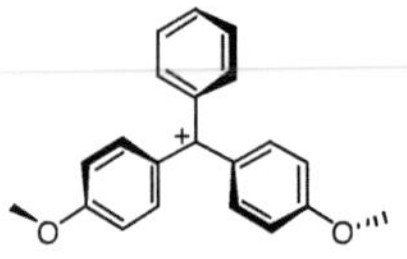

2.05 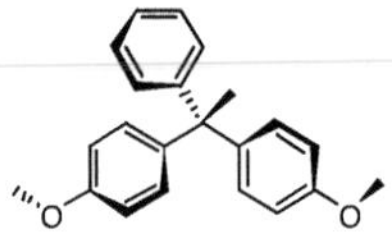        0.49

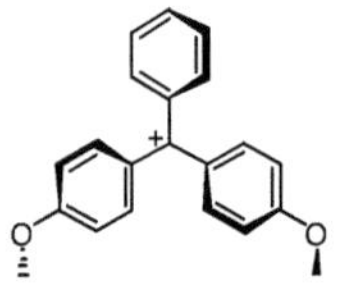

2.81 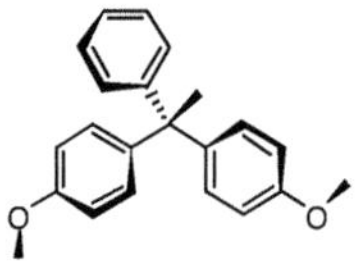        1.07

Conformations of $(MeO)_2Tr^+$ (left) and its methyl anion adduct (right). Relative energies ($G_{298}$ in kJ mol$^{-1}$) are given for the MP2(FC)/6-31+G(2d,p)//B3LYP/6-31G(d,p) level.

Experimental Part

Energy data of tritylium ions, in Hartree.

| system (conformer) | $E_{tot}$ B3LYP/ 6-31G(d,p) | $E_{tot}$ MP2(FC)/ 6-31+G(2d,p) | $H_{298}$ MP2(FC)/ 6-31+G(2d,p)// B3LYP/ 6-31G(d,p) | $G_{298}$ MP2(FC)/ 6-31+G(2d,p)// B3LYP/ 6-31G(d,p) |
|---|---|---|---|---|
| $(mF)_6Tr^+$ | -1328.20367 | -1325.008790 | -1324.75783 | -1324.82793 |
| $(mF)_5Tr^+$ | -1228.97854 | -1225.962091 | -1225.70358 | -1225.77166 |
| $(mF)_2(mF)'(mF)''Tr^+$ (1) | -1129.753268 | -1126.915187 | -1126.649127 | -1126.649127 |
| $(mF)_2(mF)'(mF)''Tr^+$ (2) | -1129.753257 | -1126.915125 | -1126.649012 | -1126.649012 |
| $(mF)_2(mF)'(mF)''Tr^+$ (3) | -1129.753247 | -1126.915196 | -1126.649155 | -1126.649155 |
| $(mF)(mF)'(mF)''Tr^+$ (1) | -1030.527806 | -1027.868094 | -1027.594450 | -1027.594450 |
| $(mF)(mF)'(mF)''Tr^+$ (2) | -1030.527748 | -1027.868016 | -1027.594411 | -1027.594411 |
| $(mF)(mF)'Tr^+$ (1) | -931.3015536 | -928.819869 | -928.538700 | -928.600474 |
| $(mF)(mF)'Tr^+$ (2) | -931.3015988 | -928.819879 | -928.538652 | -928.600342 |
| $(mF)(mF)'Tr^+$ (3) | -931.3014981 | -928.819806 | -928.538645 | -928.600454 |
| $(mF)Tr^+$ | -832.07517 | -829.771452 | -829.48270 | -829.54233 |
| $(pF)Tr^+$ | -832.07922 | -829.774346 | -829.48546 | -829.54511 |
| $(pF)_2Tr^+$ | -931.30974 | -928.825746 | -928.54427 | -928.60601 |
| $(pF)_3Tr^+$ | -1030.54017 | -1027.877041 | -1027.60295 | -1027.66666 |
| $Tr^+$ | -732.8485837 | -730.72281 | -730.42649 | -730.48400 |
| $MeTr^+$ | -772.1738629 | -769.92410 | -769.59867 | -769.66267 |
| $Me_2Tr^+$ | -811.4988717 | -809.12486 | -808.77021 | -808.83767 |
| $Me_3Tr^+$ | -850.8235926 | -848.32538 | -847.94162 | -848.01281 |
| $(MeO)Tr^+$ | -847.3853854 | -844.97674 | -844.64510 | -844.70879 |
| $(MeO)_2Tr^+$ (1) | -961.92033 | -959.22874 | -958.86177 | -958.93053 |
| $(MeO)_2Tr^+$ (2) | -961.92023 | -959.22885 | -958.86187 | -958.93132 |
| $(MeO)_2Tr^+$ (3) | -961.92006 | -959.22835 | -958.86138 | -958.93025 |
| $(MeO)_3Tr^+$ (1) | -1076.45365 | -1073.47957 | -1073.07735 | -1073.15237 |
| $(MeO)_3Tr^+$ (2) | -1076.45359 | -1073.47934 | -1073.07707 | -1073.15242 |
| $(Me_2N)(MeO)Tr^+$ (1) | -981.38158 | -978.58552 | -978.17590 | -978.24990 |
| $(Me_2N)(MeO)Tr^+$ (2) | -981.38134 | -978.58529 | -978.17573 | -978.24987 |
| $(Me_2N)Tr^+$ | -866.8485838 | -864.33559 | -863.96128 | -864.02886 |
| $(Me_2N)_2Tr^+$ | -1000.8406328 | -997.93892 | -997.48673 | -997.56495 |
| $(Me_2N)_3Tr^+$ | -1134.8282639 | -1131.53802 | -1131.00815 | -1131.09664 |

Energy data of methyl anion and hydroxide, in Hartree.

| | $E_{tot}$ B3LYP/ 6-31G(d,p) | $E_{tot}$ MP2(FC)/ 6-31+G(2d,p) | $H_{298}$ MP2(FC)/ 6-31+G(2d,p)// B3LYP/6-31G(d,p) | $G_{298}$ MP2(FC)/ 6-31+G(2d,p)// B3LYP/6-31G(d,p) |
|---|---|---|---|---|
| $OH^-$ | -75.7262924 | -75.63187 | -75.62059 | -75.64019 |
| $Me^-$ | -39.7960283 | -39.69078 | -39.65913 | -39.68109 |

Energy data of 1,1,1-triarylethanes, in Hartree.

| system (conformer) | $E_{tot}$ B3LYP/ 6-31G(d,p) | $E_{tot}$ MP2(FC)/ 6-31+G(2d,p) | $H_{298}$ MP2(FC)/ 6-31+G(2d,p)// B3LYP/ 6-31G(d,p) | $G_{298}$ MP2(FC)/ 6-31+G(2d,p)// B3LYP/ 6-31G(d,p) |
|---|---|---|---|---|
| (mF)$_6$Tr–Me | -1368.38140 | -1365.079557 | -1364.78775 | -1364.86238 |
| (mF)$_5$Tr–Me (1) | -1269.150567 | -1266.027038 | -1265.727810 | -1265.801424 |
| (mF)$_5$Tr–Me (2) | -1269.150707 | -1266.027069 | -1265.727910 | -1265.800519 |
| (mF)$_2$(mF)'(mF)''Tr–Me (1) | -1169.919833 | -1166.974423 | -1166.667842 | -1166.738864 |
| (mF)$_2$(mF)'(mF)''Tr–Me (2) | -1169.919896 | -1166.974392 | -1166.667644 | -1166.737938 |
| (mF)$_2$(mF)'(mF)''Tr–Me (3) | -1169.919789 | -1166.974383 | -1166.667710 | -1166.738165 |
| (mF)(mF)'(mF)''Tr–Me (1) | -1070.688702 | -1067.921505 | -1067.607508 | -1067.676403 |
| (mF)(mF)'(mF)''Tr–Me (2) | -1070.688820 | -1067.921541 | -1067.607457 | -1067.676015 |
| (mF)(mF)'(mF)''Tr–Me (3) | -1070.688874 | -1067.921529 | -1067.607553 | -1067.676147 |
| (mF)(mF)'(mF)''Tr–Me (4) | -1070.688776 | -1067.921542 | -1067.607442 | -1067.676053 |
| (mF)(mF)'Tr–Me (1) | -971.456748 | -968.8674448 | -968.545973 | -968.613344 |
| (mF)(mF)'Tr–Me (2) | -971.456816 | -968.8673164 | -968.545812 | -968.612253 |
| (mF)(mF)'Tr–Me (3) | -971.456791 | -968.8674134 | -968.545891 | -968.612474 |
| (mF)Tr–Me (1) | -872.224634 | -869.8129312 | -869.484002 | -869.548745 |
| (mF)Tr–Me (2) | -872.224831 | -869.8130343 | -869.484064 | -869.548493 |
| (pF)Tr–Me | -872.22437 | -869.812581 | -869.48369 | -869.54831 |
| (pF)$_2$Tr–Me | -971.45603 | -968.866482 | -968.54503 | -968.61241 |
| (pF)$_3$Tr–Me | -1070.68772 | -1067.920257 | -1067.60617 | -1067.67491 |
| Tr-Me | -772.9925127 | -770.75850 | -770.42221 | -770.75850 |
| MeTr-Me | -812.3131836 | -809.95531 | -809.58978 | -809.95531 |
| Me$_2$Tr-Me | -851.6338799 | -849.15245 | -848.75765 | -849.15245 |
| Me$_3$Tr-Me | -890.9545138 | -888.34939 | -887.92537 | -888.34939 |
| (MeO)Tr-Me (1) | -887.51760 | -885.00222 | -884.63076 | -884.69911 |
| (MeO)Tr-Me (2) | -887.51763 | -885.00220 | -884.63073 | -884.70086 |
| (MeO)$_2$Tr-Me (1) | -1002.04275 | -999.24569 | -998.83909 | -998.91343 |
| (MeO)$_2$Tr-Me (2) | -1002.04234 | -999.24524 | -998.83870 | -998.91321 |
| (MeO)$_2$Tr-Me (3) | -1002.04278 | -999.24580 | -998.83917 | -998.91362 |
| (MeO)$_3$Tr-Me (1) | -1116.56750 | -1113.48880 | -1113.04701 | -1113.12757 |
| (MeO)$_3$Tr-Me (2) | -1116.56722 | -1113.48847 | -1113.04662 | -1113.12708 |
| (MeO)$_3$Tr-Me (3) | -1116.56756 | -1113.48900 | -1113.04719 | -1113.12796 |
| (MeO)$_3$Tr-Me (4) | -1116.56785 | -1113.48946 | -1113.04762 | -1113.12908 |
| (Me$_2$N)(MeO)Tr-Me (1) | -1021.49070 | -1018.58849 | -1018.13960 | -1018.21862 |
| (Me$_2$N)(MeO)Tr-Me (2) | -1021.49072 | -1018.58860 | -1018.13970 | -1018.21866 |
| (Me$_2$N)Tr-Me | -1174.9112698 | -904.34514 | -903.93148 | -904.34514 |
| (Me$_2$N)$_2$Tr-Me | -1040.9387255 | -1037.93185 | -1037.44073 | -1037.93185 |
| (Me$_2$N)$_3$Tr-Me | -1174.9112698 | -1171.51796 | -1170.94938 | -1171.51796 |

Experimental Part                                                                    303

Energy data of triarylmethanols, in Hartree.

| system (conformer) | $E_{tot}$ B3LYP/ 6-31G(d,p) | $E_{tot}$ MP2(FC)/ 6-31+G(2d,p) | $H_{298}$ MP2(FC)/ 6-31+G(2d,p)// B3LYP/ 6-31G(d,p) | $G_{298}$ MP2(FC)/ 6-31+G(2d,p)// B3LYP/ 6-31G(d,p) |
|---|---|---|---|---|
| $(mF)_6$Tr–OH | -1404.28453 | -1400.950830 | -1400.68354 | -1400.75831 |
| $(mF)_5$Tr–OH (1) | -1305.053713 | -1301.898156 | -1301.623401 | -1301.695942 |
| $(mF)_5$Tr–OH (2) | -1305.053946 | -1301.898352 | -1301.623539 | -1301.695765 |
| $(mF)_2(mF)'(mF)''$Tr–OH (1) | -1205.823322 | -1202.845877 | -1202.563623 | -1202.633482 |
| $(mF)_2(mF)'(mF)''$Tr–OH (2) | -1205.823524 | -1202.846299 | -1202.564107 | -1202.634243 |
| $(mF)_2(mF)'(mF)''$Tr–OH (3) | -1205.823318 | -1205.561569 | -1202.846023 | -1202.563842 |
| $(mF)(mF)'(mF)''$Tr–OH (1) | -1106.592218 | -1106.322152 | -1103.793160 | -1103.503547 |
| $(mF)(mF)'(mF)''$Tr–OH (2) | -1106.592447 | -1106.322368 | -1103.793389 | -1103.503769 |
| $(mF)(mF)'(mF)''$Tr–OH (3) | -1106.592770 | -1106.322543 | -1103.793539 | -1103.503837 |
| $(mF)(mF)'(mF)''$Tr–OH (4) | -1106.592581 | -1106.322499 | -1103.793624 | -1103.504021 |
| $(mF)(mF)'$Tr–OH (1) | -1007.360459 | -1007.08209 | -1004.739220 | -1004.442164 |
| $(mF)(mF)'$Tr–OH (2) | -1007.360501 | -1007.08209 | -1004.739206 | -1004.442127 |
| $(mF)(mF)'$Tr–OH (3) | -1007.360653 | -1007.082292 | -1004.739418 | -1004.442377 |
| $(mF)$Tr–OH (1) | -908.128562 | -907.84193 | -905.685060 | -905.380579 |
| $(mF)$Tr–OH (2) | -908.128684 | -907.84205 | -905.685112 | -905.380648 |
| $(pF)$Tr–OH | -908.12880 | -905.685168 | -905.38063 | -905.44488 |
| $(pF)_2$Tr–OH | -1007.36043 | -1007.08202 | -1004.739007 | -1004.44191 |
| $(pF)_3$Tr–OH | -1106.59220 | -1106.32203 | -1103.792876 | -1103.50319 |
| Tr–OH | -808.897003 | -808.602090 | -806.63153 | -806.31962 |
| MeTr–OH | -848.2177824 | -847.895537 | -845.82873 | -845.48757 |
| Me$_2$Tr–OH | -887.5384286 | -887.188812 | -885.02565 | -884.65522 |
| Me$_3$Tr–OH | -926.859006 | -926.481928 | -924.22274 | -923.82308 |
| (MeO)Tr-OH (1) | -923.42217 | -923.09469 | -920.87525 | -920.52815 |
| (MeO)Tr-OH (2) | -923.42214 | -923.09464 | -920.87522 | -920.52812 |
| (MeO)$_2$Tr-OH (1) | -1037.94734 | -1037.58739 | -1035.11877 | -1034.73655 |
| (MeO)$_2$Tr-OH (2) | -1037.94730 | -1037.58733 | -1035.11876 | -1034.73653 |
| (MeO)$_2$Tr-OH (3) | -1037.94716 | -1035.11879 | -1034.73657 | -1034.81097 |
| (MeO)$_3$Tr-OH (1) | -1152.47249 | -1149.36238 | -1148.94493 | -1149.02483 |
| (MeO)$_3$Tr-OH (2) | -1152.47245 | -1149.36219 | -1148.94479 | -1149.02476 |
| (MeO)$_3$Tr-OH (3) | -1152.47230 | -1149.36215 | -1148.94471 | -1149.02479 |
| (MeO)$_3$Tr-OH (4) | -1152.47261 | -1149.36254 | -1148.94511 | -1149.02485 |
| (Me$_2$N)(MeO)Tr-OH (1) | -1057.39544 | -1054.46175 | -1054.03723 | -1054.11526 |
| (Me$_2$N)(MeO)Tr-OH (2) | -1057.39578 | -1054.46212 | -1054.03755 | -1054.11540 |
| (Me$_2$N)Tr-OH | -942.8703779 | -940.21829 | -939.82897 | -939.90110 |
| (Me$_2$N)$_2$Tr-OH | -1076.8434858 | -1073.80482 | -1073.33791 | -1073.41964 |
| (Me$_2$N)$_3$Tr-OH | -1210.8167096 | -1207.39162 | -1206.84741 | -1206.94010 |

## 5.2. Organocatalytic Activity of Cinchona Alkaloids: Which Nitrogen is more Nucleophilic?

The conformational space of quinine (**1a**), hydroxymethylquinuclidine (**1k**) and naphthylmethylquinuclidine (**1f**) as well as their cationic adducts has first been searched using the MM3 force field and the systematic search routine in the TINKER program.

In the case of **1k**, **1f** and their cationic adducts, the best three conformers were optimized at the B3LYP/6-31G(d) level of theory. Thermochemical corrections (B3LYP/6-31G(d)) to 298.15 K were combined with single-point energies on the MP2(FC)/6-31+G(2d,p) level.

For **1a** and its adducts the twenty energetically most favorable conformers according to the force field energies were submitted to single point calculations (B3LYP/6-31G(d)). The seven best conformers according to quantum mechanical energies were then taken as starting structures for geometry optimizations on the B3LYP/6-31G(d) level. Thermochemical corrections to 298.15 K have been calculated for all minima from unscaled vibrational frequencies, and have been combined with single-point energies calculated at the MP2(FC)/6-31+G(2d,p)//B3LYP/6-31G(d) level to yield enthalpies $H_{298}$ at 298.15 K.

When two force-field conformations turned into a single conformer during quantum mechanical geometry optimization, one was discarded, so that in each case seven different conformations were taken into account.

The other five smaller and therefore less flexible systems (lepidine, hydroxymethylquinoline, methoxyquinoline, methoxylepidine and quinuclidine) have not been submitted to conformational analyses but the structures were simply drawn in that manner, which was assumed to be the best. Care was only taken of the direction into which the methoxy group in methoxylepidine and -quinoline showed. Both possibilities have been calculated.

Solvation effects in dichloromethane have been calculated on the HF/6-31G(d) level of theory using the united atom for Hartree-Fock/polarizable continuum model PCM/UAHF. Gibbs free energies of solvation were combined with the MP2(FC)/6-31+G(2d,p)//B3LYP/6-31G(d) data.

# Experimental Part

Energy data for quinine (1a), in Hartree if not noted otherwise.

| FF-number | $E_{tot}$ (sp) B3LYP/6-31G(d) | $E_{tot}$ (opt) B3LYP/6-31G(d) | Thermochemical correction to enthalpy B3LYP/6-31G(d) | $E_{tot}$ MP2(FC)/ 6-31+G(2d,p) | $H_{298}$ MP2(FC)/ 6-31+G(2d,p)// B3LYP/6-31G(d) | Δ to best conformation (kJ/mol) | Dipole moment (Debye) | $\Delta G_{solv}$ (kcal/mol) |
|---|---|---|---|---|---|---|---|---|
| 51 | -1036.416920 | -1036.476349 | 0.432894 | -1033.656662 | -1033.223768 | 3.77 | 3.707 | -3.68 |
| 9  | -1036.418363 | -1036.476642 | 0.433203 | -1033.655041 | -1033.221838 | 8.85 | 1.317 | -4.66 |
| 43 | -1036.420258 | -1036.479073 | 0.433327 | -1033.658530 | -1033.225203 | 0.00 | 2.752 | -3.53 |
| 41 | -1036.416611 | -1036.474552 | 0.432806 | -1033.655987 | -1033.223181 | 5.32 | 3.816 | -3.89 |
| 53 | -1036.418357 | -1036.477154 | 0.433264 | -1033.656642 | -1033.223378 | 4.80 | 2.863 | -3.13 |
| 54 | -1036.418192 | -1036.476618 | 0.433285 | -1033.656051 | -1033.222766 | 6.41 | 3.025 | -3.58 |
| 60 | -1036.416934 | -1036.47683  | 0.43344  | -1033.658229 | -1033.224789 | 1.09 | 4.634 | -3.64 |

Energy data for quinine-benzyl-adduct ($N_{sp2}$), in Hartree if not noted otherwise.

| FF-number | $E_{tot}$ (sp) B3LYP/6-31G(d) | $E_{tot}$ (opt) B3LYP/6-31G(d) | Thermochemical correction to enthalpy B3LYP/6-31G(d) | $E_{tot}$ MP2(FC)/ 6-31+G(2d,p) | $H_{298}$ MP2(FC)/ 6-31+G(2d,p)// B3LYP/6-31G(d) | Δ to best conformation (kJ/mol) | Dipole moment (Debye) | $\Delta G_{solv}$ (kcal/mol) |
|---|---|---|---|---|---|---|---|---|
| 71  | -1307.170784 | -1307.238020 | 0.562618 | -1303.636377 | -1303.073759 | 0.00  | 6.513 | -29.46 |
| 189 | -1307.172260 | -1307.238740 | 0.562719 | -1303.635714 | -1303.072995 | 2.01  | 7.162 | -29.78 |
| 117 | -1307.170840 | -1307.237861 | 0.562495 | -1303.635777 | -1303.073282 | 1.25  | 5.552 | -30.18 |
| 123 | -1307.172552 | -1307.238785 | 0.562845 | -1303.635678 | -1303.072833 | 2.43  | 6.444 | -30.31 |
| 1   | -1307.170921 | -1307.235995 | 0.562734 | -1303.631249 | -1303.068515 | 13.79 | 8.259 | -31.3  |
| 72  | -1307.171249 | -1307.236261 | 0.562347 | -1303.635953 | -1303.073606 | 0.40  | 6.144 | -29.67 |
| 49  | -1307.171548 | -1307.23629  | 0.562305 | -1303.635627 | -1303.073322 | 1.15  | 5.210 | -30.29 |

Energy data for quinine-benzyl-adduct ($N_{sp3}$), in Hartree if not noted otherwise.

| FF-number | $E_{tot}$ (sp) B3LYP/6-31G(d) | $E_{tot}$ (opt) B3LYP/6-31G(d) | Thermochemical correction to enthalpy B3LYP/6-31G(d) | $E_{tot}$ MP2(FC)/ 6-31+G(2d,p) | $H_{298}$ MP2(FC)/ 6-31+G(2d,p)// B3LYP/6-31G(d) | Δ to best conformation (kJ/mol) | Dipole moment (Debye) | $\Delta G_{solv}$ (kcal/mol) |
|---|---|---|---|---|---|---|---|---|
| 30  | -1307.157340 | -1307.237017 | 0.563485 | -1303.644680 | -1303.081195 | 0.00 | 6.049 | -28.91 |
| 83  | -1307.156635 | -1307.236186 | 0.563546 | -1303.644500 | -1303.080954 | 0.64 | 6.154 | -28.75 |
| 410 | -1307.155712 | -1307.235065 | 0.563478 | -1303.643035 | -1303.079557 | 4.31 | 6.183 | -29.48 |
| 62  | -1307.151971 | -1307.232463 | 0.563714 | -1303.642412 | -1303.078698 | 6.57 | 6.757 | -29.14 |
| 1   | -1307.154753 | -1307.235100 | 0.563448 | -1303.643623 | -1303.080175 | 2.68 | 8.250 | -29.29 |
| 48  | -1307.153921 | -1307.234194 | 0.563414 | -1303.643298 | -1303.079884 | 3.45 | 8.358 | -29.38 |
| 409 | -1307.153131 | -1307.233174 | 0.563445 | -1303.641989 | -1303.078544 | 6.97 | 8.357 | -29.91 |

Energy data for quinine-benzhydryl-adduct ($N_{sp2}$), in Hartree if not noted otherwise.

| FF-number | $E_{tot}$ (sp) B3LYP/6-31G(d) | $E_{tot}$ (opt) B3LYP/6-31G(d) | Thermochemical correction to enthalpy B3LYP/6-31G(d) | $E_{tot}$ MP2(FC)/ 6-31+G(2d,p) | $H_{298}$ MP2(FC)/ 6-31+G(2d,p)// B3LYP/6-31G(d) | Δ to best conformation (kJ/mol) | Dipole moment (Debye) | $\Delta G_{solv}$ (kcal/mol) |
|---|---|---|---|---|---|---|---|---|
| 55 | -1538.201378 | -1538.288014 | 0.64807  | -1534.044981 | -1533.396911 | 0.00  | 4.512 | -23.34 |
| 65 | -1538.202616 | -1538.288829 | 0.648429 | -1534.044313 | -1533.395884 | 2.70  | 4.988 | -23.79 |
| 74 | -1538.201508 | -1538.287847 | 0.648007 | -1534.044259 | -1533.396252 | 1.73  | 3.872 | -23.79 |
| 64 | -1538.203332 | -1538.288888 | 0.648394 | -1534.044257 | -1533.395863 | 2.76  | 4.429 | -24.06 |
| 1  | -1538.201605 | -1538.286083 | 0.648187 | -1534.039759 | -1533.391572 | 14.04 | 6.440 | -24.98 |
| 88 | -1538.201446 | -1538.286146 | 0.647789 | -1534.04401  | -1533.396221 | 1.82  | 4.229 | -23.65 |
| 67 | -1538.202001 | -1538.286243 | 0.64786  | -1534.04424  | -1533.39638  | 1.40  | 3.572 | -23.94 |

Energy data for quinine-benzhydryl-adduct ($N_{sp3}$), in Hartree if not noted otherwise.

| FF-number | $E_{tot}$ (sp) B3LYP/6-31G(d) | $E_{tot}$ (opt) B3LYP/6-31G(d) | Thermochemical correction to enthalpy B3LYP/6-31G(d) | $E_{tot}$ MP2(FC)/ 6-31+G(2d,p) | $H_{298}$ MP2(FC)/ 6-31+G(2d,p)// B3LYP/6-31G(d) | Δ to best conformation (kJ/mol) | Dipole moment (Debye) | $\Delta G_{solv}$ (kcal/mol) |
|---|---|---|---|---|---|---|---|---|
| 2 | -1538.171084 | -1538.273364 | 0.649477 | -1534.046466 | -1533.396989 | 1.48 | 5.214 | -22.73 |
| 41 | -1538.169975 | -1538.272110 | 0.649211 | -1534.046762 | -1533.397551 | 0.00 | 5.296 | -22.72 |
| 668 | -1538.169455 | -1538.271008 | 0.649227 | -1534.044487 | -1533.395260 | 6.02 | 5.330 | -23.21 |
| 547 | -1538.169966 | -1538.271413 | 0.649645 | -1534.043884 | -1533.394239 | 8.71 | 5.112 | -23.67 |
| 81 | -1538.169495 | -1538.272200 | 0.649606 | -1534.044614 | -1533.395008 | 6.69 | 5.079 | -22.64 |
| 12 | -1538.168166 | -1538.269343 | 0.64907 | -1534.041442 | -1533.392372 | 13.62 | 5.591 | -22.04 |
| 13 | -1538.168361 | -1538.271647 | 0.649583 | -1534.040101 | -1533.390518 | 18.49 | 5.382 | -24.09 |

Energy data for quinine-methyl-adduct ($N_{sp2}$), in Hartree if not noted otherwise.

| FF-number | $E_{tot}$ (sp) B3LYP/6-31G(d) | $E_{tot}$ (opt) B3LYP/6-31G(d) | Thermochemical correction to enthalpy B3LYP/6-31G(d) | $E_{tot}$ MP2(FC)/ 6-31+G(2d,p) | $H_{298}$ MP2(FC)/ 6-31+G(2d,p)// B3LYP/6-31G(d) | Δ to best conformation (kJ/mol) | Dipole moment (Debye) | $\Delta G_{solv}$ (kcal/mol) |
|---|---|---|---|---|---|---|---|---|
| 5 | -1076.137741 | -1076.185863 | 0.476596 | -1073.233161 | -1072.756565 | 0.00 | 9.690 | -32.66 |
| 15 | -1076.139139 | -1076.186592 | 0.477051 | -1073.232838 | -1072.755787 | 2.04 | 10.694 | -32.68 |
| 38 | -1076.137325 | -1076.183768 | 0.476896 | -1073.228343 | -1072.751447 | 13.46 | 12.535 | -33.75 |
| 57 | -1076.137689 | -1076.183157 | 0.476415 | -1073.230314 | -1072.753899 | 7.01 | 11.190 | -33.15 |
| 18 | -1076.138506 | -1076.184303 | 0.476502 | -1073.232825 | -1072.756323 | 0.64 | 9.443 | -32.82 |
| 63 | -1076.138176 | -1076.185542 | 0.477152 | -1073.231836 | -1072.754684 | 4.95 | 10.180 | -31.88 |
| 60 | -1076.137290 | -1076.184800 | 0.477064 | -1073.230916 | -1072.753852 | 7.13 | 10.497 | -32.51 |

Energy data for quinine-methyl-adduct ($N_{sp3}$), in Hartree if not noted otherwise.

| FF-number | $E_{tot}$ (sp) B3LYP/6-31G(d) | $E_{tot}$ (opt) B3LYP/6-31G(d) | Thermochemical correction to enthalpy B3LYP/6-31G(d) | $E_{tot}$ MP2(FC)/ 6-31+G(2d,p) | $H_{298}$ MP2(FC)/ 6-31+G(2d,p)// B3LYP/6-31G(d) | Δ to best conformation (kJ/mol) | Dipole moment (Debye) | $\Delta G_{solv}$ (kcal/mol) |
|---|---|---|---|---|---|---|---|---|
| 16 | -1076.125253 | -1076.185640 | 0.477506 | -1073.242470 | -1072.764964 | 0.00 | 8.744 | -33.58 |
| 6  | -1076.122430 | -1076.183547 | 0.477439 | -1073.240881 | -1072.763442 | 4.00 | 10.829 | -33.96 |
| 10 | -1076.119378 | -1076.181190 | 0.477719 | -1073.236461 | -1072.758742 | 16.36 | 9.360 | -34.85 |
| 4  | -1076.117592 | -1076.180145 | 0.477533 | -1073.2366   | -1072.759067 | 15.50 | 11.668 | -35.87 |
| 8  | -1076.118441 | -1076.174521 | 0.477288 | -1073.232967 | -1072.755679 | 24.41 | 8.899 | -35.16 |
| 18 | -1076.117717 | -1076.179631 | 0.477367 | -1073.23338  | -1072.756013 | 23.53 | 8.966 | -35.13 |
| 12 | -1076.116254 | -1076.178741 | 0.477214 | -1073.233162 | -1072.755948 | 23.70 | 11.213 | -35.99 |

Energy data for Lewis base **1f**, and its benzyl- and benzhydryl adducts, in Hartree if not noted otherwise.

| system | FF-number | $E_{tot}$ (sp) B3LYP/6-31G(d) | $E_{tot}$ (opt) B3LYP/6-31G(d) | Thermochemical correction to enthalpy B3LYP/6-31G(d) | $E_{tot}$ MP2(FC)/ 6-31+G(2d,p) | $H_{298}$ MP2(FC)/ 6-31+G(2d,p)// B3LYP/6-31G(d) | Δ to best conformation (kJ/mol) |
|---|---|---|---|---|---|---|---|
| **1f** | 4 | -753.2651427 | -753.3165755 | 0.368887 | -751.1535243 | -750.7846373 | 0.00 |
|        | 6 | -753.2642904 | -753.3154701 | 0.368842 | -751.1527055 | -750.7838635 | 2.03 |
|        | 3 | -753.2633652 | -753.3144444 | 0.368701 | -751.1512879 | -750.7825869 | 5.39 |
| **1f-benzyl** | 8 | -1023.996961 | -1024.072854 | 0.49929 | -1021.141637 | -1020.6423467 | 0.00 |
|               | 4 | -1023.996823 | -1024.072892 | 0.499323 | 1021.1404771 | -1020.6411541 | 3.14 |
|               | 2 | -1024.002144 | -1024.075018 | 0.499381 | 1021.1407586 | -1020.6413776 | 2.55 |
| **1f-benzhydryl** | 5 | -1255.017648 | -1255.112466 | 0.585263 | 1251.5419699 | -1250.9567069 | 2.83 |
|                   | 8 | -1255.011623 | -1255.109377 | 0.585095 | 1251.5426503 | -1250.9575553 | 0.60 |
|                   | 9 | -1255.018632 | -1255.113615 | 0.585369 | 1251.5431541 | -1250.9577851 | 0.00 |

# Experimental Part

Energy data for Lewis bases **1k**, **1e**, **1h**, **1g**, and their benzyl- and benzhydryl adducts, in Hartree if not noted otherwise.

| system | FF-number | $E_{tot}$ (sp) B3LYP/6-31G(d) | $E_{tot}$ (opt) B3LYP/6-31G(d) | Thermochemical correction to enthalpy B3LYP/6-31G(d) | $E_{tot}$ MP2(FC)/ 6-31+G(2d,p) | $H_{298}$ MP2(FC)/ 6-31+G(2d,p)// B3LYP/6-31G(d) | Δ to best conformation (kJ/mol) |
|---|---|---|---|---|---|---|---|
| **1k** | 6 | -443.8296309 | -443.8355403 | 0.239482 | -442.6063017 | -442.3668197 | 0.00 |
|  | 1 | -443.8229025 | -443.8281132 | 0.23892 | -442.5997007 | -442.3607807 | 15.88 |
|  | 3 | -443.8204851 | -443.8257981 | 0.238772 | -442.5972547 | -442.3584827 | 21.92 |
| **1k**-benzyl | 3 | -714.5597291 | -714.5876988 | 0.369333 | -712.5855074 | -712.2161744 | 6.19 |
|  | 6 | -714.5621748 | -714.5906121 | 0.369853 | -712.5883835 | -712.2185305 | 0.00 |
|  | 2 | -714.5597783 | -714.5880062 | 0.369735 | -712.5860423 | -712.2163073 | 5.85 |
| **1k**-benzhydryl | 10 | -945.5759989 | -945.6290945 | 0.455626 | -942.9853885 | -942.5297625 | 1.25 |
|  | 5 | -945.5787217 | -945.6285398 | 0.455506 | -942.9857450 | -942.5302390 | 0.00 |
|  | 6 | -945.5771193 | -945.6283938 | 0.455453 | -942.9838872 | -942.5284342 | 4.75 |
| **1e** | - | - | -329.3107189 | 0.203617 | -328.3435491 | -328.1399321 | - |
| **1e**-benzyl | - | - | -600.0697431 | 0.333253 | -598.3280801 | -597.9948271 | - |
| **1e**-benzhydryl | - | - | -831.11209 | 0.419896 | -828.7281023 | -828.3082063 | - |
| **1h** | - | - | -516.4545753 | 0.178973 | -515.0510466 | -514.8720736 | - |
| **1h**-benzyl | - | - | -787.208684 | 0.308684 | -785.0225453 | -784.7138403 | - |
| **1h**-benzhydryl | - | - | -1018.2599437 | 0.394112 | -1015.4318639 | -1015.0377519 | - |
| **1g** | - | - | -441.2493576 | 0.173267 | -440.0040282 | -439.8307610 | - |
| **1g**-benzyl | - | - | -712.0053033 | 0.302897 | -709.9745664 | -709.6716694 | - |
| **1g**-benzhydryl | - | - | -943.0543945 | 0.388434 | -940.3835649 | -939.9951309 | - |

Energy data for Lewis bases **1j**, **1i**, and their benzyl- and benzhydryl adducts, in Hartree if not noted otherwise.

| system | FF⁻ number | $E_{tot}$ (sp) B3LYP/6-31G(d) | $E_{tot}$ (opt) B3LYP/6-31G(d) | Thermochemical correction to enthalpy B3LYP/6-31G(d) | $E_{tot}$ MP2(FC)/ 6-31+G(2d,p) | $H_{298}$ MP2(FC)/ 6-31+G(2d,p)// B3LYP/6-31G(d) | Δ to best conformation (kJ/mol) |
|---|---|---|---|---|---|---|---|
| **1j** | – | – | -555.7731633 | 0.208592 | -554.2496712 | -554.0410792 | – |
| **1j**-benzyl | – | – | -826.5329992 | 0.338287 | -824.2244984 | -823.886211 | – |
| **1j**-benzhydryl | – | – | -1057.5819319 | 0.423731 | -1054.6336681 | -1054.2099371 | – |
| **1i** | – | – | -516.4485684 | 0.17925 | -515.0604377 | -514.8811877 | – |
| **1i**-benzyl | – | – | -787.2041789 | 0.563485 | -785.0299946 | -784.721110 | – |
| **1i**-benzhydryl | – | – | -1018.2531412 | 0.649211 | -1015.438933 | -1015.0446651 | – |

Energy data for carbocations, in Hartree if not noted otherwise.

| cation | $E_{tot}$ B3LYP/6-31G(d) | Thermochemical correction to enthalpy $H$ B3LYP/6-31G(d) | Thermochemical correction to free energy $G$ B3LYP/6-31G(d) | $E_{tot}$ MP2(FC)/ 6-31+G(2d,p) | $H_{298}$ MP2(FC)/ 6-31+G(2d,p)// B3LYP/6-31G(d) | $G_{298}$ MP2(FC)/ 6-31+G(2d,p)// B3LYP/6-31G(d) | $\Delta G_{solv}$ (kcal/mol) |
|---|---|---|---|---|---|---|---|
| Me⁺ | -39.4803877 | 0.035413 | 0.014216 | -39.3523833 | -39.3169703 | -39.3381673 | -60.08 |
| Benzyl⁺ | -270.662271 | 0.124198 | 0.088482 | -269.8592704 | -269.735072 | -269.770788 | -38.25 |
| Benzhydryl⁺ | -501.754125 | 0.211046 | 0.163807 | -500.2934716 | -500.082426 | -500.129665 | -33.3 |

Experimental Part 311

### 5.3 Carbocationic n-endo-trig Cyclizations

The conformational space has been searched using the MM3 force field and the systematic search routine in the TINKER program. In each case the best four conformers have been subjected to geometry optimizations on the B3LYP/6-311G(d,p) level. Additionally, closed ring conformers with C(1)–C(2) = 1.54 Å have been used as input for geometry opimizations on B3LYP/6-311G(d,p). The preoptimized structures were then submitted to geometry optimizations on the MP2/6-31+G(2d,p) level. Bond length scans varied the distance C(1)–C(2) from 1.40 Å to 1.88 Å with an increment of 0.02 Å. All optimized structures were confirmed as being minima by frequency calculations on the respective level of theory. No frequency calculations have been done in the case of the bond length scans.

## 6. References

[1] J. W. Ponder, *TINKER*, Version 4.2, **2004**.
[2] M. J. Frisch, G. W. Trucks, H. B. Schlegel, G. E. Scuseria, M. A. Robb, J. R. Cheeseman, J. A. Montgomery, Jr., T. Vreven, K. N. Kudin, J. C. Burant, J. M. Millam, S. S. Iyengar, J. Tomasi, V. Barone, B. Mennucci, M. Cossi, G. Scalmani, N. Rega, G. A. Petersson, H. Nakatsuji, M. Hada, M. Ehara, K. Toyota, R. Fukuda, J. Hasegawa, M. Ishida, T. Nakajima, Y. Honda, O. Kitao, H. Nakai, M. Klene, X. Li, J. E. Knox, H. P. Hratchian, J. B. Cross, V. Bakken, C. Adamo, J. Jaramillo, R. Gomperts, R. E. Stratmann, O. Yazyev, A. J. Austin, R. Cammi, C. Pomelli, J. W. Ochterski, P. Y. Ayala, K. Morokuma, G. A. Voth, P. Salvador, J. J. Dannenberg, V. G. Zakrzewski, S. Dapprich, A. D. Daniels, M. C. Strain, O. Farkas, D. K. Malick, A. D. Rabuck, K. Raghavachari, J. B. Foresman, J. V. Ortiz, Q. Cui, A. G. Baboul, S. Clifford, J. Cioslowski, B. B. Stefanov, G. Liu, A. Liashenko, P. Piskorz, I. Komaromi, R. L. Martin, D. J. Fox, T. Keith, M. A. Al-Laham, C. Y. Peng, A. Nanayakkara, M. Challacombe, P. M. W. Gill, B. Johnson, W. Chen, M. W. Wong, C. Gonzalez, and J. A. Pople, Gaussian, Inc., Wallingford CT, **2004**.
[3] C. S. Marvel, H. W. Johnston, J. W. Meier, T. W. Mastin, J. Whitson, C. M. Himel, *J. Am. Chem. Soc.* **1944**, *66*, 914-918.
[4] W. H. Saunders, Jr., J. C. Ware, *J. Am. Chem. Soc.* **1958**, *80*, 3328-3332.
[5] S. T. Bowden, T. F. Watkins, *J. Chem. Soc.* **1940**, 1249-1257.
[6] W. N. White, C. A. Stout, *J. Org. Chem.* **1962**, *27*, 2915- 2917.

[7] D. Hellwinkel, H. Fritsch, *Chem. Ber.* **1989**, *122*, 2351-2359.
[8] A. Baeyer, V. Villiger, *Ber. Dtsch. Chem. Ges.* **1902**, *35*, 3013-3033.
[9] C. Bleasdale, S. B. Ellwood, B. T. Golding, *J. Chem. Soc. Perkin Trans. 1* **1990**, 803-805.
[10] V. Gold, C. Tomlinson, *J. Chem. Soc. B* **1971**, 1707-1711.
[11] S. V. McKinley, J. W. Rakshys, Jr., A. E. Young, H. H. Freedman, *J. Am. Chem. Soc.* **1971**, *93*, 4715-4724.
[12] V. Villiger, E. Kopetschni, *Ber. Dtsch. Chem. Ges.* **1912**, *45*, 2920.
[13] R. G. Pews, Y. Tsuno, R. W. Taft, *J. Am. Chem. Soc.* **1967**, *89*, 2391-2394.
[14] H. J. Dauben, R. H. Lewis, K. M. Harmon, *J. Org. Chem.* **1960**, *25*, 1442-1445.
[15] M. Canle L., W. Clegg, I. Demirtas, M. R. J. Elsegood, H. Maskill, *J. Chem. Soc. Perkin Trans. 2* **2000**, *8*, 85-92.
[16] M. Canle L., W. Clegg, I. Demirtas, M. R. J. Elsegood, J. Haider, H. Maskill, P. C. Miatt, *J. Chem. Soc. Perkin Trans. 2* **2001**, *9*, 1742-1747.
[17] C. A. Bunton, A. Konasiewicz, *J. Chem. Soc.* **1955**, 1354-1359.
[18] G. S. Hammond, J. T. Rudesill, *J. Am. Chem. Soc.* **1950**, *72*, 2769-2770.
[19] S. Yolles, J. H. R. Woodland, *J. Organomet. Chem.* **1973**, *54*, 95-104.
[20] N. N. Lichtin, M. J. Vignale, *J. Am. Chem. Soc.* **1957**, *79*, 579-583.
[21] P. Miles, H. Suschitzky, *Tetrahedron* **1963**, *19*, 385-390.
[22] The fluorinating agent consists of AgF dispersed on $CaF_2$. It was prepared according to T. Ando, D. G. Cork, M. Fujita, T. Kimura, T. Tatsuno, *Chem. Lett.* **1988**, 1877-1878.
[23] G. A. Olah, X.-Y. Li, Q. Wang, G. K. S. Prakash, *Synthesis*, **1993**, 693-699.
[24] C. G. Swain, T. E. C. Knee, A. MacLachlan, *J. Am. Chem. Soc.* **1960**, *82*, 6101-6104.
[25] F. Bacon, J. H. Gardner, *J. Org. Chem.* **1938**, *3*, 281-286.
[26] R. E. Lovins, L. J. Andrews, R. M. Keef, *J. Org. Chem.* **1963**, *28*, 2847-2850.
[27] M. Gomberg, *Chem. Ber.* **1904**, 1626-1644.
[28] A. Mothwurf, Chem. Ber. 1904, 3153-3163.
[29] H. Sakurai, H. Umino, H. Sugiyama, *J. Am. Chem. Soc.* **1980**, *102*, 6837-6840.
[30] N. S. Nudelman, C. Carro, *J. Organomet. Chem.* **1998**, 31-36.
[31] J. Dickhaut, B. Giese, *Org. Syntheses* **1992**, *70*, 164-168.
[32] C. Marschner, *Eur. J. Inorg. Chem.* **1998**, 221-226.
[33] H. Mayr, G. Lang, A. R. Ofial, *J. Am. Chem. Soc.* **2002**, *124*, 4076-4083.
[34] R. B. Moffett, *Org. Syntheses* **1952**, *32*, 41-44.

[35]  W. Casein, P. S. Pregosin, *J. Organomet. Chem.* **1988**, *356*, 259-269.
[36]  a) T. G. Bonner, D. Lewis, K. Rutter, *J. Chem. Soc., Perkin Trans. 1* **1981**, 1807-1810; b) B. Pério, M.-J. Dozias, P. Jacquault, J. Hamelin, *Tetrahedron Lett.* **1997**, *38*, 7867-7870.
[37]  E. J. Salmi, *Chem. Ber.* **1938**, 1803-1808.
[38]  G. Barbe, A. G. Charette, *J. Am. Chem. Soc.* **1980**, *102*, 6837-6840.
[39]  J. J. V. Eynde, F. Delfosse, A. Mayence, Y. V. Haverbeke, *Tetrahedron* **1995**, *51*, 6511-6516.
[40]  B. Loev, K. M. Snader, *J. Org. Chem.* **1965**, *30*, 1914-1916.
[41]  J. Le Bras, J. Muzart, *Tetrahedron* **2007**, *63*, 7942-7948.
[42]  G. A. Olah, G. K. S. Prakash, *Synthesis* **1978**, 397-398.

# Abbreviations and Symbols

| | | | |
|---|---|---|---|
| A | acetone | quat. | quaternary |
| AN | acetonitrile | ref. | reference |
| ani | anisyl | r.t. | room temperature |
| Ar | aryl | s | second |
| arom. | aromatic | SCE | standard calomel electrode |
| a.u. | atomic unit | | |
| DMSO | dimethylsulfoxide | $T$ | temperature |
| EI | electron impact ionization | TEA | triethylamine |
| ESI | electron spray ionization | tert. | tertiary |
| eV | electron volt | THF | tetrahydrofuran |
| FF | force field | TMS | trimethylsilyl or tetramethylsilane |
| g | gram | | |
| GP | general procedure | tol | tolyl |
| h | hour | Tr | trityl |
| HMBC | heteronuclear multiple bond correlation | trop$^+$ | tropylium |
| | | UV | ultra violet |
| HR-MS | high-resolution mass spectrometry | vis | visible |
| | | vol | volume |
| Hz | hertz | vs. | versus |
| $J$ | coupling constant | W | water |
| L | liter | | |
| M | molar | | |
| max. | maximum | | |
| Me | Methyl | | |
| mg | miligram | | |
| MHz | megahertz | | |
| min | minute | | |
| mL | mililiter | | |
| Mp | melting point | | |
| MS | mass spectrometry | | |
| NMR | nuclear magnetic resonance | | |
| Ph | Phenyl | | |
| pip | piperidine | | |
| pos. | positive | | |
| ppm | parts per million | | |

VDM Verlagsservicegesellschaft mbH

Die VDM Verlagsservicegesellschaft sucht für wissenschaftliche Verlage abgeschlossene und herausragende

# Dissertationen, Habilitationen, Diplomarbeiten, Master Theses, Magisterarbeiten usw.

für die kostenlose Publikation als Fachbuch.

Sie verfügen über eine Arbeit, die hohen inhaltlichen und formalen Ansprüchen genügt, und haben Interesse an einer honorarvergüteten Publikation?

Dann senden Sie bitte erste Informationen über sich und Ihre Arbeit per Email an *info@vdm-vsg.de*.

### Sie erhalten kurzfristig unser Feedback!

VDM Verlagsservicegesellschaft mbH
Dudweiler Landstr. 99        Telefon +49 681 3720 174
D - 66123 Saarbrücken        Fax     +49 681 3720 1749
**www.vdm-vsg.de**

Die VDM Verlagsservicegesellschaft mbH vertritt

Printed by Books on Demand GmbH, Norderstedt / Germany